NMR Basic Principles and Progress 20

Editors: P. Diehl E. Fluck R. Kosfeld

G. Govil R. V. Hosur

Conformation of Biological Molecules

New Results from NMR

With 92 Figures

Springer-Verlag
Berlin Heidelberg New York 1982

Prof. Dr. Girjesh Govil
Dr. Ramakrishna V. Hosur
Tata Institute of Fundamental Research
Homi Bhabha Road
Bombay 400 005, India

ISBN-13:978-3-642-68099-1 e-ISBN-13:978-3-642-68097-7
DOI: 10.1007/978-3-642-68097-7

Library of Congress Cataloging in Publication Data. Govil, G. (Girjesh), 1940–. Conformation of biological molecules. (NMR, basic principles and progress; 20) Bibliography: p. Includes index. 1. Macromolecules. 2. Conformational analysis. 3. Nuclear magnetic resonance spectroscopy. 4. Molecular biology. I. Hosur, R. (Ramakrishna), 1953–. II. Title. III. Series. QC490.N2 vol. 20 [QP801.P64] 538.3s 81-8976. ISBN-13:978-3-642-68099-1 (U.S.) [574.19'282] AACR2

Softcover reprint of the hardcover 1st edition 1982

Typesetting: Elsner & Behrens, Oftersheim

2152/3140-543210

Contents

1	**General Theory**	1
1.1	Introduction	1
1.2	What is Conformation?	2
1.3	Conformational Theory	4
1.4	Structure of Long-Chain Polymers	8
1.5	Problems in NMR Studies of Biological Molecules	9
1.5.1	^{1}H-NMR	9
1.5.2	^{2}H-NMR	11
1.5.3	^{13}C-NMR	12
1.5.4	^{15}N-NMR	14
1.5.5	^{31}P-NMR	15
1.5.6	Other Nuclei	16
2	**NMR Techniques in Conformational Studies**	18
2.1	Coupling Constants	18
2.2	Chemical Shifts of Hydrogen-Bonded Protons	22
2.3	Magnetic Anisotropy of Chemical Bonds or Groups	23
2.4	^{13}C Chemical Shifts	25
2.5	^{31}P Chemical Shifts	26
2.6	Relaxation Times (T_1 and T_2)	26
2.7	Nuclear Overhauser Effect	29
2.8	Paramagnetic Reagents	30
2.9	Use of Liquid Crystals as Solvents	35
2.10	Deuterium Quadrupole Coupling Constants	36
2.11	Solvent Accessibility	36
2.12	From NMR Parameters to Spatial Structures	37
3	**Nucleosides, Nucleotides and Nucleic Acids**	39
3.1	Description of Nucleotide Structures	39
3.2	Glycosidic Bond Rotations	43
3.3	Sugar Ring Conformation: Pseudorotation	46
3.4	Backbone Angles	51
3.4.1	Torsion Angles ϕ'	51
3.4.2	Angles ϕ	53
3.4.3	Angles ψ	55
3.4.4	Torsion Angles ω', ω	55
3.5	NMR Studies on Nucleosides	58

3.6 Small Nucleotides 58
3.6.1 2′-, 3′- and 5′-Mononucleotides 61
3.6.2 Cyclic Nucleotides 61
3.6.3 Pyridine Nucleosides and Nucleotides 62
3.6.4 5′-Di- and Triphosphates 62
3.7 Dinucleoside Phosphates and Short Segments of Nucleic Acids 63
3.8 Random-Coil Polynucleotides 66
3.8.1 Oligo- and Polyribonucleotides 67
3.8.2 Deoxyribonucleotides 70
3.9 Helical Polynucleotides 70
3.10 t-RNA 74
3.11 Drug-Nucleic Acid Interactions 79

4 Amino Acids, Peptides and Proteins 81

4.1 Description of the Structure of Peptide Units 81
4.2 Theoretical Considerations. The ϕ, ψ Maps 87
4.3 NMR Techniques in the Study of Peptide Conformations 92
4.3.1 *cis-trans* Isomerization (Angle ω) 92
4.3.2 ϕ and ψ Angles 94
4.3.3 Side-Chain Conformations 96
4.3.4 Detection of N—H···O=C Hydrogen Bonds 97
4.3.5 Molecular Symmetry 99
4.3.6 Conformational Mobility 99
4.3.7 ^{1}H and ^{13}C Chemical Shifts 100
4.4 Conformations of Amino Acids 101
4.5 Linear Peptides 102
4.5.1 Dipeptides 105
4.5.2 Tripeptides 106
4.5.3 Tetrapeptides 107
4.5.4 Pentapeptides 108
4.5.5 Hexapeptides 108
4.5.6 Larger Peptides. Peptide Hormones 108
4.5.6.1 Angiotensin 108
4.5.6.2 Leuteinizing Hormone-Releasing Hormone (LRF) 109
4.5.6.3 Peptides Containing α-Aminobutyric Acid (A_{ib}) 109
4.5.6.4 Human Parathyroid Hormone (PTH) 110
4.5.6.5 Insulin 110
4.6 Cyclic Peptides 110
4.6.1 Dipeptides 110
4.6.2 Tripeptides 115
4.6.3 Tetrapeptides 116
4.6.4 Pentapeptides 116
4.6.5 Hexapeptides 117
4.6.6 Larger Cyclic Peptides, Peptide Hormones and Antibiotics 119
4.6.6.1 Oxytocin and Vasopressin 119

4.6.6.2 Gramicidin S . . . 121
4.6.6.3 Valinomycin . . . 122
4.7 Homopolymeric Peptides. Helix-Coil Transition . . . 124
4.8 Characterization of Protein Structures by NMR . . . 128
4.8.1 Spatial Arrangement of Atoms in the Molecules . . . 129
4.8.2 Protein Mobility . . . 130
4.8.3 Enzyme – Substrate Binding . . . 130
4.8.4 Thermodynamic Parameters . . . 130
4.8.5 Conformation of Specific Proteins . . . 133
4.8.5.1 Lysozyme . . . 133
4.8.5.2 Basic Pancreatic Trypsin Inhibitor (BPTI) . . . 134
4.8.5.3 Dihydrofolate Reductase . . . 136
4.8.5.4 Antibody Combining Site . . . 137
4.8.5.5 Myelin Basic Protein (MBP) . . . 137
4.8.5.6 Calcium-Binding Proteins . . . 137
4.8.5.7 Elastin and Tropoelastin . . . 137
4.8.5.8 Ribonuclease . . . 138
4.8.5.9 Hemoproteins and Hemoenzymes . . . 138
4.8.5.10 Collagen . . . 140
4.8.5.11 α-Chymotrypsin . . . 140
4.9 Protein-Nucleic Acid Interaction . . . 141
4.9.1 Specificity . . . 141
4.9.2 Nature of Intermolecular Forces . . . 142
4.9.3 Applications . . . 143

5 Polysaccharides . . . 144

5.1 Structures of Polysaccharides and Carbohydrates . . . 144
5.2 Conformations of Monosaccharides . . . 144
5.3 Conformations of Polysaccharides . . . 148
5.4 Glycoproteins and Peptide-Carbohydrate Interactions . . . 152

6 Lipids and Molecular Organization in Membranes . . . 155

6.1 Biomembranes . . . 155
6.2 General Properties of Phospholipids . . . 157
6.3 Conformations of Phospholipids . . . 157
6.3.1 Conformations of the Glycerol Moiety . . . 158
6.3.2 Conformations in the α-Chains . . . 160
6.3.3 Conformations in the β and γ Chains . . . 161
6.4 Membrane Organization and Fluidity . . . 163
6.4.1 Fluidity . . . 163
6.4.1.1 Rotational Motions . . . 164
6.4.1.2 Lateral Diffusion on the Surface of the Membrane . . . 165
6.4.1.3 Flip-Flop Motions . . . 165

6.4.2 Determination of Organization 165
6.4.2.1 ^{31}P-NMR 166
6.4.2.2 ^{2}H-NMR 167
6.4.2.3 Other Studies 170
6.5 Lipid-Protein and Lipid-Cholesterol Interactions 172

7 Acknowledgements 174

8 References 175

9 Appendix 200

10 Subject Index 213

1 General Theory

1.1 Introduction

The determination of the three-dimensional structure of a biological molecule is the starting point in the understanding of molecular mechanisms involved in its complex biochemical reactions. The molecular architecture of multimolecular systems such as membranes and chromosomes provides the key to the fascinating field of molecular biology. Stereochemical details of biological macromolecules and their interactions with pharmacological agents form the basis for drug design. Naturally, the study of the structure and function of biological molecules has aroused tremendous interest and investigations in this area are being carried out in a large number of laboratories. The techniques used for this purpose include both experimental methods (X-ray and neutron diffraction measurements, study of NMR, ESR, vibrational and electronic spectra, ORD, CD and dipole moment measurements, biochemical modifications etc.) and theoretical methods (quantum mechanical and classical potential energy calculations, Monte-Carlo simulations and molecular graphics).

For several years now, X-ray diffraction [1] has served as our only source of information on the three-dimensional arrangements of atoms in biopolymers. Fiber-diffraction of DNA led to the proposal of the DNA double helix. Fibers of long-chain polymers show ordering in the direction of the fibre-axis but not in the transverse plane. Accurate estimates of the dimensions of helical structures can be made using techniques on the basis of which models of biopolymers can be constructed. On the other hand, X-ray diffraction patterns from single crystals of proteins, nucleic acids and other biomolecules, contain information on the three-dimensional arrangement of atoms. With accurate data, positions of all heavy atoms can be obtained with a high degree of precision. Structures of almost 50 macromolecules of biological interest are known with a fair degree of accuracy, not to speak of the wealth of information on small biomolecules. However, when relating these results to biological functions, one has to be sure that the crystal structure is maintained in aqueous solutions under biological conditions of pH and ionic concentrations. Secondly, X-ray crystallography gives a static picture of the molecule constrained by lattice forces and averaged over disorders and thermal motions. As a result, biologists often look at biomolecules as if these are rigid structures though it is now known that parts, or the whole molecule, may exhibit considerable degree of internal motions and flexibility.

During the last decade, NMR has proved to be one of the most powerful tools for the study of biological structures in time and space. With the existing state of art, NMR does not provide the absolute coordinates of atoms in molecules. The advantage of NMR lies in the fact that one can observe molecular structures under actual biological conditions and even monitor structural and biochemical changes in intact cells. Several books and reviews [2–10] have appeared covering various aspects of NMR of biological molecules. In addition, the Chemical Society (London) regularly reviews the subject of biological macromolecules in its periodic reports [11].

This review is devoted to one particular aspect of biological molecules, namely their three-dimensional structure in solutions. To the best of our knowledge, there have been no previous reviews on this specific subject although different classes of molecules have been covered from time to time. The emphasis of this article is on basic principles and methodology and a review of the literature published during the period 1972–80 is given. We have assumed that the reader of this series is familiar with the techniques of NMR and present developments in this area. Therefore, only brief mention has been made of techniques which have particular significance in the context of the structure of biological molecules. Further, since the reader may not be familiar with the terminology used in biomolecular structures, a brief introduction to conformational analysis has been given. A computer search procedure was used to retrieve literature from Chemical Abstracts data files. This has been supplemented by cross references. In a review containing almost 1,250 references, some omissions are inevitable. We express our sincere apologies to authors whose work could not be retrieved by the methods mentioned above.

1.2 What is Conformation?

Let us consider a linear molecule consisting of four atoms A–B–C–D. The three-dimensional structure of such a molecule requires the knowledge of three bond lengths (A–B, B–C, C–D), two bond angles (A–B–C and B–C–D) and one torsion angle ϕ (the dihedral angle between the plane ABC and BCD). In general, for a molecule consisting of N atoms, the structural parameters are N-1 bond lengths, N-2 bond angles and N-3 torsion angles (also called conformational angles or rotational angles). Thus, the structure of the molecule can be completely defined by the above 3N-6 structural elements though sometimes it is more convenient to use the 3N atomic Cartesian coordinates of the molecule with respect to an externally fixed coordinate system. Of these parameters, the bond length and bond angles show relatively small variations when their values in similar molecular fragments are compared. The reason for this is, that the force-constants involved in the stretching of chemical bonds or deformation of bond angles are very large. On the other hand, the potential energies for rotation around chemical bonds are generally of the order of a few kcal/mol. Thus, molecules have a high degree of freedom in rotation around chemical bonds, and the problem of fixing the three-dimensional structure of the molecule often reduces to the knowledge of the N-3 torsion angles. It may be noted that the torsion angles of atoms branching out from the same atom (e.g. the three hydrogen atoms in the methyl group) are related to one another. Molecular symmetry may further reduce the number of independent parameters. These considerations are very helpful in NMR investigations, since in most cases the observables leave the structural details undetermined.

The molecule may exhibit stable structures for more than one value of the dihedral angle. Different spatial arrangements of atoms within a single classical organic structure produced by simple rotation (and twisting but not breaking of chemical bonds) are called conformations [12]. The terms rotational isomers, rotamers, conformers and secondary structures are synonyms of conformation. The term conformation should be distinguished

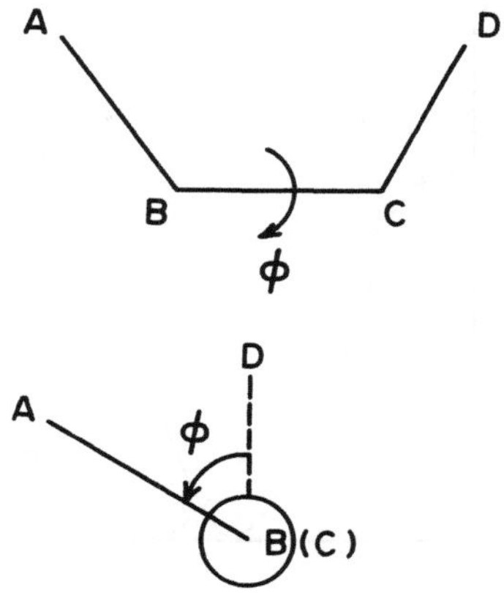

Fig. 1.2.1. The Newman projection diagram and definition of the torsion angle in the four-atom fragment A–B–C–D

from configuration (e.g. D- and L-forms of amino acids), where one can go from one structure to the other only by breaking one or more chemical bonds. In biochemistry, the term "tertiary" structure is used to denote molecular folding stabilized by long-range interactions within the same polymeric chain and "quaternary" structures to denote multichain organizations stabilized by intermolecular interactions.

The Newman projections are one of the most convenient ways to represent molecular conformations around various bonds in a molecule (Fig. 1.2.1). Several conventions exist for measuring the torsion angles. The most general one and the one followed in this review is to measure ϕ using a right-handed rotation: while looking along the bond B–C, the front atom A moves in an anticlockwise fashion relative to the far atom D. This is equivalent to saying that the far atom moves clockwise relative to near atom. The zero value of the torsion angle corresponds to the situation where the bonds A–B and C–D are *eclipsed*, and the angles are measured from 0 to 360 degrees. Such a convention has been followed in literature for most biomolecules. However, the ϕ, ψ angles in peptides are usually measured from –180 to +180 degrees (Sect. 4.1).

In organic chemistry a nomenclature which defines a family of conformations instead of precise values of torsion angles is generally used (Table 1.2.1). Each conformational state in this description covers a range of torsion angles. In solution, molecules have a greater degree of freedom for internal rotations and such a description has an obvious advantage over the specification of precise values of torsion angles.

Table 1.2.1. Classical conformational states and corresponding torsion angles

State	Abbreviations	Range of torsion angles in degrees
syn		0 ± 90
anti		180 ± 90
cis planar	*cis*	0 ± 10
trans planar	*t* (extended)	180 ± 10
gauche	g^+	60 ± 30
gauche'	g^-	300 (– 60) ± 30

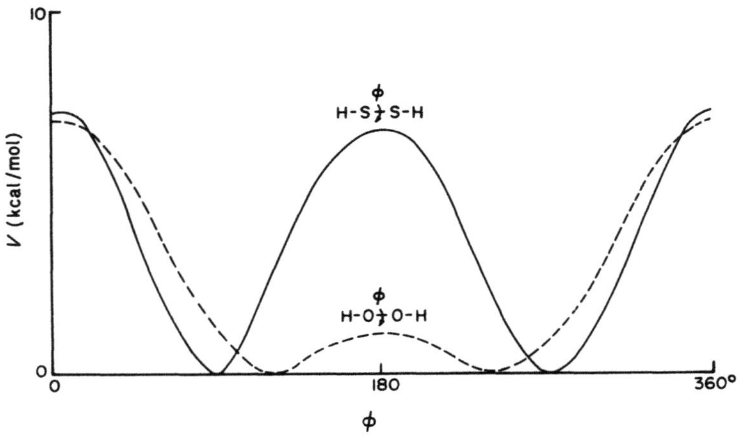

Fig. 1.2.2. Potential energy curves of internal rotation around the central bond in H–O–O–H and H–S–S–H

We can illustrate the concepts discussed above with the help of potential energy curves of two simple and related molecules H–S–S–H and H–O–O–H. The torsion energy behavior of these two molecules [13, 14] is illustrated in Fig. 1.2.2. Two minimum energy structures for each molecule are observed which are identical due to molecular symmetry. Not only the torsion angle for the minimum energy conformations but the whole potential energy curves for the two molecules are significantly different, in spite of the fact that both sulfur and oxygen belong to the same group in the Periodic Table. In fact, the conformation with torsion angles around 90° for S–S bonds are observed in proteins and peptides containing disulfide bridges [15].

1.3 Conformational Theory

It may be useful to introduce some basic principles of conformational theory which will prove useful in our later discussion. The behavior of hydrocarbon chains plays an important role in the conformational behavior of lipids, glycolipids and lipoproteins. We therefore choose *n*-butane as an example. The potential energy as a function of the torsion angle (C–C–C–C) in butane is shown in Fig. 1.3.1. There are three stable conformers for the butane molecule, g^+, t, and g^-, with the two *gauche* conformations being equivalent. The conformational thermodynamics is defined by ΔG (the difference in energy of the g and t isomers) and by ΔG^*, the energy barrier for the transition from the t to the g state. These two quantities are related to the equilibrium populations P_g and P_t of the two conformational states and the rate of interconversion $k(t,g)$ by the following expressions:

$$P_g/P_t = 2 \exp(-\Delta G/RT) \qquad (1.3.1)$$

$$k(t,g) = (kT/h) \exp(-\Delta G^*/RT) \qquad (1.3.2)$$

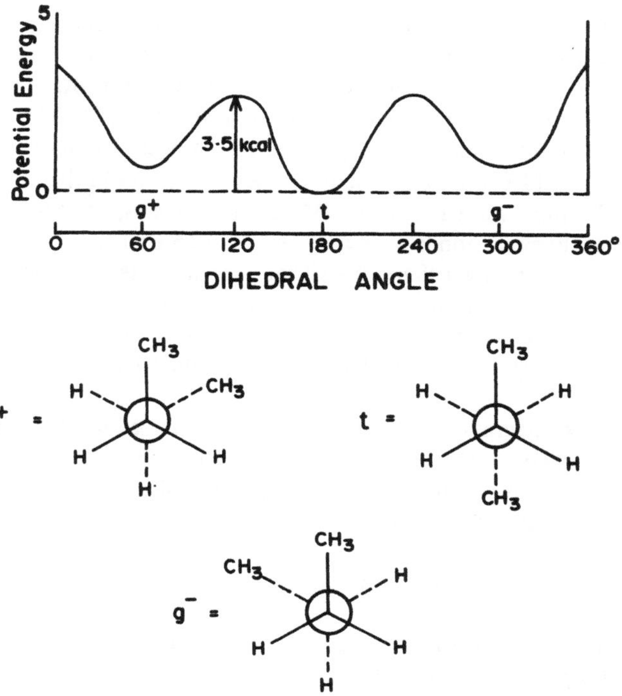

Fig. 1.3.1. Potential energy curve of rotation around the C–C–C–C bond in butane and Newman projection diagram of the minimum energy conformations g^+, t and g^-

For butane, ΔG is approx. 0.8 kcal/mol giving a value of 1 : 4 for $P_g : P_t$ and ΔG^* is 3.5 kcal/mol [16] which corresponds to a rate constant of $10^9\ s^{-1}$ at room temperature. Thus, g and t conformers interconvert at a very fast rate. For most physical techniques, the time-scale of measurements is much longer than the time for interconversion between the conformational states of butane. In such cases, one measures a time-average behavior, and the observed physical properties of the system (O) are given by:

$$\langle O \rangle = \int O\{\phi\} P\{\phi\} d\{\phi\} \qquad (1.3.3)$$

with $P\{\phi\} = \exp[-V\{\phi\}/RT]/\int P\{\phi\} d\{\phi\}$.

Here $\{\phi\}$ denotes the set of torsion angles which define the structure of the molecule, $O\{\phi\}$ is the observable under consideration (NMR chemical shift, coupling constant, relaxation time etc.), $V\{\phi\}$ the torsion potential energy, and $P\{\phi\}$ the probability of the conformational state with torsion angles $\{\phi\}$.

Both O and V are functions of torsion angles. It may be noted that what is a fast coversion rate for one physical technique may be a slow rate on the time-scale of another method. Thus, at room temperatures, one can get seperate vibration bands from t and g isomers of butane while in NMR, a time-averaged behavior is observed. The picture of the three-dimensional structure of n-butane which emerges from the above considerations is that the molecule is in a dynamic equilibrium and flips from one conformational state to another at a relatively fast rate. This freedom for internal motions is present even in several large molecules where part of the molecule may be rigid but other parts may show segmental motions because of "fast" conformational transitions. The NMR technique is ideally suited to the observation of such motions.

For NMR, the time-scale in measurements lies in the range 10^{-2} to 10^{-6} s. Under usual temperature ranges of NMR measurements, a time-averaged spectrum is observed for all chemical bonds with ΔG^* less than 15 kcal/mol. The barriers for double bonds are, however, larger than 20 kcal/mol and in these cases, spectra from individual rotamers can be resolved. In the intermediate range of ΔG^*, the NMR line shapes are sensitive to temperature variation due to transitions from "slow" to "fast" rates of interconversion. The line shapes can be analyzed to obtain the barriers for internal rotation in molecules [17].

The integration in Eq. (1.1.3) can be replaced by summation over a finite number of low-energy states. This approximation is known as rotational isomeric state model [18]. Such an assumption is usually made in the analysis of NMR data under conditions of "fast" exchange. Thus,

$$\langle O \rangle = \sum_j P_j\{\phi\} O_j\{\phi\} / \sum_j P_j\{\phi\}. \tag{1.3.4}$$

In the case of butane, the finite number of states(j) can be chosen as the g^+, t and g^- conformations of the molecule. The approximation is valid when the barrier to internal rotation is > 3 kcal/mol. However, if ΔG^* is of the order of 0–2 kcal/mol, then appreciable contribution to the observed properties may occur from states other than those corresponding to the minimum energy. If the functional form of $O\{\phi\}$ is known, then Eq. (1.3.4) and the observed NMR parameters $\langle O \rangle$ can be used to determine the relative conformer populations.

To extend the arguments to longer hydrocarbon chains, we consider the conformational behavior of *n*-pentane. This is described by two torsion angles ϕ and ψ (Fig. 1.3.2). On the basis of the conformational structure of butane, we may predict nine stable rotational isomeric states for *n*-pentane:

$$g^+g^+, g^+t, g^+g^-, tg^+, tt, tg^-, g^-g^+, g^-t, \text{and } g^-g^-.$$

These states can be divided into four groups, the members in each group being identical due to molecular symmetry.

$$tt;\ g^+g^- = g^-g^+;\ g^+t = g^-t = tg^- = tg^+;\ g^+g^+ = g^-g^-$$

A molecular model of the g^+g^- or g^-g^+ structure shows that the two terminal methyl groups come very close to each other. Such structures are clearly unstable due to short ("hard") contacts between the terminal methyl groups. This information can be diagrammatically represented by a two-dimensional isoenergy map as shown in Fig. 1.3.2. The energies in this map are from measurements based on Raman spectroscopy [19]. The *tt* conformation has the lowest energy and is called the global minimum. The four states where one of the two bonds has a *trans* conformation and the other one a *gauche* cor-

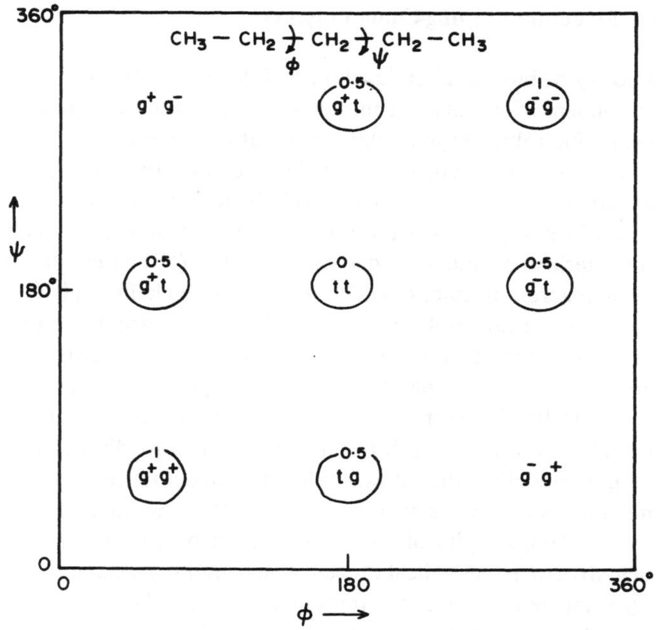

Fig. 1.3.2. Two-dimensional isoenergy map for *n*-pentane. The numbers on the map denote relative energies in kcal/mol relative to the *tt* state which corresponds the global minimum

respond to 0.5 kcal/mol and are called local minima. The g^+g^+ and g^-g^- states have energies of 1 kcal/mol while the states g^+g^- and g^-g^+ have even higher energies.

Clearly, more complex structures require the knowledge of potential energy curves in multidimensional conformational space. Fortunately, the conformational energies are dominated by nearest neighbor interactions and effects beyond two bonds become relatively unimportant except when the molecule has a possibility of folding back on itself. For example, the properties of higher alkanes and even polymethylene can be predicted fairly accurately on the basis of the isoenergy map of *n*-pentane [18, 19].

Isoenergy maps are extremely useful in the theoretical analysis of allowed conformations of biological molecules. Such maps take into account neighboring bond interactions and are thus more suitable representations of polymer conformations than the one-bond potential energy curves. They do not consider long-range interactions which are important when a long-chain molecule folds back on itself. Such polymeric structures may be stabilized further by intrachain hydrogen bonds or electrostatic interactions. A large amount of theoretical effort has gone into calculations [20–21] of such maps for all classes of biomolecules and these have proved useful in selecting the correct conformation out of the several possibilities suggested by NMR analysis.

1.4 Structure of Long-Chain Polymers

Majority of biological molecules are linear polymers (Fig. 1.4.1). They are formed by condensation between monomeric units, such as amino acids, nucleotides, and sugars. Biopolymers generally consist of a repeating backbone (B) and variable side chains (Ri). Biological polymers can be classified into three categories: proteins which contain repeating peptide units, nucleic acids (where the repeating unit is a nucleotide) and polysaccharides (repeating sugar units). Each individual molecule in these categories is thus characterized by the sequence of side chains (Ri) which determines the primary structure of the macromolecule. In addition to the side chain angle (χ_i), the three-dimensional structure of these molecules is fixed by a set of backbone torsion angles $\{\phi_i\}$ (e.g. ϕ_i, ψ_i, ω_i for polypeptides). Since a typical biopolymer consists of 100–10,000 residues, the number of structural parameters is very large. Three situations may occur:

(a) Helix: When the values of the backbone torsion angles $\{\phi_i\}$ are independent of the side chain Ri and each residue in the chain rigidly acquires the same set of values, the polymer chain describes a helix. The structure of each residue in a helical polymer is the same as that of its neighbor. Clearly, the number of independent structural parameters is much less, in spite of the large number of atoms in these molecules. Examples of such structures are helical nucleic acids and polysaccharides. Helical structures are generally stabilized by intra- or inter-chain hydrogen bonds.

(b) Ordered structures: A more general case of ordered structures is seen in globular proteins and enzymes. Here, the values of set of torsion angles $\{\phi_i\}$ depend on the nature of side chains (Ri) but do not fluctuate in time. In these cases, each peptide unit has its own unique conformation and the three-dimensional structure of the polymer is several orders of magnitude more complex than the highly symmetrical structure of helical polymers.

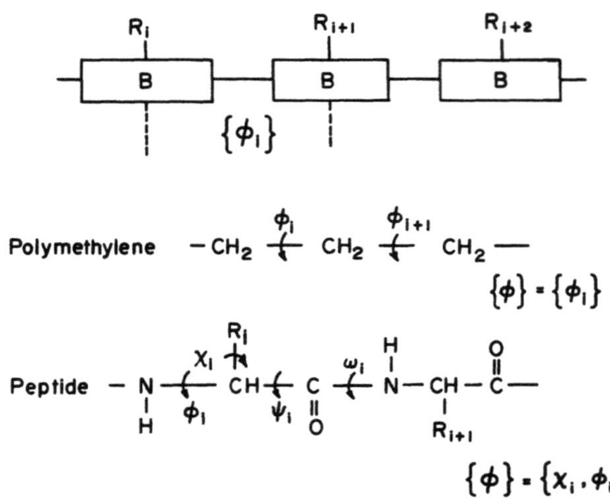

Fig. 1.4.1. Schematic representation of linear polymers. B denotes the backbone, R_i are the side chains and $\{\phi_i\}$ is the set of torsion angles which determine the conformation of the ith unit

(c) Random coil: Under the influence of heat or chemical perturbations like *p*H, solvent or ionic concentration, both types of structures can "melt" into quasirandom structures (random coils or coils). In a random coil, the structure no longer has a spatial meaning since it fluctuates both in space and time. However, a short-range order is present and the properties of a random coil are defined by time and space averages similar to Eq. (1.3.3). Thus, the observed properties of a polymer in its random coil is, to the first order, a sum of the behavior of its monomeric units. The NMR spectrum of a randomly coiled biopolymer can thus be predicted fairly accurately by summing spectra of the individual components.

In ordered (nonhelical) structures, each monomer unit has its own microenvironment and though the primary structure of a segment of the molecule may be same, small differences in chemical shifts and coupling constants may occur, due to secondary and tertiary structures of the molecule. The spectra of such molecules therefore have a large number of poorly resolved lines, particularly in the region corresponding to the backbone and the more abundant side chains. On the other hand, each unit in a helical structure or random coil has the same average environment as its neighbor and the spectra are relatively simple.

1.5 Problems in NMR Studies of Biological Molecules

In principle, each magnetic nucleus in a biomolecule can give its characteristic resonance. The resonances most commonly used in NMR studies are ^1H, ^{13}C, ^{31}P, ^{15}N and, to a lesser extent, the resonances of ions when they form part of the biological structures. Use has also been made of labelling the molecule chemically or biosynthetically with ^2H and ^{19}F and then observing these resonances. Fortunately, in most cases, labelling with these two isotopes does not alter the biological properties of the molecules. In each case, one can obtain chemical shifts (δ), coupling constants (J), relaxation times (T_1 and T_2) and, whenever necessary, the line shapes (g_ν). These parameters provide information on the microenvironment of each nuclear site under study. Alterations as a result of perturbations such as temperature, solution *p*H, metal ions, and substrates can be monitored using difference spectroscopy [2]. Table 1.5.1 summarizes properties and key applications of the above-mentioned nuclear resonances in conformational analysis of biomolecules.

For low molecular weight biomolecules, NMR investigations pose as little a problem as that for a similar organic molecule. However, in the study of a biopolymer having a well-defined structure, problems of sensitivity, resolution and assignment have to be solved before an attempt to determine biomolecular structure from NMR parameters can be made.

1.5.1 ^{1}H-NMR

Even a small protein has a fairly large number of protons. These protons resonate in a narrow spectral region and therefore give rise to an envelope of overlapping lines in the 0–10 δ region of proton NMR. The problem is further complicated by the fact that due

Table 1.5.1. Nuclear magnetic probes in the study of biomolecular conformations

Isotope	I	Relative sensitivity[a]	Resonance frequency[b]	Conformation-dependent parameter and information
^{1}H	1/2	100	100.0	δ: Position of anisotropic groups, hydrogen bonds, charged groups J: Dihedral angles, conformer populations T_1: Internal motions, paramagnetic sites Line shapes: Conformational dynamics, transition rates between rotational states
^{2}H	1	0.1	15.36	D_q: Order parameter, orientation of X–D bond T_1: Conformational dynamics
^{13}C	1/2	1.7	25.14	δ: Electronic environment T_1: Conformational flexibility J: Dihedral angles
^{15}N	1/2	0.1	10.13	δ: Molecular environment J: Dihedral angles
^{19}F	1/2	83	94.08	External probe: Environment of labelled positions
^{31}P	1/2	6.6	40.48	δ: Conformation of phosphate group Line shape: Direction of chemical shift tensor J: Dihedral angles

[a] Relative sensitivity is at constant field for equal number of nuclei
[b] Resonance frequency in MHz at a field of 2.35 tesla

to slow tumbling rates of large molecules, the correlation times are longer and the lines are broader. For this reason, one generally relies on the less crowded ^{1}H spectral region such as hydrogen bonded N–H, histidine or tyrosine protons or paramagnetically shifted resonances such as in heme proteins or metalloenzymes. The advent of superconducting NMR magnets has greatly promoted proton NMR of large molecules by providing spectral dispersion. High-resolution NMR magnets with fields up to 10 Tesla are presently available and attempts at producing even higher fields are being made.

To get meaningful results one generally works with 10 mM solutions in continuous wave (CW) proton NMR. For a 1 ml solution of a protein containing 100 amino acid residues, this corresponds to about 100 mg of sample for each experiment. Gram amounts of material are needed for less sensitive nuclei such as ^{13}C. It is difficult to obtain such large quantities of biological materials. Thus, till a few years ago, protein NMR was a difficult proposition. The advent of Fourier Transform (FT) techniques provided a big boost to NMR in biological research by reducing the minimum solution concentration to 0.1 mM for ^{1}H- and 10 mM for ^{13}C-NMR. The impact of modern NMR techniques can be judged from the growth of NMR literature (Fig. 1.5.1) on biomolecules. In addition, the more recent literature has concentrated on larger molecules such as small proteins, t-RNA, their interactions with drugs and substrates and other complicated systems.

Finally, assignment of resonances has been a major problem in getting results of biological interest. For example, the assignment of the four resonances from histidine

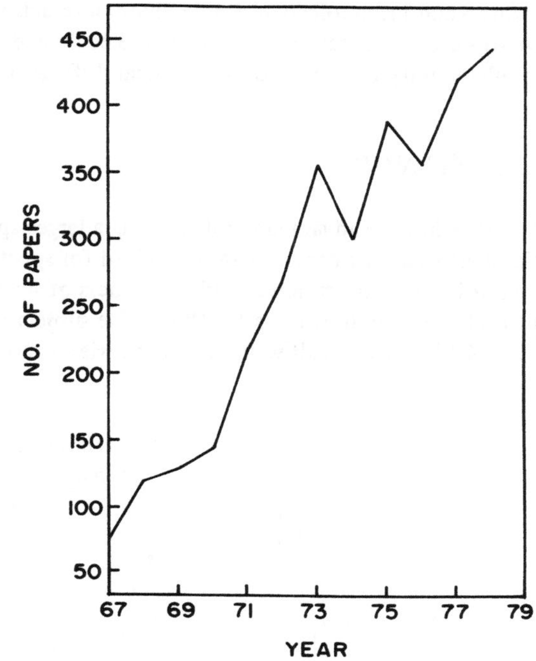

Fig. 1.5.1. Growth of NMR literature on biological molecules

residues in ribonuclease has been the subject of several investigations for over ten years of active research [22]. The use of biochemical modifications and selective deuterium labelling [23] using biosynthetic methods has been of great help in spectral assignments.

A major developement of potential use in proton NMR is the application of 2D-FT techniques to biological systems. Not only the chemical shifts and coupling constants can be seperated [24] by this method but assignments can be made more conveniently. Previously, similar experiments had to be performed using the INDOR technique which was not only time-consuming but sometimes ambiguous. It is expected that the automatic serial decoupling experiments in 2D-FT spectra will find extensive applications in the assignment of ^{1}H spectra in medium-size molecules.

1.5.2 ^{2}H-NMR

Deuterium is another nucleus of potential interest for conformational studies. Deuterium labelled molecules can be prepared by chemical or biosynthetic methods and the resultant ^{2}H-NMR spectra can be observed under proton decoupling. The major interest emerges from the effects arising due to the deuterium quadrupole moment. In partially ordered biological structures (e.g. membranes) the quadrupole coupling constants (D_q) of deuterium are not completely averaged and one observes hyperfine splittings due to residual interactions. Further, deuterium relaxation times are governed mainly by quadrupolar relaxation and are easier to interpret than proton NMR where relaxation is dominated by intra- and intermolecular dipole-dipole interactions. Both these parameters are sensitive to stereochemical features of the molecule and its internal motion [25] and act as complementary tools to proton chemical shifts and coupling constants.

1.5.3 ^{13}C-NMR

^{13}C-NMR has the advantage of displaying a larger spectral dispersion of chemical shifts. Microbiological enrichment can be utilized for spectral assignments. However, the relative sensitivity is much lower and two orders of sensitivity enhancement are required to put it at the same footing as ^{1}H-NMR. The subject of sensitivity and resolution in ^{13}C-NMR has been dealt with in recent reviews [26, 27]. A series of articles by Allerhand

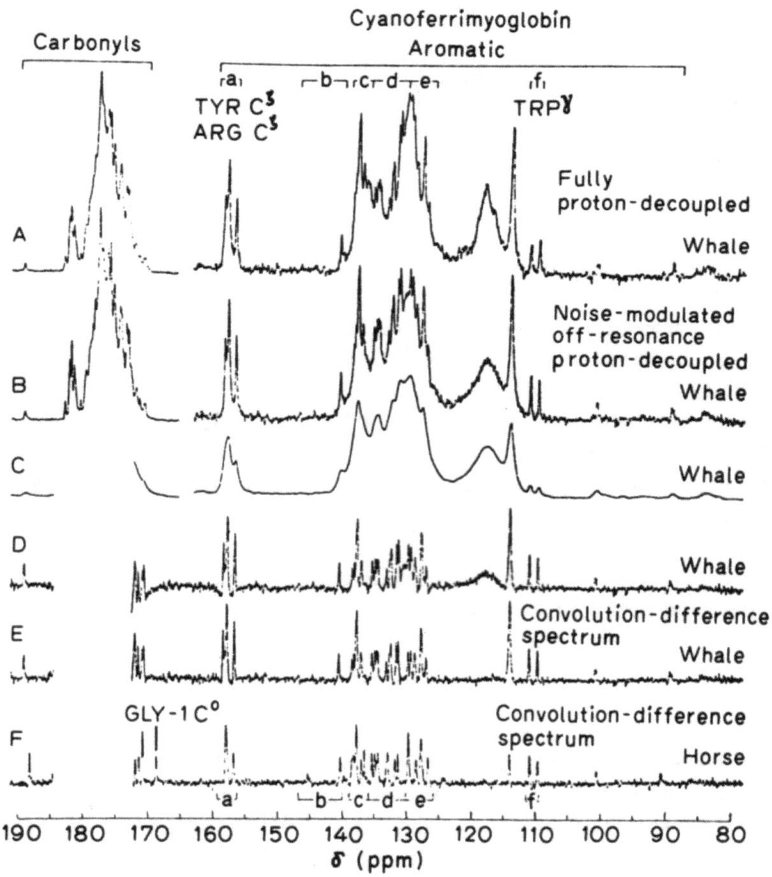

Fig. 1.5.2. Use of convolution difference technique for selecting narrow resonance lines. The spectrum A is for sperm whale cyanoferrimyoglobin (8.8 mM, 65.5 K transients, total time 20 h) (reprinted with permission from Ref. [28]). The sharp lines in the unsaturated carbon region show well-resolved resonances after the application of the convolution difference technique

and coworkers [28–31, 54, 59], provide some very useful guidelines for spectral assignments to large macromolecules and mechanisms for ^{13}C relaxation in these systems.

Two additional developements in the technological aspects of ^{13}C-NMR of biological molecules seem promising in future research on biopolymers. One of these is the convolution difference spectroscopy by which broad ^{13}C resonances (e.g. protonated carbons or those closer to a paramagnetic ion) can be filtered from the narrow resonances [28–31]. This helps not only in the assignment of ^{13}C-NMR but also in picking up selective resonance lines without interference from other bands in the same spectral region. The technique consists of weighting the FID with a large exponential factor (thus artificially broadening the FT spectrum) and then subtracting the modified FT spectrum from the original FT spectrum (Fig. 1.5.2). Only the sharp ^{13}C lines are observed in the difference spectrum.

The second technique makes use of very high decoupling fields ($\gamma H_2/2\pi \sim 75$ kHz) to narrow the proton dipolar broadening in structural proteins [32]. Collagen and fibrous proteins have relatively restricted flexibility as a result of which lines are broadened as in

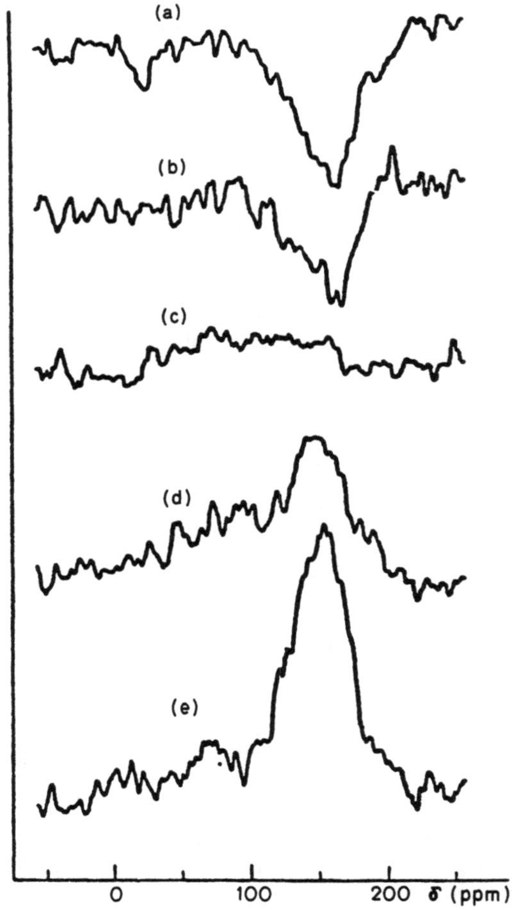

Fig. 1.5.3. Partially relaxed dipolar decoupled spectra in ^{13}C-enriched chick collagen using π-t-$\pi/2$ pulse sequences with t = (a) 12.5 ms, (b) 26 ms, (c) 78 ms, (d) 170 ms. The bottom trace shows the $\pi/2$-t-$\pi/2$ spectrum. 16–32 K transients, repetition rate 1.0 s, chemical shifts relative to CS_2 (reprinted with the permission from Ref. [32])

the solid state. By use of high-power proton decoupling, high resolution ^{13}C spectra from such samples have been observed [32, 53]. The spectra (Fig. 1.5.3), though broad, allow the estimation of NMR parameters such as chemical shifts and T_1. For example, in the case of collagen, the observed data indicate that the molecules reorient about the long axis of the helix [32].

^{13}C-NMR of biopolymers is still in its infancy and has a promising future.

1.5.4 ^{15}N-NMR

The more abundant isotope of nitrogen (^{14}N) has a spin I = 1, natural abundance of 99.64% and its chemical shifts cover a range of over 500 ppm [33]. However, lines are very broad because of the quadrupolar relaxation and ^{14}N has thus not found much use. ^{15}N has a spin 1/2 but its low natural abundance (0.36%) and a low sensitivity lead to difficulties in its detection. On the whole, the sensitivity of ^{15}N is only 2% of ^{13}C. Further, proton decoupled ^{15}N-NMR has a negative Nuclear Overhauser Effect (NOE). Thus, though nitrogen occurs in almost all biological molecules and has several potential advantages [34], its applications have been limited so far to monomeric units and oligo-

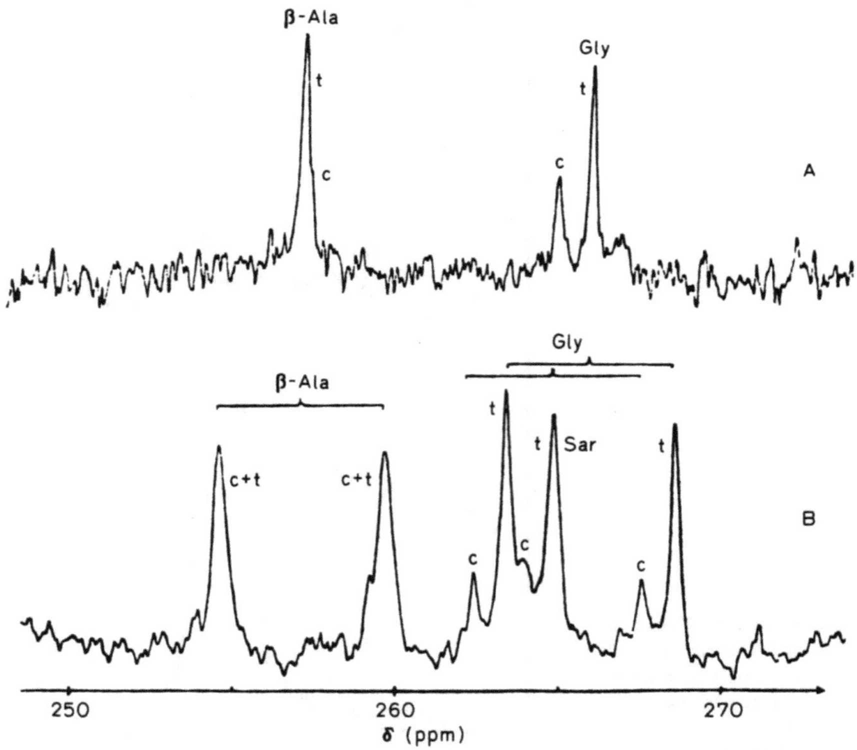

Fig. 1.5.4. ^{15}N-NMR spectra of poly(β-Ala-Sar-Gly) in H_2O at 30 °C. (A): inverse-gated proton decoupling with 24,811 transients, (B): no decoupling 27,250 transients (reprinted with permission from Ref. [36])

mers. With the current sensitivity, signals can be observed in neat liquids or relatively concentrated or enriched samples.

^{15}N-chemical shifts are sensitive both to the primary structure and to molecular conformation [35–37]. The amide chemical shifts in peptides, for example, cover a range of more than 30 ppm (Fig. 1.5.4) and depend on the sequence of the amino acid residues. In the polypeptide under discussion the sarcosine (Sar) residue can exist in *cis* and *trans* conformation and the peptide nitrogens of the neighboring glycine and alanine residues also show a splitting as a consequence of this [36].

1.5.5 ^{31}P-NMR

Unlike ^{1}H and ^{13}C, phosphorus is present in a few biological molecules. This is obviously an advantage since one has to deal with fewer resonances and problems of assignment are almost non-existent. With the spin 1/2, 100% natural abundance, moderate relaxation times, a wide range of chemical shifts and non-interference from solvent peaks, ^{31}P is an ideal probe for biological systems. The phosphorus-containing compounds such as adenosine tri- and di-phosphates (ATP and ADP) play a major role in energy storage and energy utilization while phosphate metabolites are involved in glycolitic pathways [38]. The chemical shifts are sensitive to the molecular conformation of the phosphate group whereas, in partially ordered systems, the line widths reflect the orientation of the chemical shift tensor. The advantages of ^{31}P-NMR permit the observation of NMR signals of even intact cells, live tissues and whole organs [38, 39] which, in turn, can be used to follow biochemical processes *in vivo* (see, for an example Fig. 1.5.5).

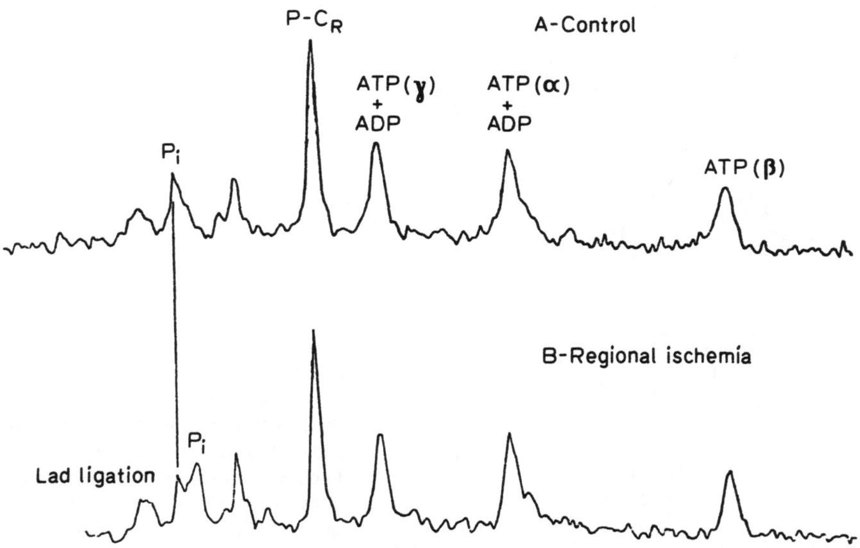

Fig. 1.5.5. ^{31}P-NMR spectra of rabit heart (A) control (B) after ligation of the left anterior descending (LAD) coronory artery (72 MHz, 37.5 °C, 5 min. accumulation) (reprinted from Ref. [39] with permission from Franklin Institute Press)

1.5.6 Other Nuclei

Among other nuclei which have been used in the study of biological structures, particular mention may be made of ^{19}F. Biological materials rarely contain ^{19}F but it can be easily incorporated through biosynthetic routes. In this way, selected sites in the molecule can be labelled. Its larger chemical shift and the fact that selective labelling does not generally alter the biological function of the molecule make ^{19}F a useful tool [40–45]. For example, the E. coli fluorotyrosine alkaline phosphatase can be prepared by *in-vivo* incorporation of *m*-fluorotyrosine and all the eleven tyrosine residues can then be monitored from their well resolved ^{19}F signals (Fig. 1.5.6). This information can then be utilized (40) to determine conformation, molecular dynamics, and effect of substrate-binding on molecular structure.

Several isotopes of biological interest have quadrupole moments. This restricts the use of nuclei such as ^{17}O, ^{23}Na, ^{25}Mg, ^{35}Cl and ^{39}K. The problem is further complicated by low sensitivity. However, in some cases the ions of these elements have been used to study their binding with macromolecules. As a result of binding, the ion tumbling rates decrease and the environment of the ion changes from almost spherical to one

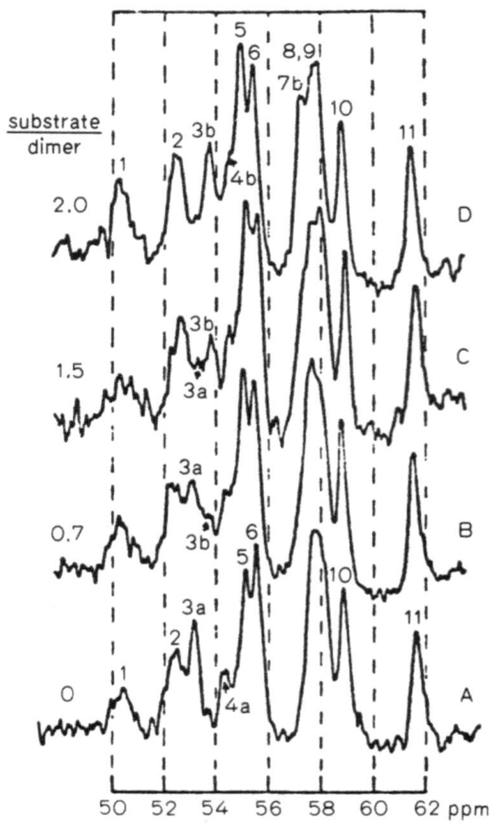

Fig. 1.5.6. ^{19}F-NMR spectra of Zn(II) fluorotyrosine alkaline phosphatase titrated with the substrate *p*-nitrophenyl phosphate (50–100 K transients, pulse interval 0.3 s). Each subunit of the macromolecule contains 11 tyrosine residues and the effect of substrate binding can be followed through the changes in ^{19}F chemical shifts. In this case, the lines No. 3, 4 and 7 significantly differ in A and D (reprinted with permission from Ref. [40])

devoid of any symmetry. This results in a broadening of the NMR lines of ionic species having quadrupole moments [46–48].

It has become increasingly apparent in recent years that the problems of understanding how biological molecules are structured, and how they function, cannot be studied by a single technique. Thus, NMR spectroscopy can be utilized most efficiently when it is applied in conjunction with other theoretical and experimental techniques such as X-ray crystallography, vibrational and electronic spectroscopy, CD, ORD, potential energy calculations, and molecular graphics. For more information on these complementary topics, the reader is referred to reviews and textbooks [49–52, 55–58].

2 NMR Techniques in Conformational Studies

In this chapter we will discuss how various NMR parameters can be used to obtain information on molecular conformation.

2.1 Coupling Constants

One of the most important NMR parameter in conformational studies of all classes of molecules is the *vicinal* or three-bond coupling constant (^3J) between nuclei A and D in the fragment A–B–C–D. Three recent reviews [60–62] have discussed the behavior of ^3J in peptides and nucleotides. These couplings show a pronounced relationship with the dihedral angle Θ between the planes A–B–C and B–C–D as was first noticed by Lemieux and coworkers [63] for carbohydrates. This relation can be represented by the general formula

$$^3J = A \cos^2 \Theta - B \cos \Theta + C \qquad (2.1.1)$$

Parameters A, B and C of Eq. (2.1.1) depend on the nature of the four atoms in the fragment as well as on the electronegativity of the other substituents in the molecule. While a power series expansion of ^3J in terms of $\cos \Theta$ is the most commonly used formula, two equivalent relations have been reported:

$$^3J = A' \cos^2 \Theta - B' \cos \Theta + C' \sin^2 \Theta \qquad (2.1.2)$$

and

$$^3J = A'' \cos 2\Theta - B'' \cos \Theta + C'' \qquad (2.1.3)$$

where

$$A' = A + C, \quad B' = B, \quad C' = C,$$

$$A'' = A/2, \quad B'' = B, \quad C'' = C + A/2,$$

and $A > B > C$. Such a relation was first derived by Karplus [64, 65] for H–C–C–H fragments. Karplus used the valence bond theory to calculate the functional dependence of the contribution of Fermi contact interactions to ^3J. Since then, the applicability of the above type of relations has been proved both theoretically and by measurements on conformationally rigid molecules for a large number of molecular fragments. Table 2.1.1 lists some of the couplings for which parameters of the Karplus type relations are known and which have potential use in conformations of biological molecules.

Table 2.1.1. Parameters for Karplus type relations

Coupling type	Parameters[a]			Area of applications	References	
	A	B	C		Theoretical	Experimental
H–C–C–H	10.5	1.2	0.0	Nucleotides, peptides, saccharides, lipids	[64–66]	[67, 68, 117]
H–C–N–H	9.4	1.1	0.4	Peptides	[69–71]	[72–74]
H–C–O–H	10.4	1.5	0.2	Saccharides, nucleosides	[71]	[75, 76]
H–C–N–C	7.4	4.4	−0.8	Peptides, nucleotides	[77]	[78]
H–C–C–C				Lipids, peptides		[79, 80, 177]
H–C–C–N	−5.3	−2.2	0.9	Peptides	[81]	[82]
H–C–C–F	54	20	2	Labelled biomolocules	[83, 84]	[83, 85]
H–C–O–P	16.3	4.6	0.0	Nucleotides, phospholipids	[86]	[87–89]
C–X–Y–C	See text			Peptides	[77]	[90–93, 116]
C–C–O–P	9.5	0.6	0.0	Nucleotides, phospholipids	[86]	[94–97, 80]

[a] These are typical parameters to be used with Eq. (2.1.1). Values for each molecular fragment should be corrected for substituent electronegativity and other long-range interactions. (For further discussions on the applications see subsequent chapters.)

Table 2.1.1 reveals that, with the exception of H–C–C–F couplings, the value of A lies in the range 7 to 20 Hz, B has values in the range 0 to 5 Hz while C is usually less than 2 Hz. The observed values of ^3J are therefore in the range 0 to 25 Hz. For some fragments, C has a negative value and in these cases, a small negative ^3J is observed when Θ is close to 90°.

In addition to the nature of the fragment under discussion and the dihedral angles, values of ^3J may be influenced by other factors. Fortunately, for most non-polar fragments, the solvent effects in conformationally rigid systems are small. However, such effects have been observed for H–C–C–F couplings [98]. A more important parameter is the electronegativity (E) of other substituents in the immediate neighborhood of the fragment A–B–C–D and the relative orientations of such groups. More electronegative substituents generally cause a decrease in the coupling constant values [99–101]. Corrections for electronegativity effects can be made using empirical procedures. For peptide H–N–C–H coupling the following correction term has been suggested [73]:

$$^3J_{obs} = {}^3J(\Theta)\,(1 - k\,\Delta E_x) \qquad (2.1.3)$$

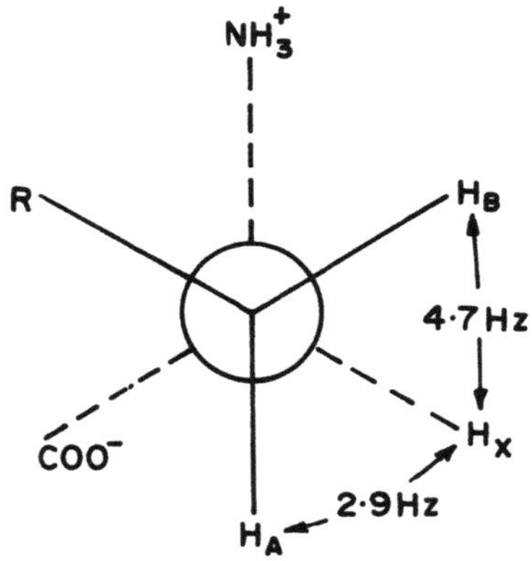

Fig. 2.1.1. ^3J(H–C–C–H) couplings of amino acids, where the large difference in the two *gauche* couplings shows the importance of the stereochemical considerations in the application of substituent corrections in Karplus type relations

where k is a constant whose value is approximately 0.1, $\Delta E_x = E_x - E_H$ is the difference in the electronegativities of the substituents and hydrogen on the Pauling scale and $^3J(\Theta)$ the coupling calculated from Eq. (2.1.1). Unfortunately, the correction term itself depends on the stereochemical disposition of the substituents. A typical example is shown in Fig. 2.1.1 [102]. Here, the coupling constants between two proton pairs both in a *gauche* conformation differ significantly. Similarly, the *cis* peptides show lower H–C–N–H coupling constants [9, 60, 103, 104] than the *trans* peptides for the same dihedral angle. For the H–C–C–H fragment, Paschler [105] has proposed the following relation to correct for the electronegativity effects as a function of dihedral angles:

$$^3J = (7.48 - 0.74\,\Sigma\,\Delta E_x) - (2.03 - 0.17\,\Sigma\,\Delta E_x)\cos\Theta + (4.60 - 0.23\,\Sigma\,\Delta E_x)\cos 2\Theta + (\pm\,0.06\,\Sigma\,\Delta E_x)\sin\Theta - (\pm 0.62\,\Sigma\,\Delta E_x)\sin 2\Theta \qquad (2.1.4)$$

Several other papers deal with the effect of substituent electronegativities on ^3J (for reviews see [60, 61]).

The values of ^3J for a particular fragment may also depend critically on the bond lengths and bond angles of the fragment. Such changes occur because of the changes in the hybridization of bonding orbitals. Thus, caution is necessary in the application of Karplus type relations. In particular, callibration for each coupling type should be made seperately and corrections for substituent electronegativity should be made. Values in Table 2.1.1 should be treated as typical values, at this stage. More accurate parameters for individual molecular fragments in biological systems will be dealt with in later sections.

Eq. (2.1.1) can be applied in two ways in conformational analysis. If it is known from theoretical or other methods that the molecule has only one possible rigid conformation, then the value of Θ can be estimated from the observed ^3J. On the other hand, when a molecule flips at a fast rate, between more than one rotational isomeric states, the values of the coupling constants corresponding to each of such states can be estimated using a Karplus type relation. The values of the component coupling constants thus obtained and the observed ^3J values, are then used for the calculation of conformer populations.

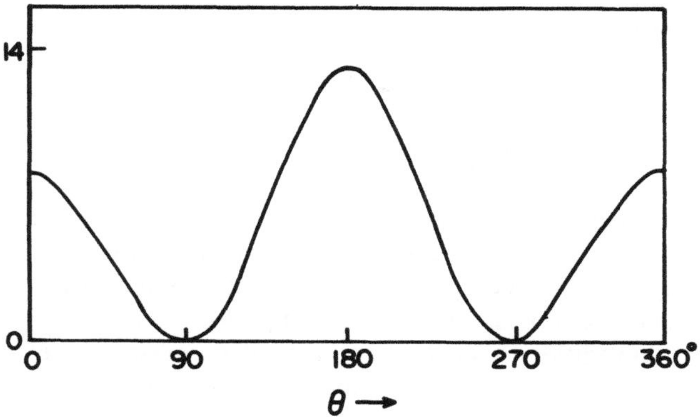

Fig. 2.1.2. Functional behaviour of the three-bond coupling constants

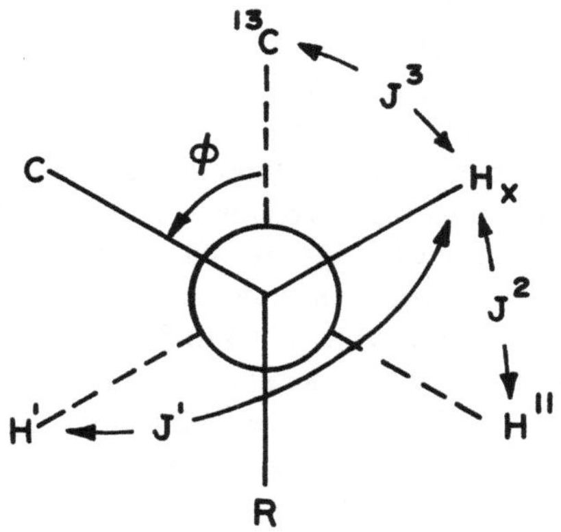

Fig. 2.1.3. Various coupling constants which can be used in the conformational analysis of hydrocarbon chains

The functional form of Karplus type relations is shown in Fig. 2.1.2. It may be observed that for each observed value of ^3J there are in general four possible solutions for Θ. In many cases however, more than one coupling constant may be available to uniquely fix the value of Θ. Consider for example, the torsion angle ϕ for rotation around the C–C–C–C bond in hydrocarbon chains. If the conformational structure is such that the two protons (H′ and H″) are chemically nonequivalent, then one has three coupling constants per proton (H_x) to measure the torsion angle (ϕ) and the conformation can be uniquely fixed (Fig. 2.1.3). Potential energy calculations have also been used to restrict the choice of possible conformational structures.

It may be noted that for rotation around C–C bonds, the rotational isomeric states can be chosen as g^+, t and g^-. Using the values of component coupling constants for these states, one can write the following equations for fractional populations P of the three states:

$$J^1_{obs} = J^1(g^+)P(g^+) + J^1(t)P(t) + J^1(g^-)P(g^-) \qquad (2.1.5)$$

$$J^2_{obs} = J^2(g^+)P(g^+) + J^2(t)P(t) + J^2(g^-)P(g^-) \quad (2.1.6)$$

$$J^3_{obs} = J^3(g^+)P(g^+) + J^3(t)P(t) + J^3(g^-)P(g^-) \quad (2.1.7)$$

$$1 = P(g^+) + P(t) + P(g^-) \quad (2.1.8)$$

It is readily seen that the information is sufficient to determine the relative conformer populations. However, problems may arise when the chemical shift between protons H′ and H″ is small (deceptively simple spectra). In these cases, only the sum of the couplings with the proton H_x can be measured [106].

In addition to the three bond coupling constants, use can be made of other coupling constants in structural determinations. One or two bond coupling constants may show a conformational dependence due to changes in electronic environment in the immediate neighborhood of the atoms involved in coupling. The one-bond couplings [^1J (A–B)] are governed, to the first order, by the nature of the orbitals localized on the bonding atoms A and B. However, long-range interactions are known to affect the values of ^1J which can be utilized for structural studies [107–109]. Similarly, ^2J (A–B–C) depend primarily on the bond angle (A–B–C) but second-order effects may arise from the orientation of neighboring groups [110, 111]. Again, use of such effects has been made in conformational studies.

Long-range couplings are generally small and may manifest themselves in high resolution NMR only in certain specific conformationally dependent situations. A relatively large value of ^4J is observed when a zig-zag (W) or *trans-trans* pathway is involved in the bonding atoms. Examples are ^4J (^{31}P–^1H) in nucleotides [112] and ^4J (H–C–C–N–H) in peptides [104]. Values of ^5J (^1H–^1H) between the protons at C_α in peptides are governed by the overall conformation of the linkages between the two α C atoms [113, 114], larger values being indicative of a W pathway. This situation can be compared with an anologous five-bond homoallylic coupling in the fragment H–C–C = C–C–H [115, 118].

2.2 Chemical Shifts of Hydrogen-Bonded Protons

As already stated for a large molecule, the normal proton spectral region is crowded due to resonances from closely related structural units in the backbone and side chains. The intra- or inter-chain hydrogen bonded N–H and O–H protons, however, resonate at low fields (10 to 15 δ). These resonances can then be used to find the number of hydrogen bonds in the molecule and the nature and environment of such bonds. To check whether the protons are locked in an intramolecular bond or bonded to the solvent, the temperature variation of the resonance is usually studied. The proton resonances from intramolecularly bonded systems show smaller temperature variation in the chemical shift [119–123, 125]. Further, while the intramolecularly bonded protons exchange slowly with H_2O or D_2O, the exchange rates for solvent-exposed protons are considerably higher.

A similar solvent dependence of ^{13}C chemical shifts is observed in the resonances of carbonyl groups locked in intramolecular hydrogen bonds [124]. Solvent titration of

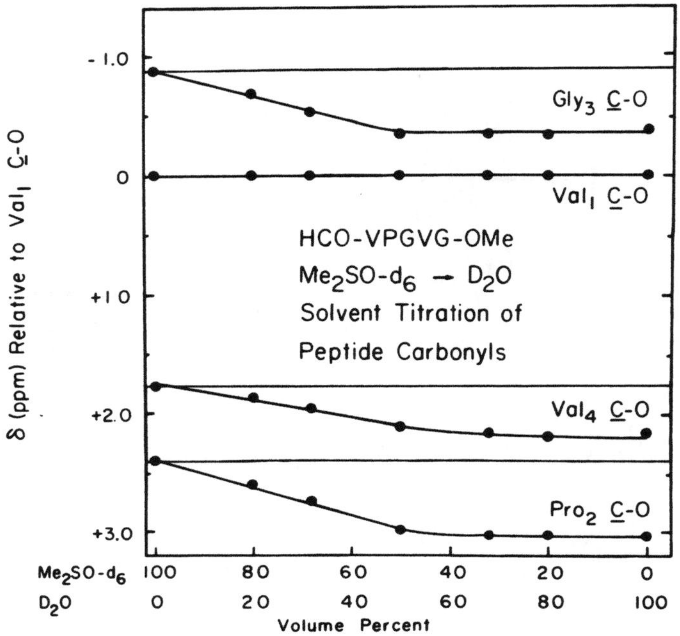

Fig. 2.2.1. Solvent titration (DMSO–D_2O) of peptide carbonyl carbon resonance in an elastin pentapeptide HCO-Val_1-Pro_2-Gly_3-Val_4-Gly_5-OMe showing the difference between the behavior of intramolecularly bonded C=O (Val_1) and solvent-exposed peptide carbonyl carbons (reprinted with the permission from Ref. [124])

the carbonyl shifts can thus be used in the same way as proton shifts for the evaluation of secondary molecular structures (Fig. 2.2.1). The methods based on solvent and exchange effects are useful but can lead to ambiguous results.

2.3 Magnetic Anisotropy of Chemical Bonds or Groups

If an anisotropic chemical bond or group is present in the vicinity of a nucleus, then its resonance frequency is altered. Suppose O (Fig. 2.3.1) is the electric center of gravity of the chemical bond or group and χ_{xx}, χ_{yy}, χ_{zz} are the components of magnetic susceptibilty tensor χ along the x, y and z axis, then the change in the electronic screening of a nucleus at point P is given by [126, 127]:

$$\sigma_p = [(1-3\cos^2\Theta_x)\chi_{xx} + (1-3\cos^2\Theta_y)\chi_{yy} + (1-3\cos^2\Theta_z)\chi_{zz}]/3r_p^3 \qquad (2.3.1)$$

For groups having axial symmetry,

$$\chi_{yy} = \chi_{zz} \quad \text{and} \quad \Delta\chi = \chi_{xx} - \chi_{zz} \qquad (2.3.2)$$

and

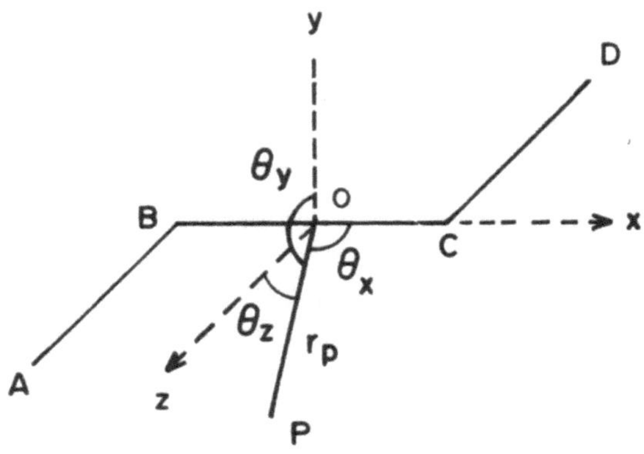

Fig. 2.3.1. Effect of magnetic anisotropic group on the chemical shift of a nucleus at point P

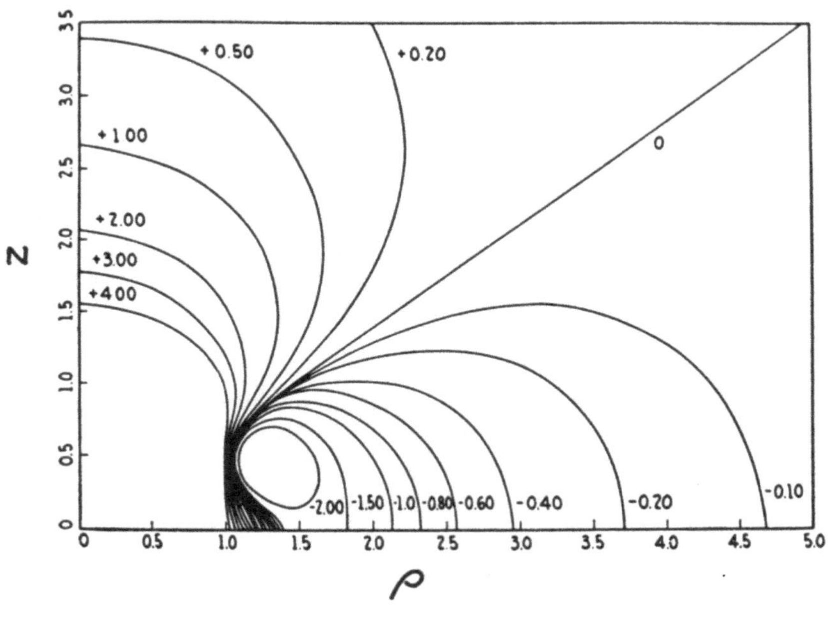

Fig. 2.3.2. Ring current effect due to the benzene ring. The lines represent shifts experienced by a nucleus due to magnetic anisotropy of the benzene ring (reprinted with the permission from Ref. [129]), z and ρ are cylindrical coordinates

$$\sigma_p = \Delta\chi\,(1-3\cos^2\Theta_x)/3r_p^3 \tag{2.3.3}$$

Some examples of such anisotropic groups are C=O bonds in peptides, aromatic side chains such as phenylalanine and tyrosine in peptides, purine and pyrimidine bases in nucleic acids, lone-pair orbitals, and double bonds in various systems.

The most common example of chemical shift alterations due to neighboring anisotropy effects is the ring-current shifts caused by magnetic anisotropy of aromatic rings.

The effects arise from the induced motions of the π electron frame of the ring in the presence of external magnetic field (Fig. 2.3.2). To a first approximation, the change in the chemical shift of a nucleus near the ring is given by [128]:

$$\delta = K(3 \cos^2 \Theta - 1)/r^3 \tag{2.3.4}$$

where K is a constant which depends on the aromaticity of the ring and its radius, r is the distance of the nucleus from the center of the ring and Θ the angle between the vector r and the normal to the base. It is seen from Fig. 2.3.2 and Eq. (2.3.4) that nuclei with $\Theta < 55°$ move upfield while downfield shifts are expected for nuclei near the plane of the ring.

Figure 2.3.2 illustrates isoshielding contours calculated theoretically for a rapidly tumbling benzene ring [129]. Extensive calculations have also been made for nucleic acid bases, because of their importance in conformational structure determinations of nucleotides and nucleic acids [130–132]. It may be mentioned that such contours have gone through several stages of refinement. At the present stage, they do provide a good qualitative idea of the disposition of the nuclear sites with respect to the aromatic rings in the molecule. However, care should be exercised in quantitative interpretations.

Among other examples of effects of close neighbors on chemical shifts, mention may be made of the effect arising from titrations of an acid-base group in the vicinity of the proton under study [133]. In polypeptides, the backbone proton titration shifts are very sensitive manifestation of the intramolecular hydrogen bonds between carboxylate groups and the amide protons.

The two protons in a methylene group can show chemical non-equivalence in conformationally rigid systems. This information can also be used in conformational studies [134].

2.4 ^{13}C Chemical Shifts

^{13}C chemical shifts cover a range of about 300 ppm and it may appear that these may be more useful than the proton chemical shifts. Unfortunately, ^{13}C shifts are dominated by the paramagnetic terms and are therefore less sensitive to environmental changes. For example, though the magnitudes of ring-current shifts in proton and ^{13}C chemical shifts are of the same order of magnitude, these are often overshadowed by other effects. Therefore, interpretations of ^{13}C chemical shifts are usually based on empirical rules rather than on theoretical equations. A typical example of ^{13}C chemical shifts in conformational analysis of peptides is the study of *cis* and *trans* proline residues which show markedly different chemical shifts [135–141] in the two conformations (Sect. 4.3). Titration of an acid-base group can lead to changes in the chemical shift of a neighboring ^{13}C nucleus and this fact can be utilized to map nuclei in close proximity of acidic or basic residues [133, 142]. This technique has special importance since acid-base groups are involved in enzyme activities. Use can also be made of neighboring anisotropy effects on chemical shifts [143]. Further, as discussed in Sect. 4.3, the larger magnitudes of ^{13}C shifts prove to be ideal for checking molecular symmetry in biological systems [144].

2.5 ^{31}P Chemical Shifts

In most biological systems, phosphorus is present in the form of phosphate groups and a great deal of effort has been devoted to the interpretation of phosphate group conformation and ^{31}P chemical shifts. From theoretical and experimental studies [145–148] the following general conclusions can be made:

(a) The ^{31}P chemical shifts are sensitive to the electronegativity of the substituent attached to phosphorus and its charge state. Thus, phosphate groups show a marked variation in shift as a function of *p*H due to the titration of the acid protons in the group. This fact has been utilized to determine the *p*H in intact cells [39].

(b) The shifts are also sensitive to the hybridization of the phosphorus orbital. Thus, in phosphates the chemical shift changes as a function of the O–P–O angle.

(c) The chemical shift changes by 3–6 ppm when the conformation of the two O–P bonds in the fragment C–O–P–O–C changes from a *gg* (the most preferred conformation for such fragments) to *gt* or *tg*. This observation is of importance in conformational studies of phospholipids, nucleotides and coenzymes since O–P bonds are relatively flexible and cannot be studied by *vicinal* coupling constants.

The ^{31}P chemical shift tensor has a large shift anisotropy and this fact tends to contribute to a significant line broadening in ordered and partially ordered system. The direction and principle components of the shift tensor in phosphate diesters have been determined and used to study the head group orientation of lipid bilayers [149].

2.6. Relaxation Times (T_1 and T_2)

For several years, coupling costants and chemical shifts were the only NMR parameters which had found routine applications in conformational studies. With the availability of FT techniques, accurate measurements of relaxation times have become possible. The methodology of measurements of T_1 and T_2 for resolved NMR lines has been established in recent years [150]. Furthermore, the physical mechanisms for nuclear relaxation are now well understood [151]. However, the theory connecting relaxation times with molecular motions is still not clearly understood. Anisotropic diffusion of large molecules, segmental mobility and seperation of various degrees of freedom for motions in a molecule are some of the topics on which detailed theoretical and experimental work is needed to put relaxation parameters on the same footing as chemical shifts and coupling constants. Nevertheless, relaxation times have already provided a wealth of information on internal motions in molecules, effect of intermolecular association on conformational rigidity and the relative flexibilities in different segments of a biopolymer.

The excited nuclei transfer their energy through interactions with fluctuating local fields having frequency components equal to the Larmor frequency of the nuclear spin under study [150–152]. Several mechanisms contribute to T_1 and T_2. For nuclei with spin 1/2 in diamagnetic samples, the dominant mechanism generally involves internuclear dipole-dipole (DD) interactions. Other mechanisms include chemical shift anisotropy (CSA), scaler coupling (SC) and spin rotation (SR). In paramagnetic systems, electron interactions with nuclear spins dominate the relaxation rates. Nuclei with $I > 1/2$ can also relax through the electric quadrupole interaction. Thus, the relaxation rate ($R1 = 1/T_1$) can be written as a sum of several contributions:

$$R1 = R1(DD) + R1(SR) + R1(SC) + R1(CSA) + \ldots \tag{2.6.1}$$

For detailed discussions on the subject of relaxation times, their theory and measurements, standard textbooks may be consulted [2–11, 150–153]. Here, we have given a brief discussion of how relaxation rates can be used in structural studies.

(a) Proton magnetic DD relaxation: In favorable cases, it is possible separate contributions from intra- and intermolecular DD interactions [154], and intramolecular R1(DD) can be utilized for conformational studies. Thus, for a system of two like spin 1/2 nuclei [153]:

$$(R1) = \left(\frac{1}{T_1}\right)_{intra} = \frac{2\gamma^4\hbar^2 I(I+1)}{5r_{ij}^6}\left(\frac{\tau_r}{1+\omega^2\tau_r^2} + \frac{4\tau_r}{1+4\,\omega^2\tau_r^2}\right) \tag{2.6.2}$$

$$(R2) = \left(\frac{1}{T_2}\right)_{intra} = \frac{\gamma^4\hbar^2 I(I+1)}{5r_{ij}^6}\left(3\tau_r + \frac{5\tau_r}{1+\omega^2\tau_r^2} + \frac{2\tau_r}{1+4\,\omega^2\tau_r^2}\right) \tag{2.6.3}$$

where τ_r is the rotational correlation time. Its value depends on a number of factors such as shape and viscosity (η) of the solvent, temperature (T), etc. For small spherical molecules, τ_r can be approximated by:

$$\tau_r \cong \frac{4\pi a^3 \eta}{3kT} \tag{2.6.4}$$

where "a" is the radius of the molecule (assumed to be a sphere) and η is the viscosity of the medium. For small molecules in non-viscous media, $\tau_r \sim 10^{-12}$ to 10^{-9} s, $\omega^2\tau_r^2 \ll 1$ and Eq. (2.6.5) applies. Such a situation is referred to as extreme motional narrowing. For high molecular weight polymers, τ_r values are much larger (typically in the range 10^{-6}–10^{-9} s), and such a simplification is not possible. Further, in these cases the relaxation rate depends on ω and thus on the magnetic field at which measurements are performed.

$$(R1) = (R2) = \frac{2\gamma^4\hbar^2 I(I+1)}{r_{ij}^6}\,\tau_r \tag{2.6.5}$$

It is seen from Eq. (2.6.5) that, if the dipole-dipole contributions can be separated, then values of $\langle r^6 \rangle$ can be estimated.

For unlike nuclei, i and j, the DD contribution to the relaxation rate is modified by the "3/2 effect" [153] and is given by

$$(R_1)_{intra}^{unlike} = \frac{3\gamma_i^2\gamma_j^2\hbar^2 I_j(I_j+1)}{r_{ij}^6}\,\tau_r \tag{2.6.6}$$

From Eqs. (2.6.2) and (2.6.3), it may be noticed that because of a larger gyromagnetic ratio, protons are usually the most important nuclei for DD relaxation. Secondly, because of r^{-6} dependence, only close neighbors make significant contributions. When more than one of such neighbors is present at the same distance (e.g. protons in CH, CH_2 and CH_3 groups), the relaxation rate is proportional to the number N of such nuclei.

An ingenious way of obtaining intramolecular dipole-dipole interaction values is the method called deuterium substitution effect on relaxation times (DESERT) [155, 156]. Here, one obtains differences in the relaxation rates of nucleus i when a proton (j) in the molecule is substituted by deuterium. Since the gyromagnetic ratio of deuterium is much lower, the contribution to relaxation rate of i due to the dipole-dipole interactions between i and j can be estimated using Eq. (2.6.7).

$$R_i(j) = 1.06\,[(R_{1i})_{j=H} - (R_{1i})_{j=D}] \tag{2.6.7}$$

where factor 1.06 has been introduced to correct for relaxation due to deuterium.

(b) $^{13}C-T_1$: Spin-lattice relaxation times corresponding to resolved ^{13}C resonance lines have found wide applications in the study of internal motions in both small and medium size biological molecules. ^{13}C-NMR measurements are generally performed with proton decoupling and the measured T_1 under these conditions depend on the speed of molecular motions relative to the precession frequency of ^{13}C [151]. The values of T_1 are therefore field-dependent and the minimum value of T_1 depends also on the size of the molecule. Medium size molecules generally lie on the decreasing side of the T_1 curve ($\omega\tau_r \ll 1$) and the value of R1(DD) is given by:

$$R_i(DD) = \hbar^2\gamma_C^2\gamma_H^2 N\tau_r/r_{CH}^6 \tag{2.6.8}$$

where N is the number of protons directly attached to carbon and r_{CH} the carbon-proton distance (usually 1.09 Å for directly bonded protons). On the other hand, very large molecules such as proteins are located on the right side of the T_1 minimum. Further, such molecules generally exhibit anisotropic motions. In these cases, T_1 and T_2 show a fairly complex relationship with molecular motions.

The application of Eq. (2.6.8) to molecular motions can again be illustrated by hydrocarbon chains. The values of NT_1 are inversely proportional to the correlation time. The behavior of T_1 in a linear alkane (decane) is shown below [157]. It is clear that the peripheral carbons reveal higher flexibility (*g* to *t* flips) than the middle ones.

$$\overset{8.74}{CH_3}-\overset{6.64}{CH_2}-\overset{5.71}{CH_2}-\overset{4.95}{CH_2}-\overset{4.36}{CH_2}-\overset{4.36}{CH_2}-\overset{4.95}{CH_2}-\overset{5.71}{CH_2}-\overset{6.64}{CH_2}-\overset{8.74}{CH_3}$$

(c) Electric quadrupole relaxation: Deuterium and some of the ionic probes (with $I > 1/2$) are often used in structural studies [158]. Such nuclei relax mainly through their electric quadrupole interactions, In the extreme narrowing case:

$$R1(Q) = R2(Q) = \frac{3(2I+3)}{40I^2(2I-1)}\left(1+\frac{\eta^2}{3}\right)\left(\frac{eqQ}{\hbar}\right)^2\tau_c \tag{2.6.9}$$

where e is the electron charge, Q the nuclear quadrupole moment, q the electric field gradient at the site of the nucleus, τ_c the correlation time for rotational motion of the quadrupolar nucleus and η the asymmetry parameter. Electric quadrupole relaxation is a very efficient process and can be utilized to study (a) correlation times and thus internal motions in a molecule and (b) binding of ions ($I > 1/2$) with biological macromolecules.

2.7 Nuclear Overhauser Effect (NOE)

An indirect method for obtaining information on T_1 is through the use of homonuclear Overhauser effect (NOE). In this method one of the nucleus (i) is saturated by applying a strong rf field corresponding to its resonance frequency. The change in the signal intensity of the other nucleus (j) is then observed [159–163]

$$f_j(i) = \frac{\text{area of line j when i is saturated}}{\text{equilibrium area of line j}} - 1 \tag{2.7.1}$$

NOE is caused by changes in populations of the nuclear energy levels when i is saturated. These changes are brought about by spin-lattice relaxation mechanisms which operate through the interactions between various nuclear spins. A detailed discussion of NOE is available in the book by Noggle and Schirmer [163] and in the review by Bell and Saunders [164]. Let us consider a two-spin system (Fig. 2.7.1) and assume that $I_i = I_j = 1/2$. Suppose further that i and j are weakly coupled ($J \ll \delta$) and spin-lattice relaxation is governed mainly by dipole-dipole interactions. Due to the strong r.f. field, populations

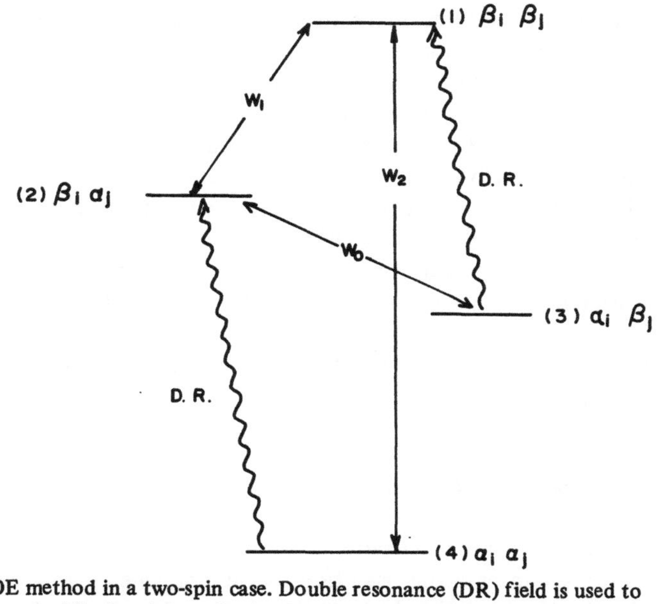

Fig. 2.7.1. Principle of the NOE method in a two-spin case. Double resonance (DR) field is used to saturate the transition of nucleus i while the changes in the signal area of spin j are monitored

of states 2 and 4 and those of 1 and 3 are equal. The system tends to return to thermal equilibrium through transitions between states which are not saturated. Let W_0, W_1 and W_2 be the transition probabilities between these levels (Fig. 2.7.1). If relaxation is due to DD interactions, the W_2 term is larger than W_0, and it will try to populate the levels 2 and 4. This in turn will lead to an enhancement of the signal of nucleus j. The NOE can be written as

$$f_j(i) = \frac{W_2 - W_0}{2W_1 + W_0 + W_2} \frac{\gamma_i}{\gamma_j} \tag{2.7.2}$$

In the case of "slow" conformational changes in a multispin system it can be shown that

$$f_j(i) = \sum_i \frac{\gamma_i}{2\gamma_j} \frac{\rho_{ji}}{R_j} - \sum_n \gamma_n \rho_{jn} f_n(i)/2\gamma_j R_j \tag{2.7.3}$$

where ρ_{ji} is the contribution to the spin-lattice relaxation rate of j due to its dipole-dipole interaction with i. R_j is the total relaxation rate of spin j and is given by

$$R_j = \sum_{i \neq j} \rho_{ji} + \rho_j^*$$

where ρ_j^* is the direct relaxation rate of j due to other relaxation paths. ρ_{ji} is given by Eq. (2.7.4).

$$\rho_{ji} = \gamma_i^2 \gamma_j^2 \hbar^2 \tau_c(ji)/r_{ji}^6 \tag{2.7.4}$$

where $\tau_c(ji)$ is the correlation time for the internuclear vector. Components of the ρ matrix can be separated by means of a series of NOE experiments and relative internuclear distances can be estimated. Since dissolved oxygen and paramagnetic ions can make appreciable contributions to R, precautions should be taken to avoid such impurities.

It should be pointed out that values of NOE depend on T_1 processes and hence on correlation times. For systems which have higher correlation times such as biopolymers, negative NOE may be observed [162]. In either case, the NOE method can be used to locate groups in close proximity of one another.

Mention may be made here of the magnetization transfer experiments which can be classified as a special case of NOE experiments. Here, the magnetization information is transferred from one set of nuclear energy levels to another through a conformational transition or a chemical exchange (see for example [165, 193]). The information can be utilized for the determination of transition rates. Another recent paper deals with the use of laser photochemically induced dynamic nuclear polarization (CIDNP) for studies on biological macromolecules [192].

2.8 Paramagnetic Reagents

In the presence of paramagnetic ions both the chemical shifts and relaxation times of nuclear resonances change significantly. In fact, when the metal ion is linked to a particular site in a biomolecule, both these parameters are dominated by shifts and relaxa-

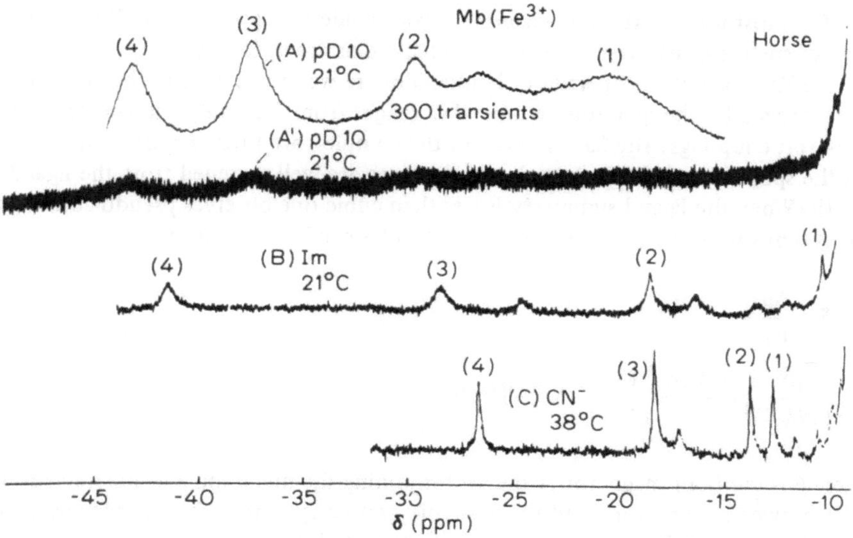

Fig. 2.8.1. ^{1}H-NMR spectrum of horse ferrimyoglobin derivatives at 220 MHz. The large chemical shifts (outside the usual 0–10 ppm region) arise from a combination of contact and pseudocontact interactions. – (A): ferrimyoglobin (pD10) at 21 °C, 300 scans, (A'), 1 scan under the same conditions. – (B): imidazole complex at 21 °C. – (C):Cyanide complex at 38 °C. – Methyl signal (1) changes into a doublet at low temperature due to hindered rotation of CH_3 groups (reprinted with the permission from Ref. [178])

tion caused by the presence of electronic moments in the molecule. The chemical shifts can change due to a direct delocalization of electronic spins at various nuclear sites in the molecule (presence of finite spin density; this effect is referred to as contact interaction) or due to anisotropy of the "g" tensor of the electron spin at the site of the paramagnetic ion which gives rise to "pseudocontact" shifts in the same way as the magnetic anisotropy of chemical groups. In favorable cases, the two effects can be seperated. However, for structural work, it is better to choose ions which contribute predominantly to the conformation-dependent pseudocontact term.

The nuclear spins in paramagnetic systems relax mainly through electron-nuclear interactions. While the nature of electron-nuclear spin interactions is very similar to nuclear-nuclear interactions, the magnitudes are much larger due to the fact that electronic magnetic moments are 10^3 times larger than nuclear magnetic moments (see for example Fig. 2.8.1). Paramagnetic ions therefore serve as ideal probes for modifying nuclear resonances and then utilizing these changes to study the conformational structure of the molecule. Frequently, a dynamic equilibrium may exist between complexed and free ligands, with only a small fraction of the ligand existing in the complex. This is often sufficient to cause detectable changes in NMR parameters and to obtain structural information. It is of course assumed in such studies that the structure does not alter as a result of metal-ion binding. In the case of metalloenzymes and heme proteins, use of the natural metal ions themselves can be made and the above restriction does not apply. Morris and Dwek [166] have reviewed this field recently.

The most useful external reagents are lanthanide ions [167–171]. The relative advantages of such reagents over other paramagnetic ions are as follows:

(a) The magnetic properties of lanthanides are well understood. The ground state is characterized by the quantum number J (J ranges from 5/2 to 8) resulting from strong spin-orbit couplings. The 2J + 1 levels of the complex are further split by the ligand fields but the splittings are small since the 4f electrons are well screened from the ligand.

(b) When the ligand symmetry is less than cubic one observes pseudo-contact shifts. For systems possessing axial symmetry the shifts can be represented by

$$\delta = \frac{\Delta\nu}{\nu_0} =$$
$$\frac{\beta^2}{60(kT)^2}\left\langle\frac{3\cos^2\Theta - 1}{r^3}\right\rangle 2A_2^0\langle r^2\rangle g^2\{J(J+1)(2J+1)(2J+3)\cdot\langle J\|\alpha\|J\rangle\}. \quad (2.8.1)$$

where β is the Bohr magneton, r the vector joining the metal ion and the nucleus, Θ the angle between the vector r and the principle symmetry axis of the complex averaged over "fast" molecular motions. Instead of applying Eq. (2.8.1) which involves several unknowns, one uses the shift ratios

$$\frac{\Delta\nu_i}{\nu_0} = D\left\langle\frac{3\cos^2\Theta_i - 1}{r_i^3}\right\rangle \quad (2.8.2)$$

$$R_i = \frac{\Delta\nu_i}{\Delta\nu_0} = \frac{(3\cos^2\Theta_i - 1)/r_i^3}{(3\cos^2\Theta_0 - 1)/r_0^3} \quad (2.8.3)$$

Thus, ratios of pseudocontact shifts provide information on molecular shapes (not sizes since we are measuring ratios). The unknowns involve the knowledge of the binding site (position of the metal ion) and the direction of the principle axis of symmetry in addition to the geometry of the ligand. Since the number of metal ions in the lanthanide series is large, one can choose those ions for shift experiments which cause an appreciable shifts but negligible broadening like Pr(III) or Eu(III) or ions such as Dy(III), Ho(III) and Yb(III) which display smaller shifts and appreciable brodening.

The contact term for lanthanides is given by

$$\Delta_c = \frac{2\pi\beta A}{3kT\gamma_N} J(J+1)g(g-1) \quad (2.8.4)$$

where A is the scaler coupling constant and g the Landé g-factor. The term contributes mainly to atoms within a few bonds from the metal-ligand bond. The contributions of this type may be used for the detection of atoms close to the metal ion. A correction for this term may be made, thus obtaining contributions from purely pseudocontact shifts.

(c) Ions such as Gd(III), Dy(III), Ho(III) and Yb(III) have large correlation times so that appreciable broadenings of spectral lines result, due to relaxation effects [172]. In the presence of paramagnetic ions the relaxation times are given by the following equations:

$$\frac{1}{T_{1M}} = \frac{2}{15}\frac{\gamma_1^2\mu_{eff}^2\beta^2}{r^6}\left(\frac{3\tau_c}{1+\omega_1^2\tau_c^2} + \frac{7\tau_c}{1+\omega_s^2\tau_c^2}\right) + \frac{2}{3}S(S+1)\left(\frac{A}{\hbar}\right)^2\left(\frac{\tau_e}{1+\omega_s^2\tau_e^2}\right) \quad (2.8.5)$$

$$\frac{1}{T_{2M}} = \frac{1}{15}\frac{\gamma_1^2\mu_{eff}^2\beta^2}{r^6}\left(4\tau_c + \frac{3\tau_c}{1+\omega_1^2\tau_c^2} + \frac{13\tau_c}{1+\omega_s^2\tau_c^2}\right)$$

$$+\frac{1}{3}S(S+1)\left(\frac{A}{\hbar}\right)^2\left(\tau_c + \frac{\tau_e}{1+\omega_s^2\tau_e^2}\right), \quad (2.8.6)$$

In these equations, the first term represents dipolar relaxation and scalar interactions contribute to the second term. If the dipolar interactions dominate, then, as in case of shifts, one can eliminate the unknowns by the use of ratios

$$\frac{(I/T_{jM})_i}{(I/T_{jM})_0} = \frac{\langle r_i^{-6}\rangle}{\langle r_0^{-6}\rangle} \quad (j = 1, 2) \quad (2.8.7)$$

(d) In both the line width and shift experiments, diamagnetic lanthanides such as La(III) and Lu(III) can be used as blanks to calculate chemical shift changes from effects such as electric fields of ions.

(e) Fairly sharp spectra are obtained with certain lanthanides and in these cases the coupling constant values can serve as additional guides in structural studies. It should be noted that due to large shifts in the presence of paramagnetic metal ions the spectra are weakly coupled and the problems of assignment and analysis are considerably simpler.

(f) Since $\Delta\chi$ changes sign as one goes to higher lanthanides, choice of metal ions leading to both positive and negative pseudocontact shifts can be made.

Computer programs to search for possible conformations of biological molecules using shift and line width ratios have been reported [168]. Frequently, such programs attempt to fit the observed data to a single family of conformations. Lanthanide reagents have been used to study the conformation of a variety of biological molecules: nucleotides, lipids, proteins, and peptides.

The main limitation of the method is that there is no guarantee that the molecular structure is not perturbed by the binding with metal ion. However, this limitation is common to all methods which utilize external probes. Secondly, fitting the average values of shift ratios and line width ratios to a single family of structures, as has been done in many cases, may be dangerous for flexible molecules since what one observes are ratios, time-averaged over possible rotational states of the molecule [173, 179].

Some other ways in which paramagnetic reagents have been used in NMR studies of biological molecules can be briefly outlined as follows:

(a) Application of spin-labelled or paramagnetic ions to perturb the resonances of nuclei in close proximity of the site of such centers. These resonances can then be detected by difference spectroscopy [166, 174].

(b) "Quenching" of the ESR signal of the spin label caused by another paramagnetic center.

(c) Water-proton relaxation rate enhancement to study complexes between paramagnetic ions and macromolecules [175, 176]. This method is based on the fact that the addition of paramagnetic ions produces an increase in T_1^{-1} (relaxation enhancement). An additional enhancement is caused upon the addition of macromolecules. This difference in proton relaxation enhancement (PRE) indicates a complex formation between the macromolecule and metal ion. The mathematical analysis of PRE involves a multiparameter fitting and the information thus derived is subject to uncertainties in the theoretical description used. A critical review of PRE has appeared recently [176] where it has been emphasised that the method empirically indicates the binding of metal ions to macromolecules. It can be used to infer structural changes from some perturbation of the system. However, it is difficult to obtain accurate data on the geometry of metal-ligand complexes.

An example of a combined application of the techniques mentioned above is the NMR study [177] of a spin-labelled complex of the enzyme alcohol dehydrogenase and the corresponding coenzyme, namely nicotinamide adenine dinucleotide (NADH). The bound paramagnetic reagent causes an enhancement of the water relaxation rate by a factor of 6. This information, along with the effects on the water relaxation rates in the absence and in the presence of the coenzymes can be utilized for the determination of the binding constants. On the other hand, changes in the relaxation of NADH protons has been used to calculate the geometry of the coenzyme bound to the enzyme (Fig. 2.8.2).

The limitations and applications of such techniques have been reviewed in detail by Morris and Dwek [166].

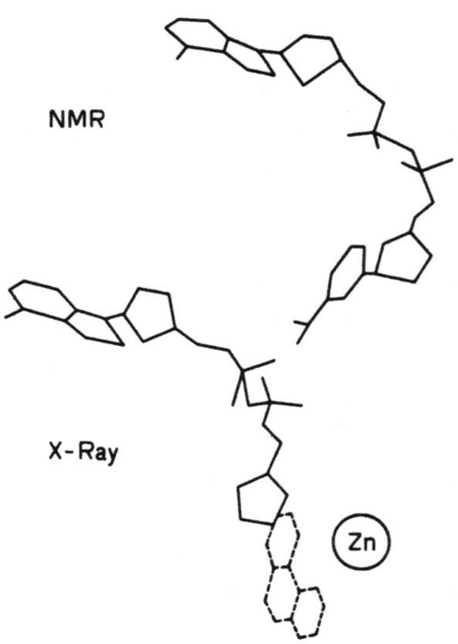

Fig. 2.8.2. Comparison of the conformation of NADH, bound to spin-labelled alcohol dehydrogenase obtained from NMR data with the structure determined by X-ray crystallography (reprinted with the permission from Ref. [177] copyright American Chemical Society)

2.9 Use of Liquid Crystals as Solvents

The direct dipole-dipole interaction [D(DD)] between two nuclear spins i and j is given by Eq. (2.9.1).

$$D(DD) = -K(1/2)\langle(3\cos^2\Theta_{ij} - 1)/r_{ij}^3\rangle \qquad (2.9.1)$$

where $K = h\gamma_i\gamma_j/4\pi^2$ and Θ_{ij} is the angle between the internuclear vector ij and the magnetic field axis, r_{ij} is the internuclear distance and γ_i, γ_j are the gyromagnetic ratios of the two nuclei. The ⟨ ⟩ indicates that the term is averaged over molecular motions in the system. The value of K is 120 067 Hz · Å^3 for a pair of protons. In solutions where a molecule is rotating rapidly and isotropically, the term $\langle 3\cos^2\Theta_{ij} - 1\rangle$ is zero due to isotropic averaging. Such DD interactions are observed directly in solids but the lines are broadened by intermolecular dipole-dipole interactions and thus the precision of measurements is low. In liquid crystalline solvents molecules are allowed to move only in certain preferred directions [180–184]. As a result, the intermolecular dipole-dipole interactions are largely averaged out but the intramolecular interactions lead to splitting of NMR lines.

In the presence of an external magnetic field, nematic liquid crystals are aligned with their optic axis (longitudinal axis), approximately parallel to the direction of the field. The motion of the solute molecules is anisotropic and the observed dipole-dipole

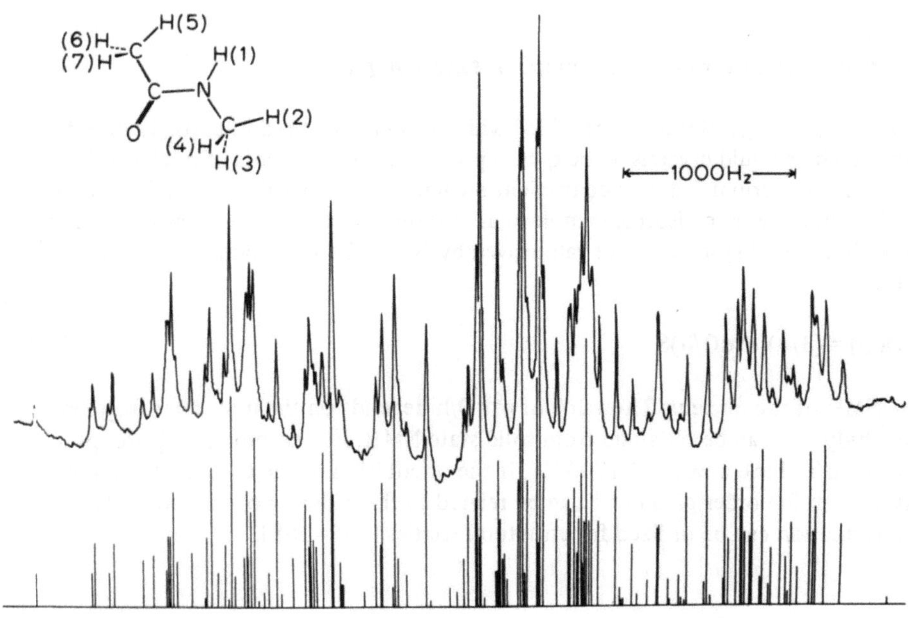

Fig. 2.9.1. 360 MHz ^{1}H-NMR of *N*-methylacetamide oriented in the nematic phase of propyl 4-amino-4-methoxybenzylidine-α-methylcinnamate. The lower trace is the calculated spectrum after an iterative fit to obtain directly dipolar coupling constants, chemical shifts and coupling constants (reprinted with the permission from Ref. [181])

interactions arise only from intramolecular interactions. If the molecule itself is rigid, then

$$D(DD) = -K\,S_{ij}/r^3 \qquad (2.9.2)$$

where $S_{ij} = \langle 3\cos^2\Theta_{ij} - 1\rangle/2$.

Here, S_{ij} is the degree of orientation of the vector ij. The order parameter is unity for a perfectly ordered system and zero for a completely disordered one. Since values of different internuclear vectors are interdependent, five independent elements of the S matrix must be determined for an unsymmetrical molecule. The remaining NMR information can be used to determine relative values of internuclear distances. The number of independent elements of the S matrix is less for molecules possessing symmetry. The amount of structural information from the spectrum of a molecule oriented in the nematic phase therefore depends on the number of independent couplings, which can be observed in a given spectrum, and on molecular symmetry.

The relative interatomic distances (not their absolute values) can be measured by this technique. Thus, the shape but not the size of the molecule can be established.

So far, the method has been applied mainly to small rigid molecules. In the field of biological systems, an interesting application of this method is the determination of the planarity of peptide group [181, 183] (Fig. 2.9.1). Using isotopically labelled *N*-methylformamides and *N*-methylacetamide it has been concluded that the amide group may show deviations of 10–15° from a planar structure.

2.10 Deuterium Quadrupole Coupling Constants [D(Q)]

Another interesting NMR parameter [25] which can be measured in partially oriented systems such as liquid crystals is the quadrupole coupling-constants of nuclei with spin > 1/2. Deuterium is a very convenient nucleus for biological studies of this type. In molecules undergoing isotropic molecular motion, this parameter is again averaged to zero. In ordered systems, a net value given by Eq. (2.10.1) is manifested through splitting

$$D(Q) = (3/4)(e^2qQ/h)S \qquad (2.10.1)$$

in the NMR of the nucleus. The value of e^2qQ/h depends on the molecular fragment under study and can be obtained from solid-state NMR. For example, CD_2 groups in paraffin chains have a value of 170 kHz. If the orientation of the C–D vector can be guessed, then the order parameter can be related to the ordering of the molecule. Again, this information can be utilized for structural studies [185–188].

2.11 Solvent Accessibility

One of the techniques applied to check the folding of biological molecules involves determination of the accessibility of different parts of the molecule to solvent [119,

125, 195–199]. Labile and hydrogen bonding groups are specially suited to NMR studies of this type because of the spectral alterations with a change in solvent or temperature. The following techniques have been used (also see Sect. 2.2).

(a) Temperature-induced changes in chemical shifts of N–H or O–H protons when the molecule is dissolved in hydrogen-bonded solvents. If these parts of the molecules are exposed to the solvent, then their temperature variation is different from the behavior shown by intramolecularly bonded protons.

(b) Exchange rates of labile protons with deuterium when the molecule is dissolved in D_2O.

(c) Changes in the chemical shifts of solvent-exposed nuclear resonances with change in solvent. The protons which are not exposed to solvent show relatively minor variations in chemical shifts when the solvent is changed. The ^{13}C chemical shifts of carbonyl groups are markedly influenced by protonic solvents such as trifluoroethanol, trifluoroacetic acid and water.

Aromatic solvents can induce an anisotropic solvent shift on proton and ^{13}C chemical shifts [191]. This can be utilized for studies on interactions of solvent with solute molecules.

(d) Saturation transfer from solvent peak to a proton in the molecule. This technique again depends on the rate of proton transfer between solute and solvent [189, 190, 193].

(e) Relaxation behavior of water protons as a result of a change in its order near a macromolecule [194].

2.12 From NMR Parameters to Spatial Structures

It may be inferred from the above discussion that the NMR method, in contrast to X-ray crystallography, does not give absolute coordinates of atoms in the molecule. In fact, except for small molecules, the determination of the complete three-dimensional structure by NMR techniques is impossible. On the other hand, NMR provides answers to certain vital structural questions which cannot be answered by other techniques. For example: Is the crystallographic conformation preserved in solution? If not what parts of the molecule show structural changes? Which part of the molecule is exposed to the solvent? What groups are hydrogen-bonded? What are the structural changes accompanying the binding of a substrate or a drug? Which part of the molecule exhibits structural rigidity when present in solutions? How does the molecular conformation change when an acid-base group in the molecule is titrated? And, finally, NMR can be used to study internal motions in the molecules and to the molecular structure in solutions under conditions very close to physiological environment which is not possible by X-ray crystallography.

Another important difference lies in the various time-scales of measurements of the two techniques. The electron density map in X-ray crystallography is obtained over a period of several hours on crystals containing a large number of molecules. If the molecules exist in different conformations in the crystal, then the information in electron-density maps is space-averaged. Further, the observations are also averaged with respect to time for any molecular motions present in the crystal. The observation time in NMR is much shorter (less than a few milliseconds). Thus, several motions which are averaged out by X-ray techniques can still be monitored by NMR.

However, it will be futile to claim that NMR can replace X-ray crystallography. A more meaningful claim is that NMR takes over at the stage where X-ray crystallography leaves the problem of biomolecular conformation and function unresolved.

3 Nucleosides, Nucleotides and Nucleic Acids

3.1 Description of Nucleotide Structures

From a chemist's viewpoint, DNA (deoxyribonucleic acid) and RNA (ribonucleic acid) are polymers composed of nucleotide units (Fig. 3.1.1). (For standard textbooks on physical aspects of nucleic acid chemistry see [200–202].) The side chains in these polymers are purine and pyrimidine bases. The common bases found in DNA are guanine (G), adenine (A), cytosine (C), and thymine (T), while the common RNA bases are G, A, C and uracil (U). RNA also differs from DNA in that it bears a hydroxy group in the 2′ position. The five-membered furanose ring (ribose or deoxyribose) has a well-defined configuration by which we mean that the relative spatial arrangements of bonds in the 1′, 2′, 3′ and 4′ positions are fixed. The sugar-base combination is called a nucleoside and the phosphates esters are called nucleotides. The linkage in DNA and RNA structures is always 3′–5′. Molecules with short segments of 2′–5′ linkages are known but they play a minor role in biology.

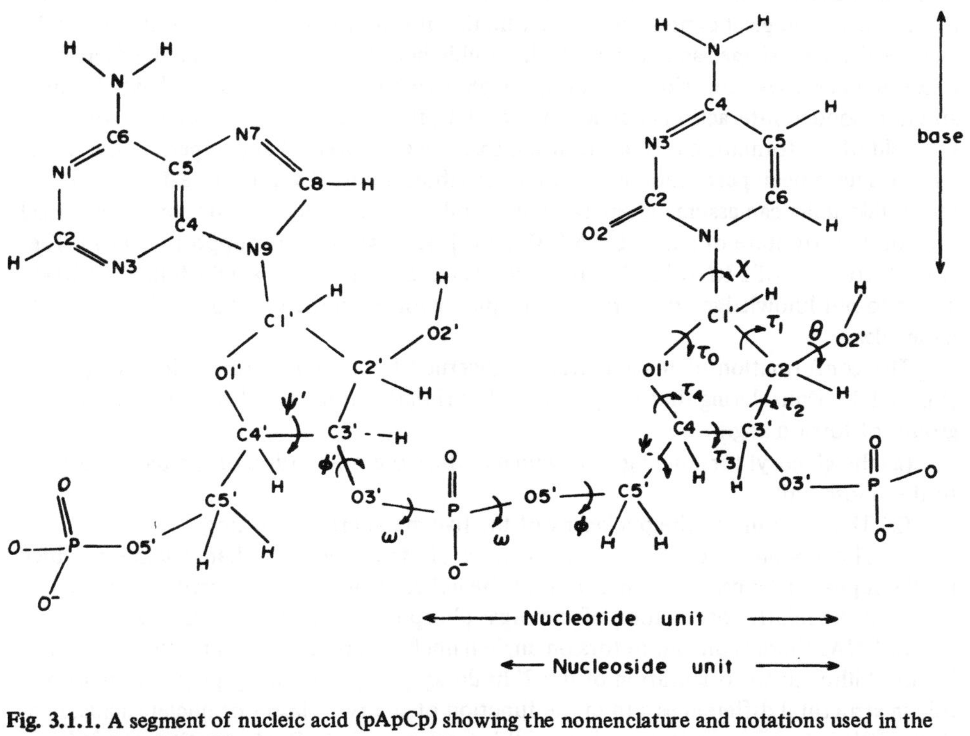

Fig. 3.1.1. A segment of nucleic acid (pApCp) showing the nomenclature and notations used in the structure of nucleic acids

In addition to the common bases listed above, DNA and RNA are known to contain some modified bases. Transfer-RNA (t-RNA) for example, is known to contain bases such as β-pseudouridine, dihydrouridine, thioanalogs, and inosine. Aza-purines and primidines are used as pharmacological agents. 7-methyl-G occurs at the 5′ position of m-RNA and acts as a primer in protein synthesis. Thus, variations in nucleic acid bases are widespread in character, but the frequency of such variations is rather low.

The importance of nucleic acids in the control of life processes is well known. From the viewpoint of a biologist, nucleic acids display three important properties. Under suitable conditions, a DNA molecule can synthesize its replica (replication). It also acts as a template in the synthesis of RNA (ribonucleic acids), a process called transcription. The third property is translation, a term used for RNA controlled synthesis of proteins which, in turn, are required to carry out various metabolic activities in a living cell. All these processes are monitored by the sequence of bases in the polynucleotide chains. The information about the biological control processes can thus be traced out ultimately to the primary structure of nucleic acids. The conformational structure plays a dominant role in each of the three basic processes listed above. For several years, the double helical structure of DNA, which is based primarily on its fiber-diffraction pattern [203], served as a useful model in the field of molecular biology. According to this model, two strands of DNA coil around each other, with the two chains running in the opposite sense. The bases of the two strands are interlocked through hydrogen bonds. In this way, the genetic code, which is determined by the sequence of the bases, is kept locked. During replication and transcription, the bases interact with the medium by unwinding of the double helix. Thus, biological expression of the genetic information demands that nucleic acids assume alternate shapes and structures. In double helical structures, it may be desirable to know the relative contributions of hydrogen-bonding interactions, base-base stacking and non-bonded interaction energies. After 1970, there has been a renewed activity in the field of conformation of nucleic acids. Extensive theoretical and experimental investigations have been performed in a number of laboratories. It was soon realized that nucleotide units can assume alternate shapes and structures, and doubts have been raised even on the structure of native DNA [204, 205], see also [206] for opposite views). The crystal structure of t-RNA [207–210] and several low molecular weight fragments have added to our knowledge on the basic principles involved in the folding of DNA and RNA molecules.

The conformation of nucleic acids is governed by a number of torsion angles (Fig. 3.1.1). Considering that the purine and pyrimidine rings are planar one has three groups of torsion angles:

(a) the glycosyl torsion angle (χ) which defines the orientation of the base relative to the ribose unit.

(b) The structure or the puckering of the five-membered sugar ring.

(c) The torsion angles ϕ', ω', ω, ϕ, ψ and ψ', in the sugar phosphate backbone which fix the repeat distances and dimensions of the helical structure in ordered polynucleotides and the relative orientation of the sugar-phosphate backbone and the bases.

In RNAs there is one more torsion angle namely the rotation around the C2′–O2′ bonds. Although the orientation of the 2′-hydroxy group very likely plays an important role in structural differences and in the function of the two classes of nucleic acids, relatively little is known about the values of this torsion angle in RNA constituents. It is,

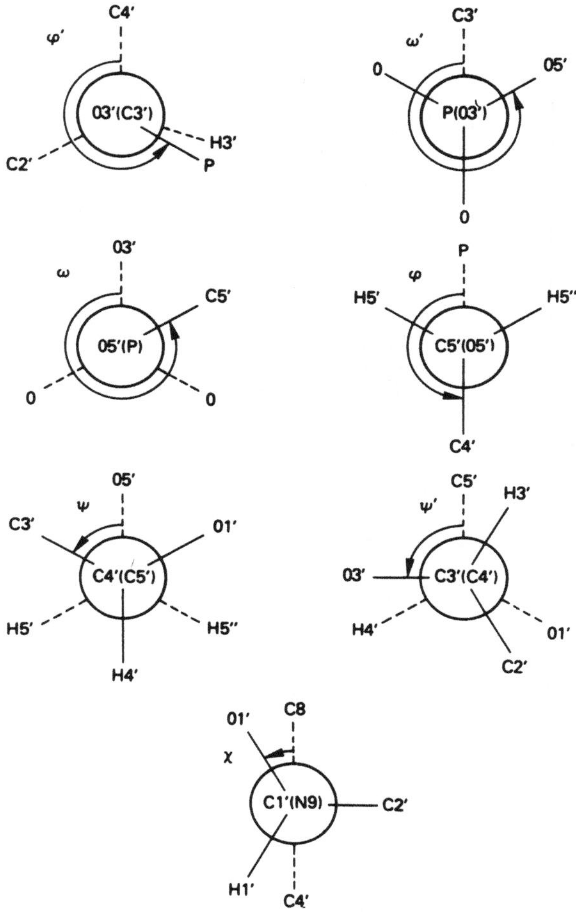

Fig. 3.1.2. Projection diagrams of the various torsion angles which fix the conformational structure of nucleosides, nucleotides and nucleic acids. The values used in the figure are the most preferred conformations although alternative conformations exist in nucleic acid anaalogs

however, known to be involved both in intra- and intermolecular hydrogen bonds and seems to play an important role in the process of translation of nucleic acid sequences into the protein structure via t-RNA [200–202].

Figure 3.1.2 represents the projection diagrams of the torsion angles involved in the description of nucleotide structures. While efforts have been made to provide unified description of the torsion angles in nucleotides, no agreement has been reached in this regard, and each working group tends to follow its own nomenclature and notations [211–216]. Table 3.1.1 selectively compiles some of these alternative descriptions [211]. To add to the confusion, some groups have changed either their notations or their nomenclature. In this chapter we follow the recommendations of an international committee [212]. We should like to point out that at this stage, none of the conventions dis-

Table 3.1.1. Torsion angles involved in the nucleic acid structures and the nomenclature proposed by various authors

Backbone: –O–P→O5′→C5′→C4′→C3′→O3′→P–O–

	–O–P→O5′	O5′→C5′	C5′→C4′	C4′→C3′	C3′→O3′	O3′→P–O–	Backbone nomenclature by	Side chain
1.	φ(O5′–P)	φ(C5′–O5′)	φ(C4′–C5′)	φ(C3′–C4)	φ(O3′–C3′)	φ(P–O3′)	Pullman and Saran [211]	χ = O1′–C1′–N1–C6 for pyrimidines or O1′–C1′–N9–C8 for purine [213]. Alternatively, χ can be measured using the positions of C2 or C4 respectively [212]
2.	ω	ϕ	ψ	ψ'	ϕ'	ω'	Sundaralingam [212, 213]	
3.	ψ	θ	ξ	σ	ω	ϕ	Arnott [214]	
4.	ψ	$\theta 1$	$\theta 2$	σ	$\theta 3$	ϕ	Sasisekharan [215]	
5.	ψ''	ϕ'	ϕ''	ω'	ω''	ψ'	Olson and Flory [216]	
6.	γ	δ	ϵ	ζ	α	β	Seeman et al. J. Mol. Biol. 104, 142 (1976)	
7.	α	β	γ	δ	ϵ	ζ	Davies [218]	

cussed in Table 3.1.1 have been accepted by international unions. Further suggestions with regard to nomenclature in NMR studies of polynucleotides and nucleic acids have been reported in another paper [218]. As regards abbreviations and symbols followed in chemical nomenclature of nucleic acids, reference [217] should be consulted.

Several earlier reviews on the NMR aspects of the conformation of nucleic acid components are available [217, 219–223].

3.2 Glycosidic Bond Rotations (χ)

The glycosidic bond angle defines the orientation of the pyrimidine or purine base relative to the sugar ring. It can be measured either with respect to the position at C6 in pyrimidines and C8 in purines as suggested in Table 3.1.1 or in terms of positions at C2 and C4, respectively. We choose the former convention. In this respect, we deviate from the recommendations of the committee referred to earlier [212] but follow [213]. The torsion angle χ(C–N) is related to angle ϕ(C–N) of Donohue and Trueblood [225] in the sense that ϕ(C–N) = $-\chi$(C–N). The classical *anti* and *syn* conformations for this torsion angle correspond to χ(C–N) angles of $0 \pm 90°$ and $180 \pm 90°$.

The *syn-anti* equilibrium is one of the most extensively studied problem using NMR. Several approaches [223, 226–228] have been applied:

(a) Use of three-bond coupling constants. Though no proton-proton ^3J couplings are available, use can be made of the coupling between H1′ and C4 and C8. Data on conformationally rigid pyrimidine nucleosides and nucleotides have enabled the calibration of the curve between χ(C–N) and ^3J(H1′–C1′–N–C) [78]. The functional form of the ^3J(H1′–^{3}C) can be represented by a Karplus-type relation of the form

$$^3J(\mathrm{H1'{-}C{-}N{-}C}) = 7.0 \cos^2 \Theta - 1.5 \cos \Theta \qquad (3.2.1)$$

Such calibration curves are not available for purine nucleotides although a very similar relationship is expected. It has been proposed that the value of parameter A in the Karplus type relation is somewhat larger and B is somewhat smaller so that the curve for purine nucleosides and nucleotides is more symmetrical around $\Theta = 90°$ than that for pyrimidine systems [229].

The knowledge of coupling constants for both C4 and C8 in purines and C2 and C6 in pyrimidines allows a complete structural determination of the torsion angle around the glycosidic bond [253]. However, it should be noted that, in general, the *anti* conformations in nucleotides have a χ(C–N) angle of approximately 30° while the corresponding angle for *syn* nucleotides is around 250° (Fig. 3.2.1). The values of dihedral angles (Θ) and ^3J corresponding to these torsion angles are as follows: for $\chi = 30°$, Θ(H1′–C4) = 30°, J = 4.7 Hz, Θ(H1′–C8) = 30°, J = 4.7 Hz, while for the *syn* conformation with $\chi = 250°$, Θ(H1′–C4) = 170°, J = 8 Hz and Θ(H1′–C8) = 10°, J = 5.5 Hz (for pyrimidine replace C4 by C2 and C8 by C6). Thus, the differences of J in these two situations are relatively small and the time-averaged values of J may not be very sensitive to conformer populations.

(b) Extensive use of proton homonuclear NOE has been made. Generally, the ribose protons are saturated and the signal enhancement of the base protons is detected. Cau-

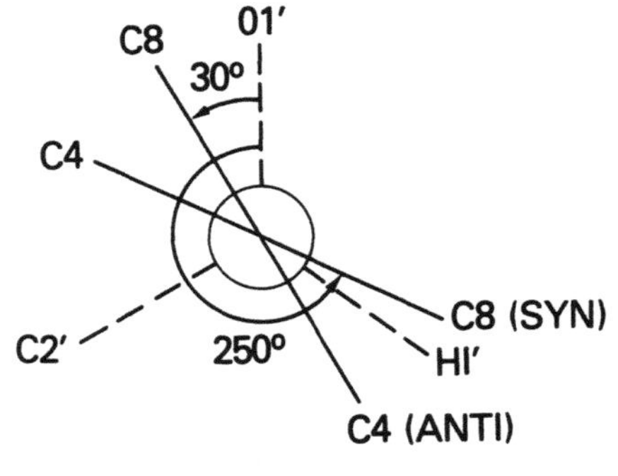

Fig. 3.2.1. Projection diagram for *syn* and *anti* conformations of nucleotides and the relationship to various dihedral angles involved in three-bond coupling constants

tion is needed in the interpretation of NOE results. An interesting example in this connection is that of nucleosides containing uridine, cytosine and their 6-methyl analogs. X-ray crystallography indicates that the first two exist in the *anti* (χ(C–N) = 20°) while the latter exist in the *syn* (χ(C–N) = 253°) conformation [230]. Similar conclusions were reached on the basis of the effect of chemical shift anisotropy of the carbonyl group at C2 and three bond coupling constants [231, 232]. NOE enhancement on H6 as a result of the irradiation of the ribose protons results in significant enhancements on the saturation of H1′, H2′ and H3′. It has been argued that the large NOE on the saturation of H1′ proton indicates that the conformation is *syn* since only in this conformation H6 and H1′ are very close to each other [233]. However, it has now been realized [160, 234] that such qualitative arguments need more careful analysis. When both *syn* and *anti* conformer coexist in finite populations, the relative NOE depends on the function

$$\mathrm{NOE}\,(\chi) \propto r^{-6}(\chi) \exp\left(-V(\chi)/RT\right) \tag{3.2.2}$$

where r is the interatomic distance between nuclei i and j involved in the NOE experiment and $V(\chi)$ the potential energy as a function of angle χ. Since both the factors in Eq. (3.2.2) are sensitive functions of the torsion angle χ, significant NOE may occur from states which do not correspond to minimum energy conformations. For β-pseudouridine, typical plots of Eq. (3.2.2) are shown in Fig. 3.2.2. It is noticed that even a small amount of *syn* conformer with a χ value close to 240° can make an appreciable contribution to NOE. For *anti* conformations with χ around 30°, NOE from saturation of H2′ and H3′ will be larger than H1′. On the other hand, all protons contribute equally if χ is near 15°.

The problem is less serious in cases where more than one nucleus on the base can be simultaneously monitored. Interesting examples are pyridine nucleotides [235] where both sides of the nicotinamide base have a proton. Here, one observes equal NOE at the two base protons as a result of the saturation of H1′, indicating almost equal probabilities of the existence of *syn* and *anti* conformers.

(c) Use of relaxation times: Use of proton T_1 has been made of to study the *syn-anti* equilibrium [155, 156, 223, 224, 236–241]. The methodology involves separation of intramolecular DD contributions to T_1 which in turn are related to interatomic dis-

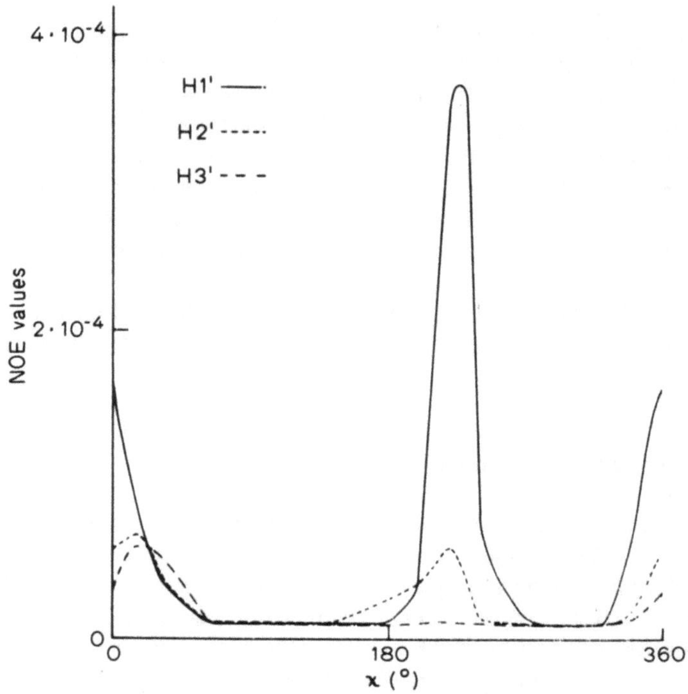

Fig. 3.2.2. Factor $r^{-6}(\chi) \exp(-V(\chi)/RT)$ (Relative NOE values in an arbitrary scale) as a function of torsion angle χ in β-pseudouridine. The energies $V(\chi)$ have been calculated by the CNDO method using a C3′ *endo* conformation (reprinted with the permission from Ref. [234])

tances. As in the case of NOE, the method involves careful examination of the r^6 dependence as a function of the glycosidic torsion angle and the conformation of the furanose ring.

(d) Magnetic anisotropy of the chemical groups: Purine and pyrimidine rings can influence the chemical shifts of the nuclear resonances in the furanose ring through several mechanisms. The more important effects include the ring current, local diamagnetic susceptibility anisotropy and polarization. In addition, the N3 lone pair orbitals of purine and the C2 = O2 carbonyl group in pyrimidine make appreciable contributions to the chemical shifts. These effects depend both on the angle χ and the conformation of the furanose ring. Theoretical estimates for the proton chemical shifts as a function of various conformational parameters have been made [242–245]; these are in general agreement with the experimental findings [228, 229, 231, 232, 246, 247]. For example, for a C2′ *endo syn* purine nucleotide, the N3 lone pair orbital points into the direction of the C2′–H2′ bond and causes downfield shift on the proton and an upfield shift on the carbon. In C3′ *endo syn* conformation, a similar effect occurs on C3′ and H3′ resonances, while the neighbor anisotropy effects on C2′ and H2′ are considerably less pronounced in this state.

The results of X-ray crystallography, energy calculations and NMR are consistent with a model in which the *anti* conformational state with angle χ in the range 0–70° is generally preferred. However, some contributions from the *syn* conformer cannot be ruled out. An *all-syn* conformation is observed in the following situations:

(a) When a bulky substituent (Br or methyl) is present at C8 of the purine base, or at C6 in pyrimidines, the steric interactions between the substituents and ribose cause the molecule to flip to a *syn* structure [228, 229, 248, 249].

(b) In nucleosides, an intramolecular hydrogen bond between the base and the 5′-hydroxy group stabilizes the *syn* and C2′ *endo* conformation in certain cases. The corresponding 5′-nucleotide, however, may assume an *anti* conformation in these cases [250, 251, 328].

(c) Certain cyclonucleosides and cyclonucleotides where the ring closure forces the molecule to adopt a conformation intermediate between normal *anti* and normal *syn* [143, 252].

3.3 Sugar Ring Conformation: Pseudorotation

The furanose ring is centrally located in the nucleotide fragment. Conformations of cyclic systems introduce some constraints – due to ring closure – which are not present in linear non-cyclic molecules. In ribose we have five atoms in the ring and we need N–3 = 2 torsion angles to define its three-dimensional geometry [215, 254, 255]. Thus, out of the five torsion angles $\tau_0 - \tau_5$ in Fig. 3.1.1, two are enough to define the geometry of the ring. For example, let us consider a plane passing through C1′, O1′ and C4′. We should now like to know the positions of C2′ and C3′ with respect to this plane. This is given by values of τ_0 and τ_1. However, even τ_0 and τ_1 are not independent, since bond lengths and bond angles are expected to show relatively minor changes as a result of ring closure. If $\tau_0 = 0°$, i.e. atom C2′ lies in the plane described by the other three atoms, then the two positions for C3′ dictated by the bond lengths (C3′–C4′ and C2′–C3′) and bond angles (C1′–C2′–C3′) and C2′–C3′–C4′) are 0.6 Å above the plane (when C3′ is on the same side of the plane described by the remaining atoms as C5′ and the base: such a conformation is described as C3′ *endo* or ^{3}E) or 0.6 Å below the plane (a conformation labelled as C3′ *exo*). The values of τ_1 corresponding to these conformations are ± 25.5°. In a similar way, one finds two values of τ_1 for each value of τ_0. Further, there is a maximum value τ_m which can be assigned to τ_0 and τ_1 in order to allow a ring closure to take place. The conformational freedom of a ring system is thus more restricted. For a five-membered ring, we need just the value of one of the five torsion angles and the sign (or the phase) of the second angle to completely define the three-dimensional structure or pucker of a five-membered ring. Such arguments can be extended to other ring systems in biological systems such as proline and hydroxyproline rings in peptides and six-membered rings in saccharides.

It has been observed quite early by X-ray crystallography [213, 254] that the more abundantly observed conformations of the furanose ring correspond closely to the classical envelope forms (Fig. 3.3.1) namely, C3′ *endo* (^{3}E) and C2′ *endo* (^{2}E). However, twisted non-envelope conformations (denoted by T) where both C2′ and C3′ are displaced from the plane defined by C4′, O1′ and C1′ may also occur.

Altona and Sundaralingam [256] have introduced a general two-parameter description of the conformation of the ribose ring. In this description, the conformation of the sugar ring is described by two parameters P and δ. The five torsion angles $\tau_0 - \tau_5$ are related to these parameters by the equation:

$$\tau_j = \tau_m \cos [P + (j - 2)\delta] \qquad (3.3.1)$$

with

$$\tan P = \frac{[(\tau_4 + \tau_1) - (\tau_3 + \tau_0)]}{2\tau_3(\sin 36^\circ + \sin 72^\circ)} \qquad (3.3.2)$$

For a five-membered ring, the approximate values of τ_m and δ are 38° and 144°, respectively; j in Eq. (3.3.1) can be 0, 1, 2, 3, or 4. As P goes through the complete pseudorotation cycle (0 to 360°), values of $\tau_0 - \tau_5$ range from $-\tau_m$ to $+\tau_m$ through 0°.

Figure 3.3.1 illustrates classical structures of ribose as a function of P, ranges of various torsion angles, their interrelationship and energies calculated by the molecular orbital method PCILO [211, 256]. From the potential energy curve shown in Fig. 3.3.1, it is clear that the energy shows a minimum for two conformational states. One of the states has a P value around 20° and the other a P value around 160°. These two isomeric states are denoted by N and S to indicate that they cover the north and south parts of the conformational wheel in Fig. 3.3.1. It should be noted that the state N covers ranges of conformations including the classical C3′ *endo* and C2′ *exo* while the S state includes C2′ *endo* and C3′ *exo* puckering of the ribose ring. An energy barrier of about 4 kcal/mol separates these two rotational states.

NMR parameters prove ideal for studying ribose conformations [224, 256–263, 272–274]. Use can be made of the proton-proton three-bond couplings J(H1′–H2′), J(H2′–H3′), J(H3′–H4′) and in the case of deoxyribose J(H2″–H1′), J(H2″–H3′). Since we actually have only one unknown (P), the problem is overdetermined. In solution, the sugar ring may exist in more than one conformational forms and rotational isomeric states N and S can be utilized. The relative conformer populations can then be estimated from the coupling constants listed above. For ribonucleotides, the relation

$$^3J(\text{H–C–C–H}) = 10.5 \cos^2 \Theta - 1.2 \cos \Theta \qquad (3.3.3)$$

has been proposed [259] for the ^3J values of the ribose ring. However, Davies and Danyluk [264] have recommended A = 9.8 and B = 0.9 Hz. The parameters for deoxyribose are slightly different because of the replacement of 2′–O by H″. Here, A ~ 11.7 and B ~ 0.4 [264].

For the states N and S, the values of ^3J(H2′–H3′) and sum of J(H1′–H2′) + J(H3′–H4′) are very close to one another. It is however clear from the Newman projection diagram for τ_1 and ψ' shown in Fig. 3.3.2 that the values of J(H1′–H2′) and J(H3′–H4′) are very sensitive to N $\rightleftharpoons$ S equilibrium. Thus, for an N family of conformers ^3J(H1′–H2′) is around 0 Hz ($\Theta \sim 95^\circ$) while ^3J(H3′–H4′) is around 9.5 Hz ($\Theta \sim 155^\circ$). For S type conformers, these values are interchanged. Thus, a variation of 1 Hz in coupling constants correspond to approximately 10% difference in conformer populations. It has been shown that this information, in conjunction with values of other ^3J in the ribose ring,

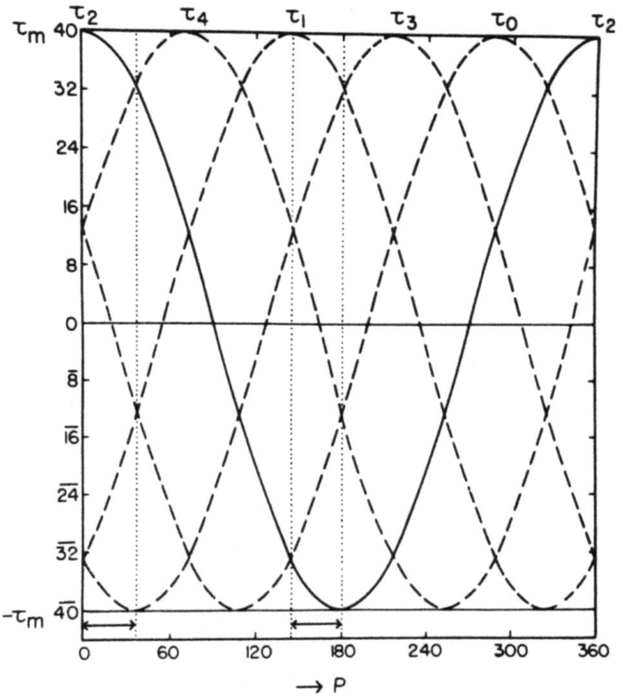

a

Fig. 3.3.1 (a–c). Pseudorotational description of the furanose ring. (a) Values of the five torsion angles $\tau_0 - \tau_4$ in the furanose ring as P goes from 0 to 360°. (b) Classical conformations of the furanose ring as a function of P. (c) Potential energy of ribose as a function of P as calculated by the PCILO method. (Figs. 3.3.1(a) and 3.3.1(b) reprinted with the permission from Ref. [256]; copyright American Chemical Society)

is sufficient not only to determine the conformer populations but also the maximum puckering angles (τ_m^N and τ_m^S) and pseudorotational angles (P^N and P^S) for the two states, N and S. Graphical plots for determining the pseudorotational parameters and populations of the two conformer families have been reported [263, 264].

Values of ^3J(H1′–H2″) and ^3J(H3′–H4′) for nucleosides, nucleotides and randomly coiled polynucleotides are generally in the range 3–7 Hz. Thus, the solution conformations of nucleic acid fragments generally coexist as a mixture of N and S type conformers both in DNA and RNA systems. The free energy difference between the two states is less than 1 kcal/mol. The deoxynucleotides generally favor S-type conformers to a greater extent but "conformational purity" is not attained in most cases. A detailed analysis of 5′-ribo- and 5′-deoxyribonucleotides [263] carried out in this way has also shown that the pseudorotational parameters do not change significantly between nucleotides and are generally within the ranges of values observed in crystalline states.

A correlation between χ_{C-N} and the sugar puckering has been observed both by NMR and X-ray crystallography on nucleosides [265–268]. Thus, a *syn* conformer is generally associated with a larger population of C2′ *endo* conformer while *anti* conformation is associated with a C3′ *endo* structure. Exceptions to this correlation have been

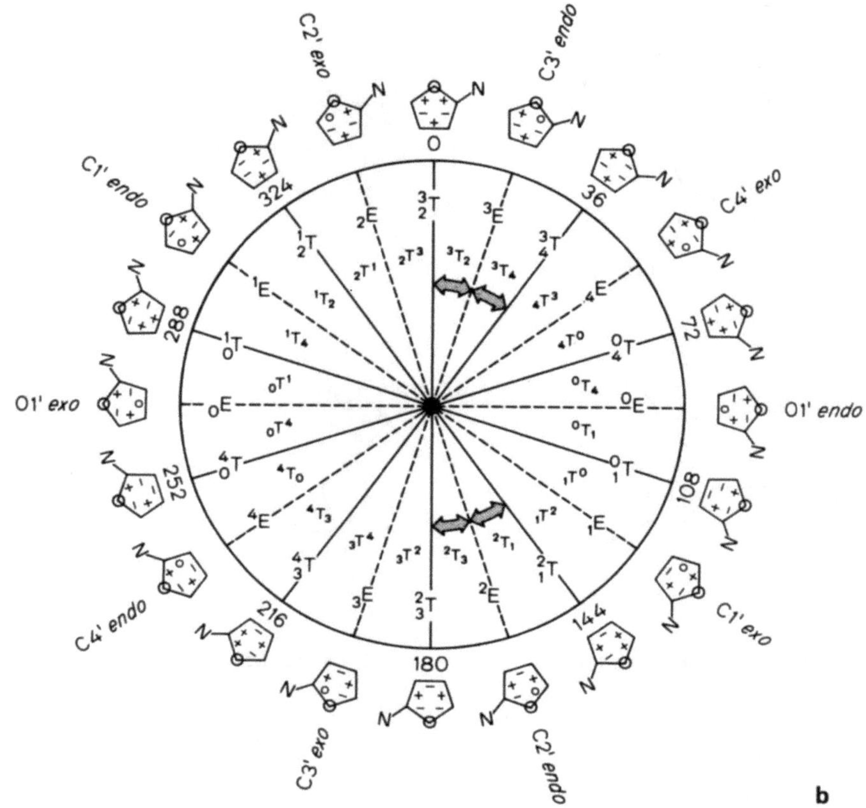
C2' exo
C3' endo
C4' exo
O1' endo
C1' exo
C2' endo
C3' exo
C4' endo
O1' exo
C1' endo
0
36
72
108
144
180
216
252
288
324

b

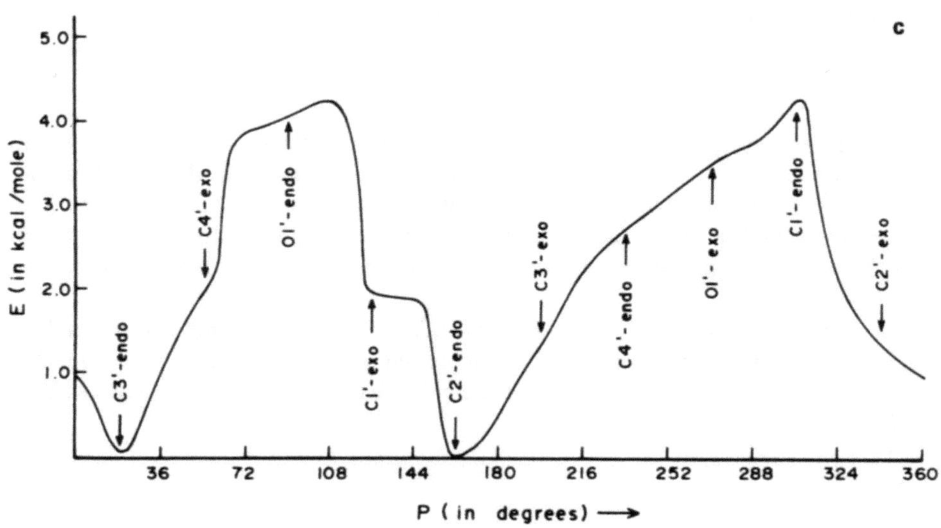
E (in kcal/mole)
P (in degrees) ⟶
C3'-endo
C4'-exo
O1'-endo
C1'-exo
C2'-endo
C3'-exo
C4'-endo
O1'-exo
C1'-endo
C2'-exo

c

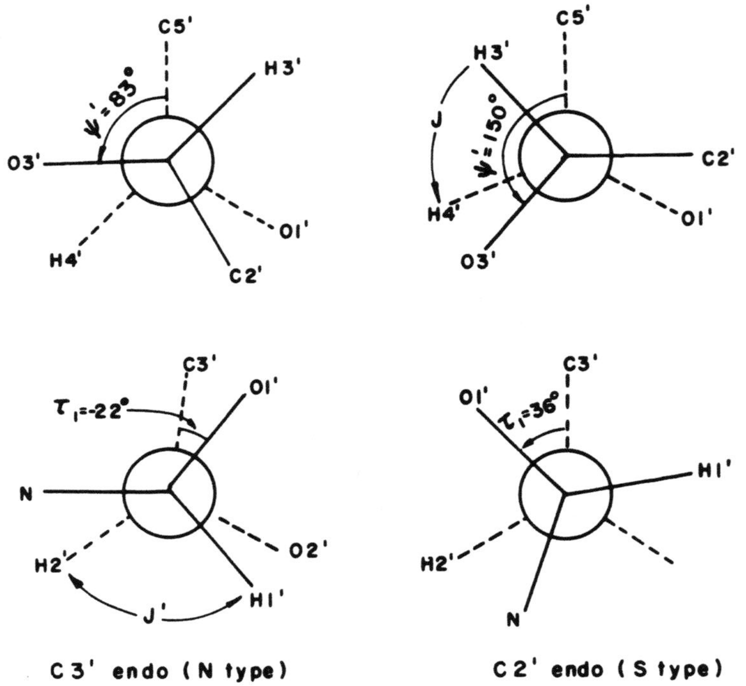

	C3' endo (N type)	C2' endo (S type)
J (H3'- H4')	~ 9·5	0
J'(H1'- H2')	~ 0	9·5

Fig. 3.3.2. Newman projection diagrams with respect to angles ψ and τ_1 in the furanose ring and their relationships with the dihedral angles involving proton-proton ^3J coupling

observed recently [229, 269, 270], and it now appears that such a correlation holds only for certain nucleosides where a *syn*-C2, *endo* conformation is stabilized by an intramolecular hydrogen bond between the 5′-hydroxy group and the base N3. A further correlation in the *anti* range of rotational states is that an N conformer has a χ angle in the range 0 to 40° while an S type conformer is associated with χ angles ranging from 40 to 80° [264].

In deoxyribonucleotides, it is sometimes difficult to resolve 2′ and 2″ protons. In these cases, population analysis can be accomplished by taking the sum of the two ^3J involving the above protons and H1′. The use of lanthanide shift reagents may also be helpful in such cases [271].

3.4 Backbone Angles

The sugar-phosphate backbone conformation is fixed by the values of six conformational angles. Of these the value of ψ' is fixed by the conformation of the sugar ring (ψ' is 85° for ^3E conformation and 150° for ^2E conformation).

3.4.1 *Torsion Angles* ϕ'

The torsion angle ϕ' around the C3′–O3′ bond can be verified by the three-bond coupling constants between H3′ and P and the two ^{13}C–^{31}P coupling constants involving C2′ and C4′. The parameters for the Karplus-type relationships for these coupling constants have been obtained [95, 97, 258, 264, 275–279]. These can be represented by

$$^3J(H\text{–}C\text{–}O\text{–}P) = 16.3 \cos^2\Theta - 4.6 \cos\Theta \quad (3.4.1)$$

$$^3J(C\text{–}C\text{–}O\text{–}P) = 9.8 \cos^2\Theta - 0.5 \cos\Theta \quad (3.4.2)$$

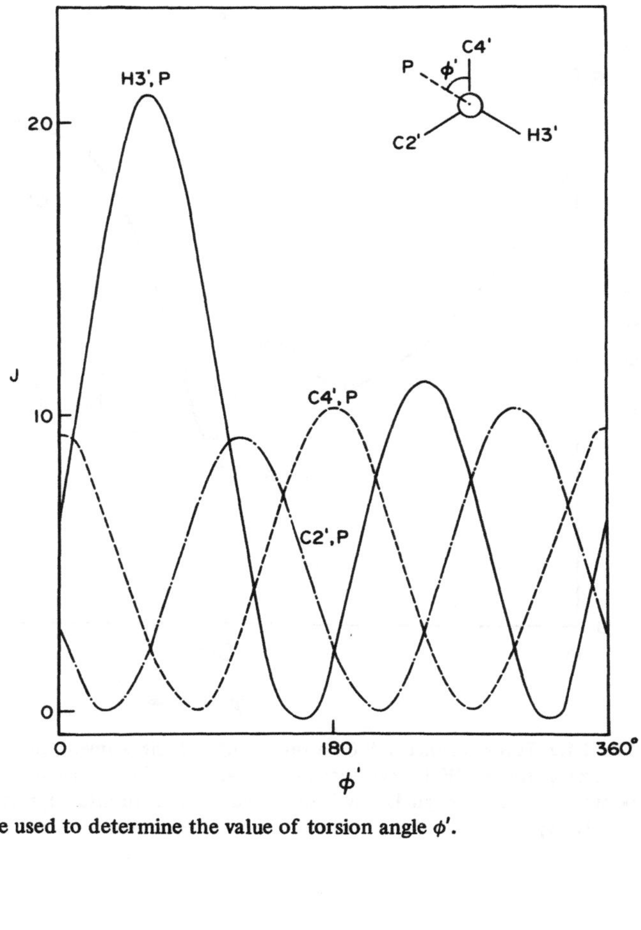

Fig. 3.4.1. ^3J values which can be used to determine the value of torsion angle ϕ'.

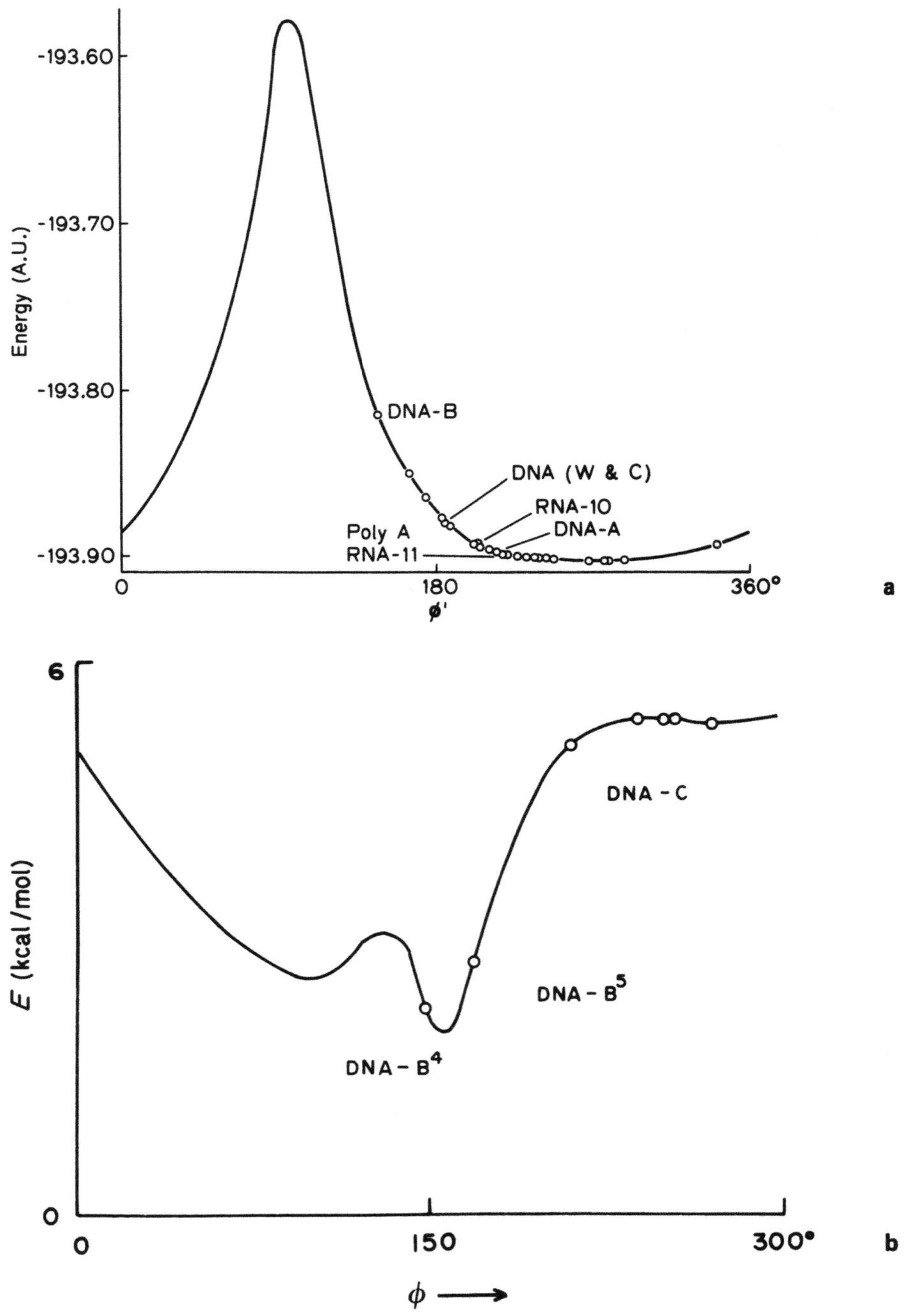

Fig. 3.4.2. Potential energy E (in atomic units) of the segment diribose monophosphate as a function of angle ϕ' for (a) ^{3}E deoxyribonucleotide and (b)^{2}E deoxyribonucleotide. A strong correlation between ϕ' and sugar puckering is suggested by these results (reprinted with the permission from Ref. [278])

and can be used to estimate values of ϕ'. While the relation for $^1H-^{31}P$ is presently well-founded, the behavior of $^{13}C-^{31}P$ couplings is based on a limited number of data sets. The functional behavior of these coupling-constants as a function of angle ϕ' is shown in Fig. 3.4.1.

In earlier NMR works on ribonucleotides [280, 281], attempts were made to interpret the conformational structure around this bond in terms of a three-state model involving three rotational isomeric states g^+, t and g^-. However, such a model is inconsistent with potential energy calculations involving only one shallow minimum around ϕ' value of 230° (Fig. 3.4.2) [278, 279]. The actual minimum energy value of ϕ' depends, to some extent, on the sugar pucker and the presence or absence of 2′–O. Likewise, the known crystal structures of all ribonucleotides, have ϕ' values in the range 200–260° [207–211, 213, 214, 278]. For deoxyribonucleotides in the S state, on the other hand, these results indicate a value of 120–160° [211, 213, 214, 278, 282]. An S state with $\phi' \sim 140°$ is precluded due to hard contacts in ribonucleotides. It has also been argued that solvent stabilization may be responsible for the exclusive occurrence of the high ϕ' values observed in ribonucleotides [283]. The fact that NMR results fit better to a model of a conformation intermediate between classical t and g states was first pointed out in connection with ^{13}C-NMR of poly (U) [97, 279, 280]. Since then, most investigators have dropped the three-state model.

The values of ^3J(H3′–P) are generally around 8–9 Hz in ribonucleotides which gives a value of ±30° or ±125° for the dihedral angle H3′–C3′–O3′–P. In view of the potential energy and crystal structure results discussed above, the only acceptable solutions out of these correspond to $\phi' = 210°$ or $\phi' = 270°$. Alternatively, the ribonucleotides may show a dynamic equilibrium over the range of $\phi' = 200–270°$ consistent with energy calculations. The N-state conformer may prefer the lower value of ϕ'(210°), while the S state may be associated with $\phi' = 270°$ This view is supported by two independent evidences. First, a four-bond coupling constant between H2′ and P(J ~ 1–3 Hz) has been observed in ribonucleotides [112, 284]. Such a coupling indicates that a substantial proportion of conformational states provides a W pathway for this coupling constant. Second, it may be noticed from Fig. 3.4.1 that for a ϕ' value of 210°, J(C2′–P) = 0 and J(C4′–P) = 7.8 Hz, while for $\phi' = 270°$, J(C2′–P) = 7.8 Hz and J(C4′–P) = 0 Hz. The observed values of these two $^{13}C-^{31}P$ coupling constants lie in the range 3–5 Hz and the actual values depend on the N ⇌ S equilibrium.

For deoxyribonucleotides, ^3J(H3′–P) values are much lower (3–8 Hz) (see for example [285]) and ^4J(H2′–P) is not observed. It is possible that the DNA fragments are locked at $\phi' = 200°$ in both the N and S state [285]. However, fibre-diffraction results suggest that deoxyribonucleotides are more flexible and the S state in these cases prefer a ϕ' angle around 130°. Thus, the results can also be reconciled in these cases with a model where the two conformers correspond to N-210° and S-130°.

3.4.2 Angles ϕ

NMR results on nucleosides and nucleotides have been analyzed using a three-state model consisting of staggered conformations with respect to C5′–O5′ bonds (Fig. 3.4.3). The three states, g^+, t, and g^- are also called $g't'$, $g'g'$ and $t'g'$ because of the orientation of

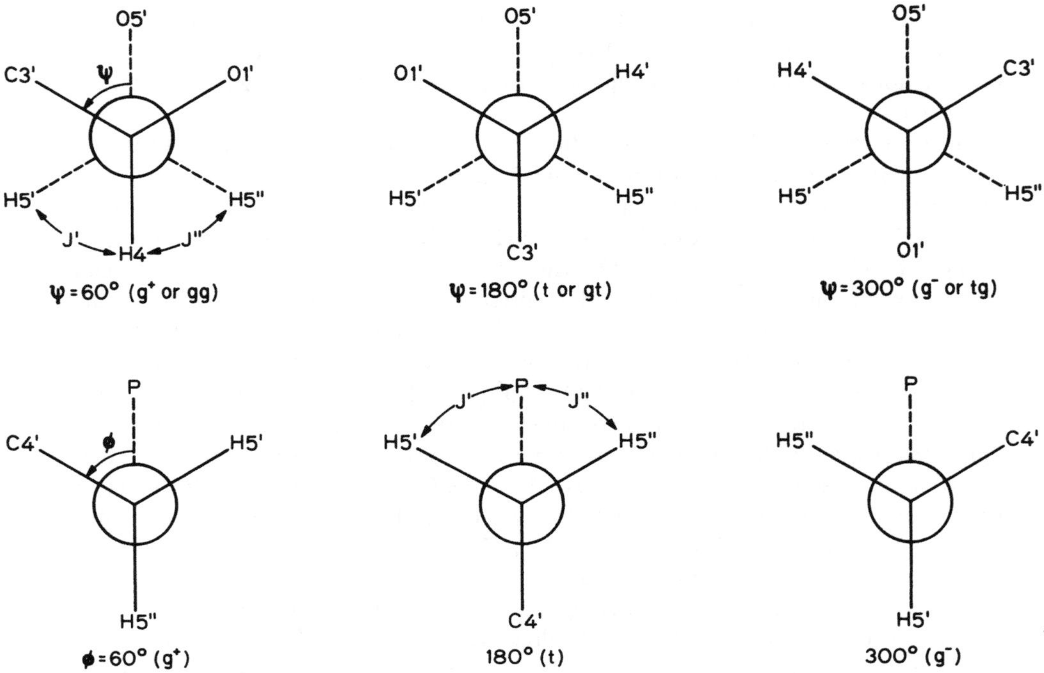

Fig. 3.4.3. Newman projection diagrams for staggered conformations around the C5′–O5′ bond (angle ϕ) and C4′–C5′ bond (angle ψ)

protons relative to the ^{31}P. While the three-bond ^{13}C–^{31}P coupling constants are also indicative of conformational preference, quantitative estimates are generally made using the coupling constants involving ^1H and ^{31}P. For ^1H–^{31}P coupling constants in nucleotides the following values of component coupling constants have been suggested. (See, for example [264, 279, 280, 281, 285]): J(g) = 1.8, J(t) = 21 Hz. With these values, the populations of the three conformers with respect to each bond can be established using the relations

$$J(g)[p(g^+) + p(t)] + J(t)p(g^-) = J(H5'\text{–}P) \qquad (3.4.3)$$

$$J(g)[p(g^-) + p(t)] + J(t)p(g^+) = J(H5''\text{–}P) \qquad (3.4.4)$$

and

$$p(g^+) + p(t) + p(g^-) = 1 \qquad (3.4.5)$$

with $p(t) = [24.6 - (J(5'\text{–}P) + J(5''\text{–}P))]/21.0$.

Values of $^3J(H5'\text{–}P)$ and $^3J(H5''\text{–}P)$ usually lie in the range of 3–5 Hz. This indicates that the *trans* conformer with respect to C5′–O5′ bond is preferred to an extent which is greater than 70%.

3.4.3 Angles ψ

Like in the case of torsion angle ϕ the rotational states for torsion angles around the bond C4′–C5′ are also chosen as the three staggered conformations $g^+(gg)$, $t(gt)$ and $g^-(tg)$ (Fig. 3.4.3). The values of $^3J(H4'–H5')$ and $^3J(H4'–H5'')$ can be utilized for the analysis of conformer populations. The values of component coupling constants generally used in the analysis are: J(g) = 1.5 Hz, J(t) = 10 Hz (see for example [264, 279–281, 285]).

$$J(g)[p(gg) + p(tg)] + J(t)p(gt) = J(H4'–H5') \quad (3.4.6)$$

$$J(g)[p(gg) + p(gt)] + J(t)p(tg) = J(H4'–H5'') \quad (3.4.7)$$

$$p(gg) + p(gt) + p(tg) = 1 \quad (3.4.8)$$

with $p(gg) = [13 - (J(4', 5') + J(4', 5''))]/10$

The observed values of both coupling constants generally lie in the range 2–5 Hz, showing that the g^+ (or gg) state is populated to the extent of 60% or more.

Assignment of H5′ and H5″ protons: At this stage, it may be pointed out that though in general both protons H5′ and H5″ are well resolved, there is controversy about their assignments. Son and Guschelbauer [286] used the effect of *p*H changes on chemical shifts and rotamer populations and argued that the statistically low field H5′ proton is closer to H3′. This leads to the assignment as shown in Fig. 3.4.3.

The alternative assignment is supported by Remin and Shugar [287] who have used arguments based on the effect of 3′-phosphorylation on the 5′ and 5″ protons in uridine and β-pseudouridine. The assignment is further supported by a comparative study of the chemical shifts of 5′ and 5″ protons in 2′- and 3′-nucleotides [288] and the observation of a hydrogen bond between the 5′-hydroxy and 2-oxo group of *N*(3)-methyl-2′,3′–O-isopropylidene-uridine in $CDCl_3$ and CCl_4 solutions. Further, these assignments provide a basis for the observed correlation with the chemical shift difference between the two methylene protons and the conformationally dependent ^3J value of the exocyclic group [288, 289].

An inversion of assignment does not alter the estimated populations of the g^+ (or gg) conformer with respect to angle ψ and of the t (or $g'g'$) conformer with respect to angle ϕ. However, the populations of the other two states is reversed in each case. Thus, the general trend of the relative conformer populations using the first assignment is $g^+(gg) > g^-(tg) > t(gt)$ while the second assignment gives $g^+(gg) > t(gt) > g^-(tg)$.

3.4.4 Torsion Angles ω', ω

These angles are of special significance for the stereochemistry of nucleic acid components since both potential energy calculations and X-ray crystallography have revealed that there is a great degree of flexibility around these bonds [213, 279, 286, 290]. The most preferred conformation is g^-g^- but other states can contribute to the structure of nucleotides. Unfortunately, coupling constants in nucleotides do not show a direct relationship with rotation around the O–P bonds. The following approaches, however, have proved useful in conformational studies of phosphate groups:

Stacking Shifts: The principle of this method is that when values of other torsion angles in the nucleotide fragment are fixed at their preferred ranges discussed above, then a g^-g^- arrangement with respect to O–P bonds in a dinucleoside phosphate or in longer nucleic acid segments leads to a structure where the bases lie on the top of one another. This in turn causes an upfield shift of the base protons due to the magnetic anisotropy of purine and pyrimidine rings. Generally, stacking shifts can be estimated from the isoshielding contours [130, 131] or from NMR of ordered nucleotides where the bases are fully stacked [229, 291]. This observation then allows to estimate the percentage of the stacked conformer.

The theoretical ring current maps of Giessner-Prettre and Pullman [130, 131] have been extensively used in conformational studies of nucleic acid components. There are two versions of the maps both of which have been widely employed. The earlier calculations [130] were based on the estimation of the "ring current" effects following an approach very similar to that of Johnson and Bovey [129]. It was shown that the chemical shift changes due to stacking interactions at a distance of 3.4 Å from the purine rings can amount to as much as 1.3 ppm for adenine and 0.73 ppm for guanosine. The pyrimidine rings contribute relatively much lower stacking shifts (up to 0.4 ppm). Recently, these calculations have been refined to take care of the atomic diamagnetic anisotropies [131]. The atomic anisotropy contributions are significant and have led to predictions of much higher shifts arising from base stacking than predicted earlier. The shielding effects due to cytosine are higher by a factor of 4 and those due to uridine by a factor of 2 than those predicted solely on the basis of ring-current effects. The purine effects are likewise 25–40% higher.

Figure 3.4.4 shows the result of latest calculations [131]. The calculations have been performed for the shieldings in the plane at distance of 3.4 Å from the plane of the rings. It can be noticed that the observed shieldings depend critically on the orientation of the bases with respect to one another. For this reason, the ring-current contributions to the chemical shifts due to stacking from the ring above and the one below, can be different. More recently, the spatial dependences of the ring-current effects have been described by a series of graphs [132] in cylindrical coordinates. These graphs allow estimates of ring-current shifts at any point in a cylinder of radius 10 Å, extending to 8 Å above and below each ring.

In nucleic acid double helices, the shieldings of a base proton may be influenced both by the base pair above and the one below it. Calculations for such contributions in A-RNA and B-DNA geometries have been made [291].

While the calculations of Giessner-Prettre and Pullman do give a good agreement between theoretical and experimental results, it may be emphasized that there are no model systems with which the calculations can be tested rigorously. The pattern of base stacking is sensitive to torsion angles. In solution, these angles may show time-dependent fluctuations. Therefore, it is not possible to use the crystallographic data to calibrate the isoshielding contours. In fact, the older calculations enjoyed extensive applications until serious discrepancies were noticed by experimental measurements.

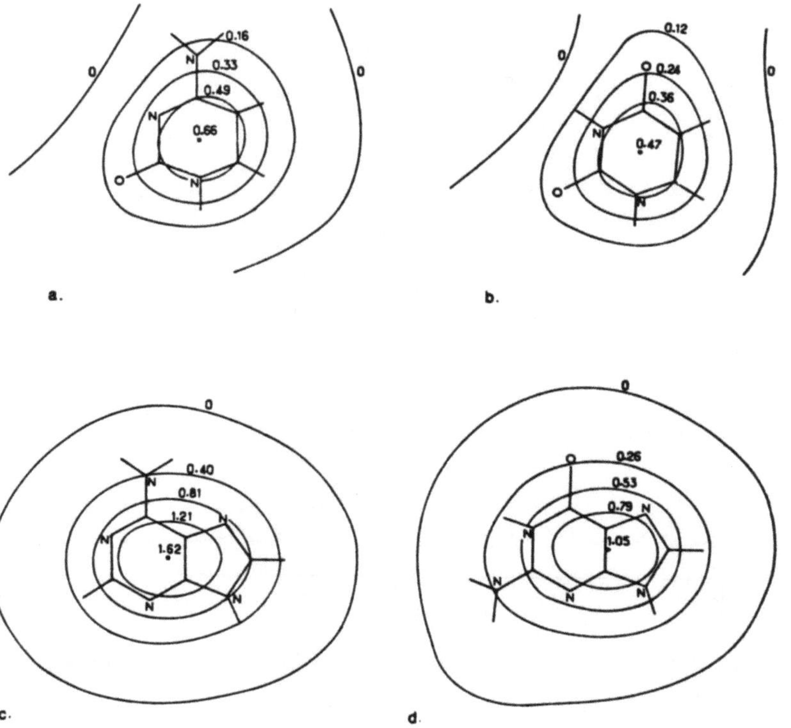

Fig. 3.4.4. Effect of base anisotropy of purine and pyrimidine rings on the chemical shift of a nucleus at a distance of 3.4 A above the plane of the base (reprinted with the permission from Ref. [131])

^{31}P Chemical Shifts: The ^{31}P chemical shifts in nucleic acids are sensitive to conformational structures [280, 292–295]. For example, when ordered polynucleotides are heated (see Sect. 3.8, 3.9), the "melting" of the ordered structure is roughly accompanied by a 1 ppm downfield shift. Considering that the major conformational change involved in this process of conversion of an ordered to a random-coil structure is a rotation around one of the two O–P bonds from a *g* to *t* state, the ^{31}P chemical shifts very likely monitor this effect. An even more interesting example is the ^{31}P-NMR spectrum [293] of t-RNA (see Sect. 3.10) where the phosphate group exists in several non-equivalent environments. The ^{31}P spectrum of t-RNA (Fig. 3.4.5) reveals several phosphate chemical shifts dispersed in a 7 ppm region [293]. While the ^{31}P chemical shifts may not be very useful to the analysis of conformer populations of *t* and *g* states, these can be fruitfully utilized to monitor conformational differences and transitions between various secondary and tertiary structures of a molecule.

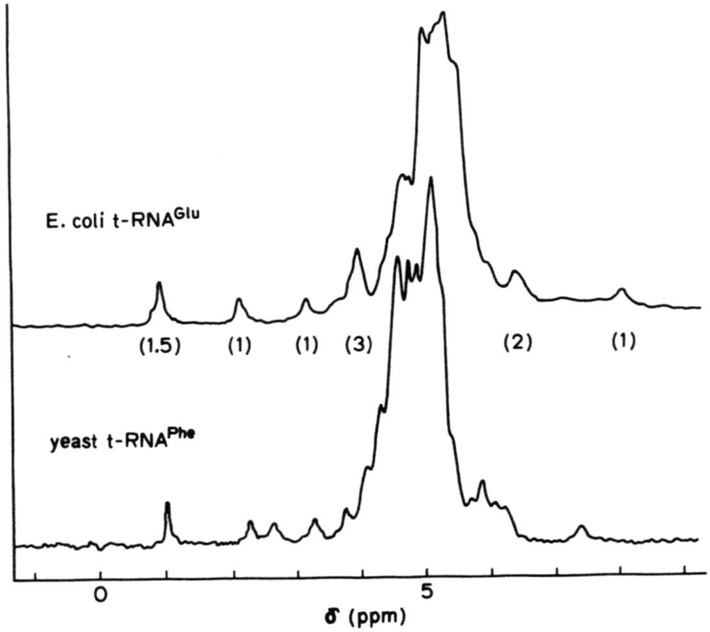

Fig. 3.4.5. ^{31}P-NMR spectrum of t-RNA (reprinted with the permission from Ref. [293])

3.5 NMR Studies on Nucleosides

Table 3.5.1 gives an overall review of the recent literature on NMR studies on the conformation of nucleosides. Extensive work on nucleoside conformations has been done before the advent of FT techniques and the earlier reviews may be consulted for this purpose [219, 220, 222]. An interesting aspect of the conformational analysis is the structure around C3'–O3' and C2'–O2' bonds which can be monitored by making use of ^{3}J(HCOH) in non-hydroxylic solvents (e.g. DMSO) [296–298]. The expected value of this coupling for a free rotation around the C–O bond is ~5.4 Hz [75, 76]. The coupling constant between OH(2') and H2' lies in the range 5.5 to 6 Hz indicating a slight preference for the *t* conformer along the H2'–C2'–O2'–H pathway. The values for the 3'(OH) group are however lower (^{3}J ~ 4–5 Hz) than the value expected for a free-rotation model indicating a preference for *g* states. On the whole, nucleosides exhibit considerable flexibility.

3.6 Small Nucleotides

Table 3.6.1 gives a review of the recent literature on nucleotides. Some general features of the structure of short segments of nucleic acids are outlined below.

Table 3.5.1. Conformational studies on nucleosides

Nucleus	System and special information	Ref.
^1H	Isopropylidinepyridine	[300]
^1H	Acetylylglucopyranosylpyridinium	[301]
^1H	Arabinosyl nucleosides: effect of hydroxyl dissociation	[302]
^1H	6-Methyl-2′-deoxyuridine: *syn-anti* equilibrium	[248]
^1H	Thiouridines	[303]
^1H	Nucleosides: bond lengths, NMR, hydrolysis, X-ray studies	[304]
^1H	Oxathianepuryl, nucleosides of hetero analogs	[305]
^1H	Thymidine: oxidation mechanism and conformation	[306]
^1H	Purine nucleosides: NOE	[160]
^1H	β-Pseudouridine: NOE, *syn-anti* equilibrium	[234]
^1H	β-D-Arabinonucleosides	[307]
^1H	Pyrimidine nucleosides: NOE, *syn-anti* equilibrium	[233]
^1H	Nucleosides, *syn, anti* conformers	[231, 233]
^1H	Purine, pyrimidine nucleosides, nucleotides: δ-^3J relation	[289]
^1H	Purine-β-nucleoside, sugar-χ correlation	[265]
^{13}C	Adenine cyclonucleosides: neighboring effect on δ	[143]
^1H	Furanose conformation in ribonucleosides	[308]
^1H	Adenine nucleosides: effect of C5′ substitution	[309]
^1H	Pyrimidine-2′-fluoro-2′-deoxyribonucleosides	[310]
^1H	Virucide: ribofuranosyltriazolecarboxamide	[311]
^{13}C	8-Substituted purine nucleosides	[228]
^1H	Anhydroisopropylidene uridine	[313]
^1H	Purine 8-cyclonucleosides	[246]
^1H	Cytidine and their O′-methyl derivatives	[314]
^1H	Uracil, 6-azauracil and corresponding nucleotides	[315]
^1H	2′-Deoxynucleosides	[316]
^1H, ^{13}C	Pyrimidine nucleosides: *syn-anti* equilibrium	[317]
^1H	Isopropylideneuridine, isopropylideneadenosin	[318]
^1H	Dihydrouridine	[319]
^1H	Guanosine derivatives: *syn* nucleosides	[320]
^1H	Purine nucleosides	[321]
^1H	Isoformycins, pyrazolo pyrimidines	[322]
^1H	6-Methyl-2-deoxyuridine: *syn* conformation	[249]
^1H	Isopropylideneadenosine	[323]
^1H	Aminodeoxynucleosides: theophylline	[324]
^1H	8-Methyladenosine	[325]
^1H	3′-Amino-3′-deoxyadenosine: puromycin	[326]
^1H	Cytidine, pseudouridine, pyrazomycin: T_1	[237]
^1H	C8 aminopurinenucleoside	[247]
^1H	Uridine: T_1	[238]
^1H	Formycin: 2-azaadenosine; nebularine	[327]
^1H	*C*-nucleosides: T_1, solvent effects	[239]
^1H	N(6)-dimethyl-2′,3′ isopropylideneadenosine: intermolecular hydrogen bonding	[328]
^1H	*S*-cyclonucleosides from nebularine	[252]
^1H	Thymidine nucleosides: substituent effects	[329]
^1H	2′,3′-O-Methyloxy-methyleneuridine	[330]
^1H	2′-O-Methyladenosine: protonation effect	[331]
^1H	Pyridine nucleosides	[332]

Table 3.5.1 (continued)

Nucleus	System and special information	Ref.
^1H	4-Thiouridine	[299]
^{13}C	Pyridine nucleosides containing sugar apiose	[333]
^1H	Purine and pyrimidine nucleosides: *syn-anti* equilibrium, T_1	[223]
^1H	α-pseudouridine	[317]
^1H	2′,3′-Isopropylidene-3′,5′-cycloguanosine	[224]
^1H	Common nucleosides, uridine, dihydrouridine	[296]
^1H	Oxy- and deoxynucleosides	[297]
^1H	1- and 3-Deazadenosine	[298]

Table 3.6.1. Conformational studies on small nucleotides

Nucleus	System and special information	Ref.
^1H	5′-Nucleotides: T_1 study	[241]
^1H	5′-Ribo- and deoxyribonucleotides	[264]
^1H	2′-and 3′-Ribonucleotides	[258]
^1H, ^{31}P	5′-β-Nucleotides, 8-aza-5′-GMP, 8-azadenosine	[266]
^1H	Nucleotides: *syn-anti* equilibrium, model systems	[335]
^{31}P	Mononucleotides	[280]
^1H, ^{31}P	C8-substituted adenine nucleotides: correlation χ–N/S	[267]
^1H	Pyrimidine ribonucleotides: substituent effects	[336]
^1H	*syn* Nucleotides, 5′-GMP derivatives	[320]
^{13}C	5′-Amino-5′-deoxythymidine derivatives	[337]
^1H	Uridine-5′-phosphate: lanthanide shifts	[338]
^1H	Thymidine nucleotides: substituent effects	[329]
^1H	6-Thiopurineribonucleoside-5′-phosphate	[339]
^1H	6-Azauridine-5′-monophosphate	[340]
^1H	Guanosine-2′-phosphate	[341]
^1H	Nucleotidyl units: cooperativity	[342]
^1H	Pyridone nucleotide	[343]
^{13}C	8-Substituted purine nucleotides: Phosphate effect	[344]
^1H	5′-ribo and deoxyribonucleotides: lanthanide shifts	[345]
^1H	benzoadenine nucleotide, T_1	[346]
^1H	adenosine-5′-phosphate: conformational flexibility	[347]
^1H	adenosine 2′,5′- and 3′,5′-diphosphates	[348]
^{13}C	Cyclonucleosides and nucleotides: neighboring anisotropic effect on δ	[143]
^1H	Nucleoside diphosphohexose	[302]
^1H	8-(6Aminohexyl)aminoadenosine-5′-phosphate	[349]
^{31}P	Mononucleotides, dinucleotides	[350]
^1H	Diruridyl phenyl phosphates	[351]
^1H	2′,5′-Cyclic nucleotides	[352]
^1H	2′,3′-Cyclic nucleotides	[353]
^1H	Cyclic nucleotides	[94, 275, 354–357]
^1H	Cyclic 2′-deoxynucleoside-3′,5′-phosphate: lanthanide shift	[271]
^{13}C	Cyclic nucleotides	[94, 276, 277, 334]

Table 3.6.1 (continued)

Nucleus	System and special information	Ref.
^1H	Cyclic nucleotides, lanthanide shifts	[358]
^1H	3′,5′-Cyclic nucleotides: T_1	[240]
^1H	Cyclic nucleotides: water bridge with the 2′-OH-group	[283]
^1H	Cyclic nucleotides	[359]
^1H	Nicotinamide xyloside: Oxidized and reduced forms	[360]
^1H	NADH, NADPH: intramolecular association	[361]
^{13}C	NAD: *p*H-dependence of ^{13}C-NMR	[362–364]
^1H	NAD: adenine-nicotinamide interaction	[365]
^1H	NAD, NMN: lanthanide shift, conformational dynamics	[366, 367]
^1H	NADH: $Co^{2\oplus}$ complex	[368, 369]
^1H, ^{31}P	Pyridine nucleotides	[370–376]
^1H	NMN, UpU: lanthanide shift reagents	[377]
^1H	Pyridine coenzymes	[378, 379]
^1H	Pyridine nucleotides: NOE, *syn-anti* equilibrium	[235]
^1H	Chiral models of NAD	[380]
^1H	Pyridine nucleotides, intramolecular association	[332]
^1H	NAD: solution dynamics	[381]
^1H	NAD: cycloadduct	[382]
^1H	Guanosine mannose diphosphate	[383]
^1H, ^{13}C	ATP: Palladiumglycyl.histidine complex	[384]
^1H	ADP: Interaction with lanthanides, Mn, Cu	[385]
^1H	Nucleoside di- and triphosphates: T_1	[236]
^1H	Adenosine, ionosine 5′-triphosphates: metal complexes	[386]
^1H	ATP: catechol complex	[387]
^1H	2′(*R*), 3′(*S*), 5′-trihydroxypentyl derivatives of adenine and cytosine, non-glycosidic analogs	[388]
^1H, ^{13}C	Nucleic acid analogs	[281]

3.6.1 2′-, 3′- and 5′-Mononucleotides

These nucleotides do not markedly deviate from the general principles of nucleic acid conformations discussed above. They usually show almost equal preference for N and S furunose states and a general preference for *anti*, *gg*, *g′g′* conformations but in no case is a "conformational purity" reached.

3.6.2 Cyclic Nucleotides

A certain degree of conformational rigidity is imparted to the mononucleotide structures when the phosphate group is involved in cyclization. Thus, a series of compounds with different patterns of cyclization i.e. 2′,3′, 3′,5′, 2′,5′ can be achieved. The cyclic nucleotides are thus bicyclic compounds. Biologically, the most important compound in this series is 3′,5′-cyclic-AMP(*c*-AMP) which is involved in certain neural processes. The cyclic nucleotides have been extensively studied using ^1H chemical shifts and coupling-constants (Table 3.6.1), ^{13}C chemical shifts, lanthanide shifts, and ^1H relaxation rates. Both, the structure of the ribose ring and the glycosidic bond torsion angles show more pronounced conformational preferences than the corresponding uncyclized mononucleotides. For example, *c*-AMP is locked in a conformation in the N range (lying between the ^{3}E and $_{4}T^{3}$ conformations). The six-membered cyclic phosphate ring in this case adopts a somewhat flattened chair conformation.

In recent years, nucleosides and nucleotides where cyclization between the 2′ position of the ribose and the C8 position of the purine ring is achieved through a bridging oxygen or sulfur atom, have been synthesized. These compounds have a fixed conformation around the glycosidic bond. Cyclonucleosides and nucleotides serve as good models for the elucidation of the effect of angle χ on the spectroscopic properties of nucleotides and polynucleotides [495].

3.6.3 Pyridine Nucleosides and Nucleotides

Nicotinamide-adenine-dinucleotide (NAD) acts as a natural coenzyme for pyridine-linked dehydrogenases. Therefore, NAD, its reduced form NADH, nicotinamide mononucleotide (NMN), reduced NMN (NMNH), nicotinamide ribonucleoside (NR), reduced NR (NRN), and other analogs have been extensively studied by NMR. A variety of parameters have been employed to study the microdynamic behavior of pyridine nucleotides, i.e. ^{1}H chemical shifts and coupling constants (Table 3.6.1), ^{13}C chemical shifts, intramolecular NOE, lanthanide shifts and ^{31}P-NMR. From these studies the following general conclusions can be drawn:

(a) The solution conformation of pyridine nucleotides has a high degree of conformational flexibility.

(b) The nicotinamide moiety exist as an equilibrium mixture of *syn* and *anti* conformers with a slight preference for the *anti* (60:40) conformer. The adenine moiety on the other hand most likely adopts an *anti* conformation with a small contribution from the *syn* state.

(c) At concentrations of < 5 mM, the intermolecular association is negligible.

(d) The sugar and backbone angles show the familiar pattern of conformational preferences discussed in previous cases.

(e) The dinucleotides show an equilibrium between several states with "stacked" and "unstacked" bases.

3.6.4 5′-Di- and Triphosphates

Adenosine di- and triphosphates (ADP and ATP) play a major role in the processes of energy storage, utilization and transfer. ADP and ATP have two and three (α, β and γ) phosphate groups, respectively, linearly attached to the 5′ end of the adenosine moiety. NMR studies have been carried out using ^{1}H chemical shifts and coupling constants (Table 3.6.1), ^{13}C and ^{31}P chemical shifts, relaxation rate measurements, and lanthanide shifts. The metal ion complexes of nucleoside triphosphates have also been studied. The furanose ring in these molecules has a slight preference for the S state (60:40), the bases show a *syn* $\rightleftharpoons$ *anti* equilibrium while the phosphate groups assume a somewhat extended conformation. The metal ions bind to the phosphate chains and in some cases can also chelate to the N1 and N7 positions of the base.

3.7 Dinucleoside Phosphates and Short Segments of Nucleic Acids

We have pointed out earlier, that the O–P bonds are relatively flexible and can adopt a variety of conformations. Figure 3.7.1 shows the isoenergy contours for the phosphate O–P bond rotations in some nucleic acids fragments [394]. It is clear from these maps that the conformational freedom of nucleic acid substructures becomes increasingly restricted as the substructural length increases. For this reason, a clear understanding of the principles governing the folding of nucleic acids can only be reached if larger fragments are studied. In recent years, a great deal of efforts has been devoted to NMR studies of such fragments (Table 3.7.1).

Table 3.7.1. Dinucleoside monophosphates and short segments of nucleic acids

Nucleus	System and special information	Ref.
^{1}H	Dipurine, dipyrimidine	[395]
^{1}H	Dinucleoside monophosphates	[396, 397]
^{1}H, ^{31}P	Dinucleoside monophosphate: chain flexibility	[294]
^{1}H	Dideoxynucleoside ethylphosphotriesters	[398]
^{1}H	6, 6-*N*-dimethyladenylyl-3′,5′-uridine: stacking	[399]
^{1}H	Deoxyribotrinucleoside diphosphates	[400]
^{1}H	Dinucleoside monophosphate: left-handed helix	[401]
^{1}H	Adenine thioxylonucleoside monophosphate	[402]
^{1}H	Dinucleoside monophosphate: lanthanide shifts	[403]
^{1}H	Adenine dideoxynucleoside monophosphate	[260]
^{1}H	Dinucleoside diphosphates: methylation effect	[404]
^{1}H	3′,5′- and 2′,5′-Uridyl-2′-deoxythymidine	[405]
^{1}H	Conformation relation: stability of phosphodiester bond	[406]
^{1}H	ApU: substituent effects	[407]
^{1}H	Modified Dinucleoside diphosphate, 1,N6 ethanoadenosine and 2′-O-methylcytidine	[408]
^{1}H	Deoxyribodinucleotide	[285]
^{1}H	CpC: 2′-O-methylation	[409]
^{1}H	GpU, UpG: T_1	[410]
^{1}H	Dinucleoside monophosphates: ionization	[411]
^{1}H	Dinucleoside monophosphates containing 8-methyladenine	[412]
^{1}H	Dinucleoside monophosphates: influence of furanose ring on conformation	[413]
^{1}H	ApApA: assignment and structure	[414]
^{1}H	d(TpTp)	[415]
^{1}H	d(T-T-A)	[416]
^{1}H	Oligoinosinates: chain-length dependence	[417]
^{1}H	Oxy- and deoxyadenyl dinucleosides: backbone flexibility	[418]
^{31}P	Dinucleotides	[280]
^{1}H	Dimer segments: stereodynamics	[419]
^{1}H	2′,5′ -and 3′,5′-Cytidalcytidine	[421]
^{1}H	Conformational equilibria on 2′,5′- and 3′,5′-nucleotides and their influence on DNA stability	[390–393]
^{1}H	Oligonucleotides: solution conformations	[422]
^{1}H	Vertically stabilized double helices	[423]
^{1}H	3′-Acceptor end of t-TNA: stereodynamics	[424]

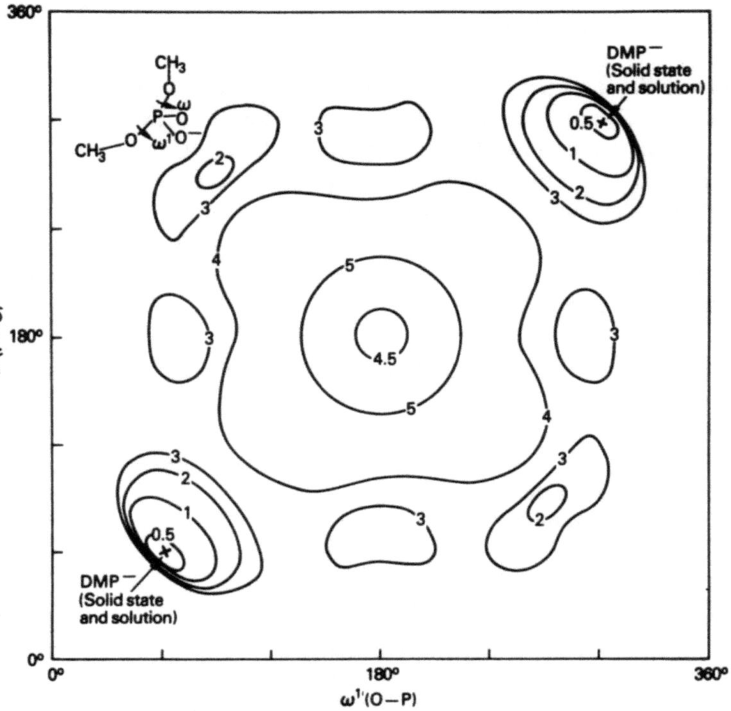

360°
CH3
O
ω
P
O
O
ω1
O−
CH3
DMP−
(Solid state
and solution)
0.5
1
2
3
4
5
4.5
ω (P—O)
180°
DMP−
(Solid state
and solution)
0°
0°
180°
360°
ω1'(O—P)

a

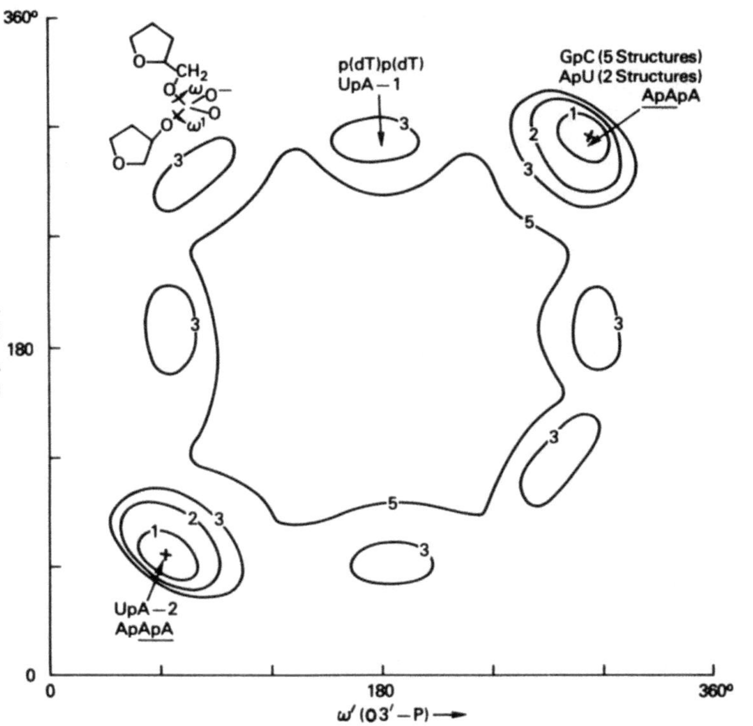

360°
O
CH2
O
ω
O−
O
O
ω1
O
p(dT)p(dT)
UpA − 1
GpC (5 Structures)
ApU (2 Structures)
ApApA
1
2
3
5
ω (P—O5') ⟶
180
UpA − 2
ApApA
0
0
180
360°
ω' (O3' − P) ⟶

b

c

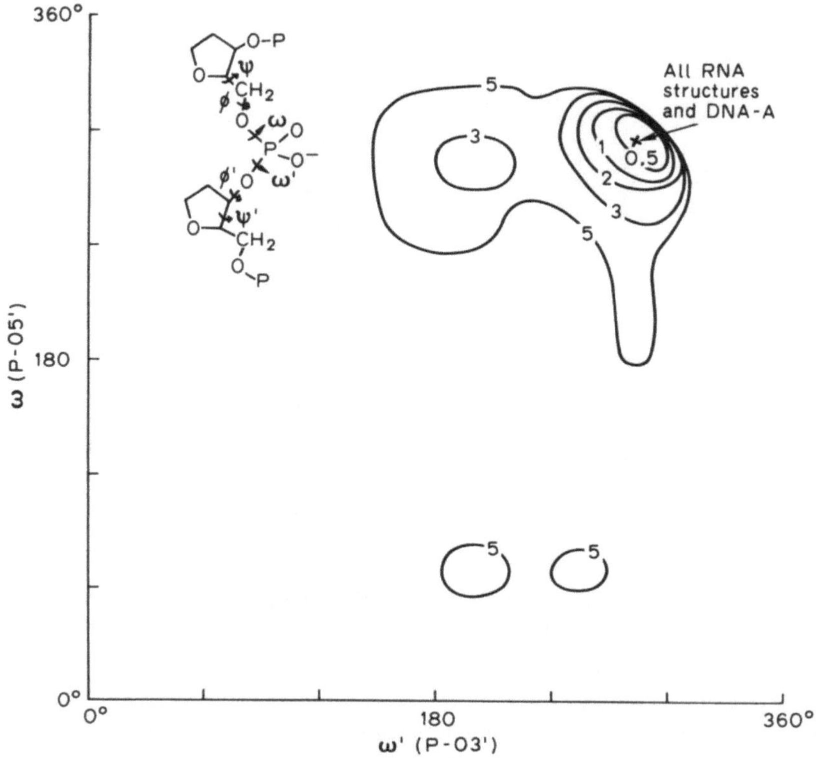

Fig. 3.7.1 (a–c). Potential energy calculations by the CPF method of (a) dimethyl phosphate, (b) diribose monophosphate and (c) diribose triphosphate as a function of torsion angles around the O–P bonds (reprinted with the permission from Ref. [394]). Energy is in kcal/mole

One of the problems which arises in such studies is the resolution and assignment of the large number of spectral lines. In case of ApApA [414], the assignments were achieved by selective deuteration using enzymatic methods. The H2 chemical shifts in the trinucleotides suggest that Ap- is more closely stacked with -pAp- than is -pA, in agreement with the crystal structure of this compound [420]. The coupling constants of H1′ indicate that the ribose conformation favors an N conformer in all the three nucleotyl units.

An analogous example of deoxynucleotide is d(TpTpA) [416]. The complexity of the proton spectrum for such systems can be gauged from Fig. 3.7.2. In this molecule, the ribose conformation is shifted predominantly to the direction of the S state [416]. The conformation around the bond C5′–O5′ (angle ϕ) is almost 80% in favor of the $t(g'g')$ state while that around the bond C4′–C5′ (angle ψ) is about 70% in favor of the $g^+(gg)$ state.

Extensive experimental data on dinucleoside phosphates have become available in recent years (Table 3.7.1). The conformational purity for the preferred states of ribo- and deoxyribonucleotides is somewhat lower in these cases compared to trinucleotides

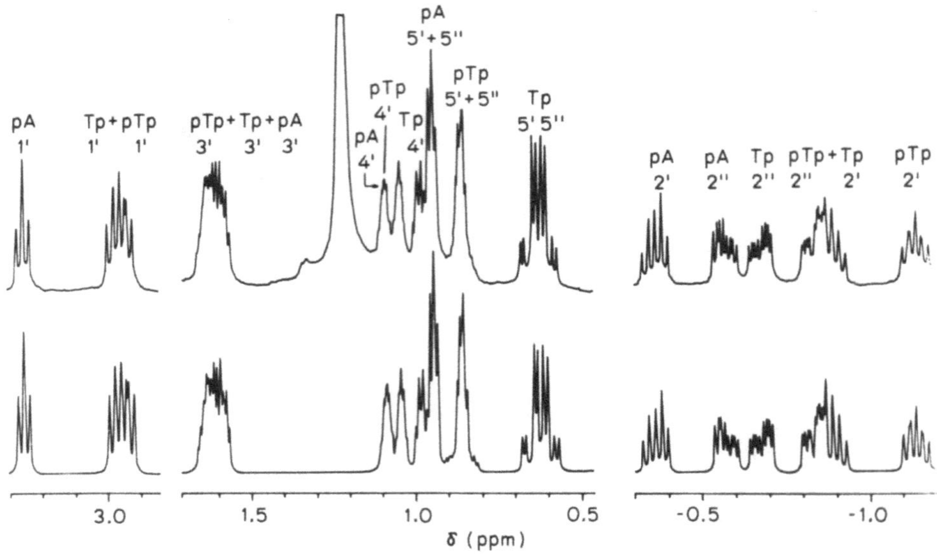

Fig. 3.7.2. 360 MHz H^1-NMR spectrum of d(TpTp) and assignment of the spectrum. Lower trace is computer-simulated spectrum (reprinted with the permission from Ref. [416])

but the evidence of marked preferences for the various states is obvious. Further, the difference in the solution conformations of DNA and RNA fragments is evident even at this level of chain length.

3.8 Random-Coil Polynucleotides

Most polynucleotides and naturally occurring nucleic acids exist as random coils at higher temperatures. The structures of such coils have been of interest since the investigations have thrown light on the importance of backbone interactions concerning the stabilization of nucleotide chains since the base-base hydrogen bonding is absent in such systems. These results also allow an assessment of the extensive theoretical calculations and views to be made on the structure of nucleotide backbones which has been a matter of some controversy. In this connection, two opposing views exist. Sundaralingam [213, 268] has advanced the concept of a "rigid" nucleotide backbone with the torsion angles: $\phi' \sim 220°$, $\phi \sim 180°$, $\psi \sim 60°$, $\omega' \sim 290°$, $\omega \sim 290°$, χ in the *anti* range and a preference for C3' *endo* or C2' *endo* sugar pucker. While Sundaralingam himself pointed out some of the alternative conformations which are possible in polynucleotides [268], the concept has often been taken to imply a 100% rigidity. On the other extreme is the view that the intrinsic preference for the so called "rigid" nucleotide backbone angles is very low (< 5%), and that the random coils in nucleotides acquire a conformation very different from that found in ordered structures stabilized by base-base hydrogen bonding [425, 426]. Finally, studies of this type have enabled to delineate the conformations, which may occur during biological processes involving opening of ordered structures, and to follow conformational changes during ordering and nucleation of DNA and RNA [430–443].

3.8.1 Oligo- and Polyribonucleotides

Figures 3.8.1 and 3.8.2 show some typical NMR spectra of high molecular weight polyribonucleotides. It is immediately clear that the spectra are very sharp (line widths 2–3 Hz) thus enabling a complete analysis of the chemical shifts, coupling constants and relaxation times [97, 229, 279–281, 427, 429–433]. Table 3.8.1 compiles some typical data on model polynucleotides [229].

The analysis of such data in terms of population analytical methods outlined in previous sections has provided very interesting information. The first point to be observed is that while poly(A), poly(U) and other polynucleotides are expected to predominantly acquire an *anti* conformation for the base, poly(8-BrA) has an *all-syn* conformation due to the bulky substitution at the 8-position. This is fully confirmed by NMR chemical shift and coupling constant data in Table 3.8.1. The ribose conformation in all cases consists of an equilibrium between the N and S type conformer with some preference for N states. It was believed that while *anti* conformations are associ-

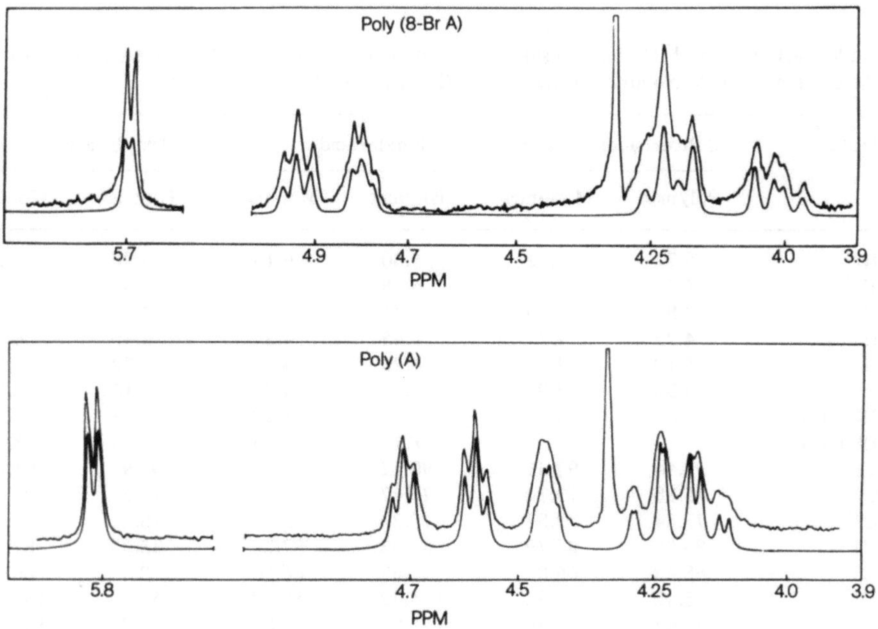

Fig. 3.8.1. ^{31}P decoupled proton NMR of poly(8BrA) and poly(A) in the ribose region at 71 °C. The bottom traces are simulated spectra with parameters of Tables 3.8.1 and 3.8.2 (reprinted with permission from Ref. [229])

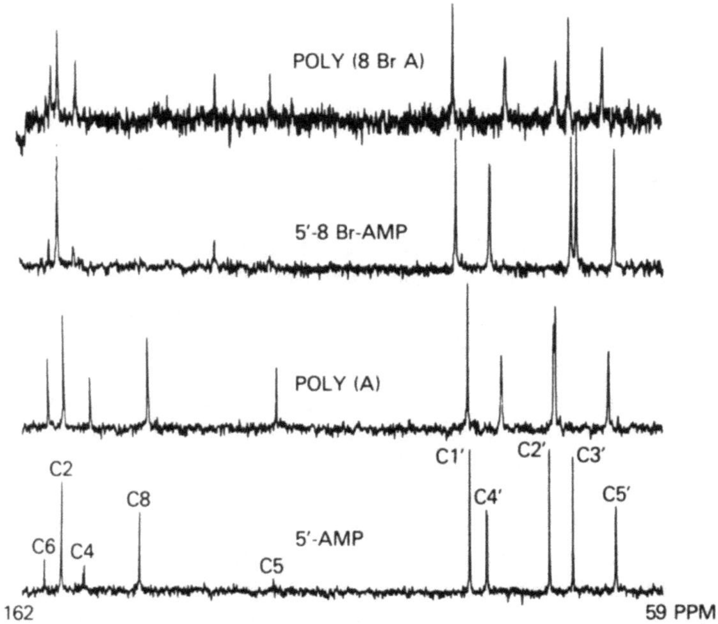

Fig. 3.8.2. ^{13}C-NMR of randomly coiled polyribonucleotides and the corresponding 5' mononucleotides (reprinted with the permission from Ref. [229])

Table 3.8.1. ^{1}H- and ^{13}C chemical shifts of randomly coiled poly (8BrA), poly(A), poly(U) and the corresponding 5'-mononucleotides at 70 °C (relative to TSP)

Proton	8-Bromoadenylic acid		Adenylic acid		Uridylic acid	
	Polymer	Monomer	Polymer	Monomer	Polymer	Monomer
H1'	5.72	6.12	5.80	6.13	5.96	5.98
H2'	4.79	5.30	4.58	4.78	4.43	4.41
H3'	4.91	4.60	4.71	4.52	4.67	4.34
H4'	4.23	4.27	4.45	4.37	4.46	4.25
H5'	4.19	4.09	4.26	4.04	4.22	4.02
H5"	4.02	4.02	4.15	4.04	4.15	3.97
H2 (H5)	7.85	8.25	7.94	8.28	5.93	5.98
H8 (H6)	-	-	8.13	8.56	7.85	8.04
C1'	92.44	92.31	90.42	90.13	91.9	91.6
C2'	75.88	73.74	76.47	77.20	77.2	77.0
C3'	73.84	72.88	76.20	73.44	76.2	73.3
C4'	83.98	86.79	84.93	87.25	85.6	87.1
C5'	68.42	66.86	67.68	66.50	68.5	66.4
C2	155.42	155.92	155.29	155.79	155.2	155.3
C4	152.44	153.38	151.09	152.01	169.0	169.6
C5	121.56	122.20	121.18	121.61	106.2	105.7
C6	156.48	157.31	157.80	158.49	144.8	145.2
C8	130.32	130.95	141.89	143.14	–	–

Table 3.8.2. Conformation-dependent NMR parameters of poly(8BrA), poly(A), and poly(U) at 70 °C and corresponding data on 5′-mononucleotides (from 229)

Parameter	8-Bromoadenylic acid		Adenylic acid		Uridylic acid	
	Polymer	Monomer	Polymer	Monomer	Polymer	Monomer
Glyocosidic bond (angle χ)						
J[H(1′)–C(4)]	5.2	4.5	4 ± 1	4 ± 1	–	–
J[H(1′)–C(8)]	5.4	5.4	3 ± 1	3 ± 1	–	–
Δ[H(2′)][a]	0.21	0.52				
Δ[C(2′)][a]	−0.59	−3.46				
Δ[H(3′)][a]	0.20	0.08				
Δ[C(3′)][a]	−2.36	−0.56				
Ribose conformation						
J[H(1′)–H(2′)]	3.5	6.1	4.5	5.9	4.5	5.1
J[H(3′)–H(4′)]	6.3	4.8	4.4	3.6	4.3	4.1
J[H(2′)–H(3′)]	5.3	6.3	4.9	5.0	5.2	4.8
J[H(1′)–H(2′)] + J[H(3′)–H(4′)]	9.8	10.9	8.9	9.5	9.1	9.2
Angle φ′						
J[H(3′)–P]	8.7	–	8.7	–	7.8	–
J[C(2′)–P]	5 ± 1	–	5 ± 1		5	–
J[C(4′)–P]	6 ± 2	–	4 ± 1	–	3	–
Angle φ						
J[H(5′)–P]	4.5	5.6	4.0	4.3	3.9	3.8
J[H(5″)–P]	6.0	6.0	4.0	4.3	3.4	5.2
J[C(4′)–P]	12 ± 2	7.6	8 ± 1	8.5	7	8.5
Angle ψ						
J[H(4′)–H(5′)]	1.7	4.8	2.0	3.2	2.4	2.3
J[H(4′)–H(5″)]	7.4	6.3	4.2	3.2	4.0	2.8
J[H(5′)–H(5″)]	−11.8	−11.8	−11.7	–	−11.8	−11.8
Angles (ω′, ω)[b]						
Δ_1[C(2)]	0.50		0.50		0.1	
Δ_1[C(4)]	0.94		0.92		0.6	
Δ_1[C(5)]	0.64		0.43		−0.7	
Δ_1[C(6)]	0.83		0.69		0.4	
Δ_1[C(8)]	0.63		1.25		–	
Δ_1[H(2)]	0.40		0.34		–	
Δ_1[H(8)]	–	0.43	0.43		–	
Δ_2[H(2)]	0.80		0.61		–	
Δ_2[H(8)]	–		0.89		–	

[a] $\Delta = \delta$(8-bromoadenine derivative)$-\delta$(adenine)

[b] $\Delta_1 = \delta_{monomer} - \delta_{polymer}$ (70 °C); $\Delta_2 = \delta_{monomer} - \delta_{polymer}$ (20 °C)

ated with both S and N type conformers, *syn* conformations are associated only with ^{2}E geometry. The results listed in Table 3.8.2 evidence against such a hypothesis. In fact, poly(8-BrA) has a larger population of N conformer [i.e. J(H1′–H2′) = 3.5 Hz] than poly(A) and poly(U) [J(H1′–H2′) = 4.5 Hz]. Finally, the contribution of the overall rigid nucleotide fragments is almost 50% irrespective of the base orientation, as estimated from ring-current effects. The preferences for individual bonds of the "rigid" nucleotide backbone angles is much larger. Clearly, the structures found in ordered DNA and RNA (Sect. 3.9) are not very different than those predominantly present in random coils and are dictated to a large extent by backbone non-bonded interactions. Finally, the alternative conformations of polynucleotides may arise primarily from possible rotations around O–P bonds, different ribose conformations and, to a somewhat lesser extent, due to the orientation of the C4′–C5′ bond.

3.8.2 Deoxyribonucleotides

Extensive investigations on randomly coiled deoxyribonucleotides and helix-coil transitions in these systems have been made [434, 441, 443]. These systems differ from RNA coils in the conformation of the furanose ring. The more populated state in this case is the S family of conformers. There may also be a difference in the ϕ' angle, since the fiber diffraction indicates that the ^2E conformers are associated with a ϕ' value of 130–155° [442]. However, the NMR evidence on this point is not quite clear. The bases are considerably stacked even in random coils.

To summarize these results, both RNA and DNA fragments in solutions at room temperature, and randomly coiled oligo- and polynucleotides at high temperature exist as unassociated molecules, at least when present in low concentrations. Both deoxyribo- and ribo-nucleotides show a preference for a *t* conformation for torsion angle ϕ and for a $g^+(gg)$ conformation for angle ψ to an extent of 70% or more. The bases generally assume an *anti* conformation and exhibit a varying degree of right-handed base stacking. The preferred sugar conformation is S for deoxyribonucleotides and N for ribonucleotides, the preferred conformer having a population > 60%. The values of NMR parameters, which depend on angle ϕ', are different for the two series and very likely reflect various conformations of DNA and RNA analogs observed for C3′–O3′ bond by other physical and theoretical methods.

The evidences presented in this and previous sections reveal that the nucleotide unit is not totally rigid but displays a strong preference for "A" family structures in RNA and "B" family structures in DNA segments. The conformational freedom becomes increasingly restricted with chain length. The NMR results thus support the concepts developed from X-ray crystal structure analysis [213, 214] and potential energy calculations [11, 279, 290].

3.9 Helical Polynucleotides

Around room temperatures, many polynucleotides and almost all naturally occurring nucleic acids form ordered helical structures. While most natural nucleic acids acquire a helix formed from two strands (with the exception of t-RNA discussed in Sect. 3.10), synthetic polynucleotides have been reported to order with 1, 2, 3, 4, and 6 strands. In all cases, base-base hydrogen bonds and "stacking interactions" act as additional stabilization forces to prevent any randomness in structures. The structures of both natural and

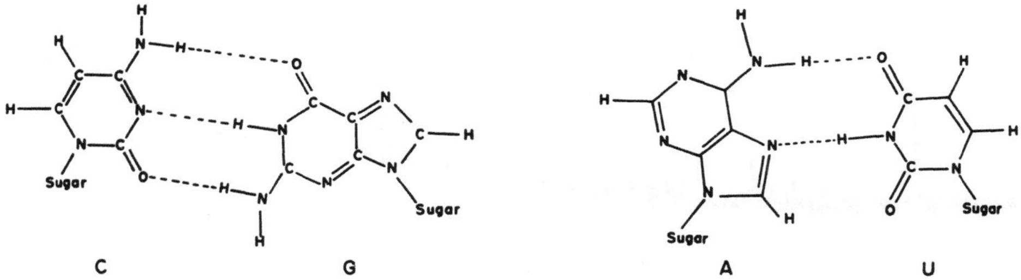

Fig. 3.9.1. Hydrogen-bonding schemes for standard Watson-Crick base pairs

synthetic polynucleotides have been extensively studied by fiber-diffraction techniques [214, 442]. The torsion angles of the helical structures fall into two categories. The "A" type structures have a ^3E geometry and ϕ angle around 220°. Such structures have been observed for RNA, DNA-RNA hybrids and DNA systems. The B type structures have been observed only in deoxyribonucleotides; they display ^2E geometry and possess an angle ϕ' of about 150°. The values of other torsion angles lie in the preferred ranges discussed above, namely t for ϕ, g^+ for ψ and g^-g^- for ω, ω'. The bases assume an *anti* conformation. Thus, the basic difference in the A and B type helices is only in the values of two adjacent angles in the sugar phosphate backbone ϕ' and ψ' which differ by almost 70° but in opposite sense in the two classes of structures. In DNA, the transition from the A to the B form can be brought about by minor perturbations such as changes in concentration of counter ions, pH, humidity etc.

The hydrogen-bonding arrangement in helical nucleic acids is base-specific. G binds to C and A bind to U or T by a scheme proposed by Watson and Crick in their original DNA model and shown in Fig. 3.9.1. Theoretically, other hydrogen-bonding schemes are possible, and some of these have been observed in synthetic polymers and t-RNA.

The conformation of ordered polynucleotide helices and mini-helices have been studied by several groups [436–440, 444–455]. Several earlier reviews on the subject are available [452–455]. Of particular interest is the work on short segments of DNA [437–439, 443] such as d-ApTpGpCpApT, RNA [431, 444, 446] and ApApGpCpUpU which form mini-helices. These molecules provide useful models for the investigation of hydrogen bonding, base stacking and helix-coil transitions in nucleic acids and polynucleotides.

The NMR spectrum in Fig. 3.9.2 reflects a typical behavior of poly(rU) as one goes from a random-coil to an ordered structure. Two facts are immediately obvious from such investigations. The base protons (with the exception of those involved in hydrogen bonding) and some of the furanose nuclear resonances move upfield because of the greater stacking in ordered structures. All the resonance lines broaden due to aggregation and loss of segmental motions. Because of such broadening, the proton and ^{13}C spectral lines from these regions merge into a relatively broad envelope and do not provide useful information. However, the base protons can still be monitored and used in favorable cases to determine the ring stacking patterns.

It is interesting that the resonances which provide the bulk of information for ordered systems are those which are of relatively little use in the study of nucleic acid

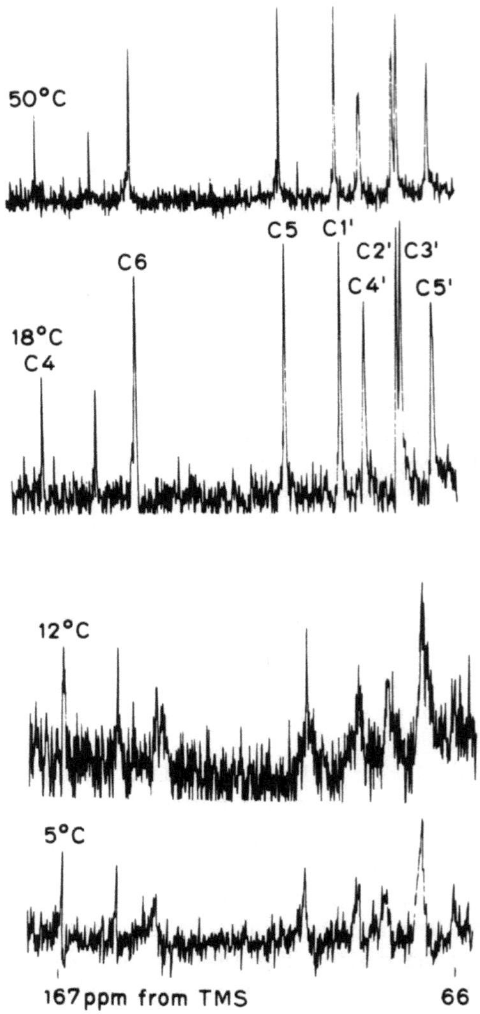

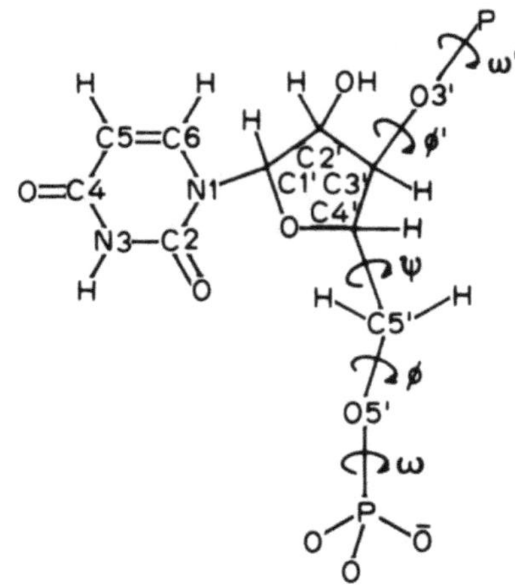

Fig. 3.9.2. ^{13}C-NMR of Poly(rU) at various temperatures (reprinted with the permission from Ref. [97])

fragments and random coils. These are the imino protons and, to a lesser extent, the amino protons in the bases. In the systems discussed so far, these protons exchange with water at rates such that the line widths are of the order of more than hundred Hz. As a result of hydrogen bonding, the resonance lines move downfield (to about 8 ppm for amino and 11–15 ppm for imino protons). Further, the exchange with solvent is considerably slower which results in fairly sharp proton lines [456]. It may be noticed that each WC base pair contributes to one imino proton resonance. By integrating the 11–15 ppm region, the number of hydrogen-bonded base pairs can be immediately established.

The line widths of the exchangeable base protons is a measure of the life-time of the hydrogen-bonded state [456, 457]. For example, in poly(dA-dT) duplexes, the thymidine H3 proton exhibits a line width of about 90 Hz between 30 and 60 °C. Just below the melting temperature of the helix, the exchange of the base proton with solvent causes a gradual broadening of the proton line to an extent that it cannot be observed in the NMR spectrum [457].

Attempts have been made to estimate the chemical shift contributions to imino protons arising from hydrogen bonding and base stacking. It may be noted that though these two are the major factors, contributions from other mechanisms are insignificant.

Table 3.9.1. Shielding of the nucleic acid base-pairs in A-RNA and B-DNA double helices (from [291])

of	A-RNA				B-DNA			
by	AU	UA	GC	CG	AU	UA	GC	CG
First Order					First Order			
R–H1′	0.01	0.05	0	0.01	0	−0.01	0.01	−0.01
*R–H1′	−0.01	0.03	0.01	0.03	0.01	0.16	0.08	0.17
Pu–H8	0.62	0.04	0.36	0.12	0.15	−0.01	0.09	0.02
A–H2	0.06	0.65	0.11	0.35	0.04	0.78	0.04	0.31
G–NH	0.09	0.73	0.17	0.42	0.21	0.89	0.14	0.38
*Py–H6	−0.03	−0.06	−0.02	−0.03	−0.02	−0.01	0	0
*Py–H5	−0.03	−0.06	−0.03	−0.04	−0.02	−0.01	−0.01	0
*U–NH	0.06	0.54	0.19	0.36	0.11	0.83	0.15	0.40
Py–H6	0.46	0.03	0.27	0.08	0.08	−0.02	0.06	0
Py–H5	1.47	0.14	1.02	0.42	0.64	0.06	0.49	0.20
U–NH	0.30	0.61	0.16	0.27	0.78	0.32	0.32	0.15
*Pu–H8	−0.03	−0.06	−0.02	−0.03	−0.02	−0.01	0	0
*A–H2	0.06	0.97	0.21	0.59	0.04	1.00	0.10	0.46
*G–NH	0.14	0.87	0.15	0.41	0.41	0.68	0.18	0.27
Second Order					Second Order			
R–H1′	0.04	0.11	0.03	0.06	0.02	0.06	0.02	0.03
*R–H1′	0	0	0	0	0.01	0.05	0.02	0.03
Pu–H8	0.12	0.10	0.07	0.06	0.05	0.03	0.03	0.02
A–H2	0.03	0.09	0.03	0.06	0.06	0.23	0.07	0.13
G–NH	0.04	0.08	0.03	0.05	0.10	0.18	0.07	0.10
*Py–H6	0	−0.01	0	−0.01	0.01	0.02	0.02	0.01
*Py–H5	0	−0.01	0	−0.01	0.03	0.02	0.03	0.02
*U–NH	0.03	0.05	0.03	0.03	0.09	0.15	0.07	0.09
Py-H6	0.10	0.10	0.06	0.05	0.04	0.03	0.02	0.02
Py-H5	0.22	0.09	0.13	0.08	0.12	0.04	0.06	0.03
U–NH	0.07	0.13	0.06	0.08	0.13	0.18	0.08	0.09
*Pu–H8	0	−0.01	0	−0.01	0.01	0.02	0.02	0.01
*A–H2	0.02	0.08	0.03	0.06	0.06	0.20	0.07	0.12
*G–NH	0.05	0.11	0.05	0.07	0.12	0.19	0.08	0.11

Note: The convention used in the table is to consider that the strand going in the 5′ → 3′ direction is the principle strand (unstarred); protons of the complementary strand are marked by a star. The base-pairs causing the shielding are on the 5′ side of the principle strand. First order shifts are due to a base pair directly above the base under consideration, while the second order are due to the next nearest base pairs.

Even in the estimation of these two factors, there are considerable difficulties due to the absence of suitable models. Katz and Penman [458] observed large downfield shifts of imino protons when A and U or G and C were mixed in organic solvents. Since, however, the equilibrium in these systems toward the hydrogen-bonding side is far from complete, chemical shifts for hydrogen bonds cannot be estimated from such data. Results on a series of polynucleotides have been used to arrive at the following set of values for the intrinsic chemical shifts of hydrogen-bonded imino protons

$(AU)^\circ = 14.6 \pm 0.1$ ppm
$(GC)^\circ = 13.6 \pm 0.1$ ppm

The second contribution to the chemical shift of imino protons comes from the stacking interactions. This term depends on the relative geometry of the purine and pyrimidine bases. The more important contributions stem from the base pairs above and below the proton in questions. The shielding effects determined from the DNA-B structure and the 1970 isoshielding contours of Giessner-Prettre and Pullman are given in Table 3.9.1. The values may be higher by about 20% if the calculations are based on the latest maps.

If it is assumed that the two effects discussed above are sufficient to predict the imino proton chemical shifts in ordered polynucleotides, then the observed data can be used to check structures of DNA and RNA helices in solutions. Several attempts in this direction have been made (see for example [459]).

Recently, it has been suggested that the backbone conformation of DNA duplexes may depend on the sequence of bases [460, 461]. This may have important consequences in the recognition of DNA by proteins. If the model is correct, then the DNA sequence may be recognized through protein-nucleic acid interactions at the backbone level. Recent ^{1}H- and ^{31}P-NMR studies support that the backbone conformation of poly(dA-dT) depends on the sequence, i.e. the segments -dA-dT and -dT-dA- have different conformations [462, 463]. Likewise, at high salt concentrations, the ^{1}H- and ^{31}P-NMR of oligo(dG-dC) duplexes show that every other glycosidic torsion angle and phosphodiester linkage adopts a conformation different from that observed in B-DNA [464].

In recent years, studies have been extended to the secondary structures of ribosomal RNAs [465, 466]. Even at the substructural level [462], these are fairly complex systems and the fact that such systems can be fruitfully explored at present shows that NMR has come a long way since the first applications to nucleotide structures were published about 15 years ago.

3.10 t-RNA

Transfer-RNA (t-RNA) is a relatively small polymer containing between 76 and 90 nucleotides. It is involved in the recognition of amino acids during protein synthesis (translation) [467] and in regulation of other biochemical pathways [468, 469]. The primary structure of a t-RNA was first established by Holly et al. in 1965 [470]. In an attempt to form as many complementary base pairs as possible by folding of the single nucleotide chain, Holly arrived at a model for the structure of t-RNA which is commonly known as the clover-leaf model (Fig. 3.10.1). This model contains 20 base pairs

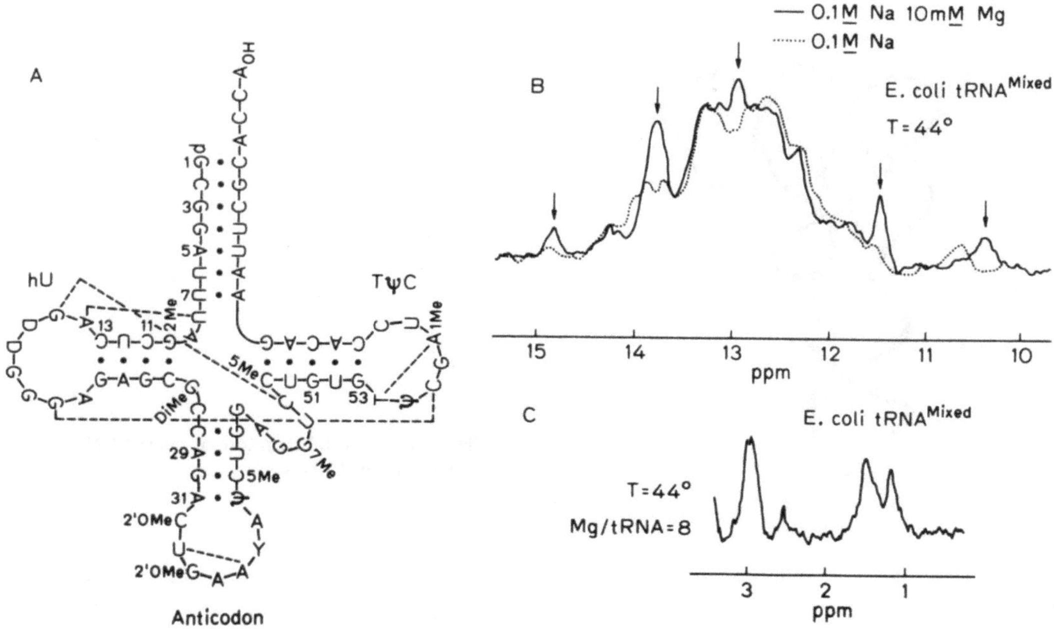

Fig. 3.10.1. Clover-leaf model of t-RNA with some of the tertiary interactions postulated by Levitt (reprinted with permission from Ref. [452]. The spectra on the right show the NMR spectra of unfractionated E. Coli t-RNA in the 10–15 and 1–3 ppm region and the sensitivity of tertiary interactions on Mg(II)

which are involved in the so-called secondary interactions of the molecule. The structure has four helical segments, three loops (the DhU, TψC, and the anticodon), and an extra arm. Levitt [471] showed that in addition to the hydrogen bonds involving secondary structures proposed by Holly, some of the bases in the loops form hydrogen bonds through so-called "tertiary" interactions. Bivalent metal ions participate in the stabilization of such interactions.

In recent years, four crystal structures of t-RNA [207–210] have been reported and though the chain folding in all the cases is very similar, the structures differ in their microscopic details. At the present stage of refinement, it is difficult to say whether these differences are real or artefacts due to inaccuracies in the structural analysis. The structures show the essential features of the models of Holly and Levitt although the overall structures are L-shaped (Fig. 3.10.2). Most of the ribose units in the crystallographic structures exhibit a C3′ *endo* conformation and the backbone torsion angles for the majority of the units correspond to the values preferred by ribonucleotides. The alternative conformation occur mostly in the loops of t-RNA.

Extensive NMR studies on the solution conformations of t-RNA [472–479] and its fragments have been carried out recently [444]. (For reviews on this subject, see [452, 453, 455, 480–483]). The resonances in the 10–15 δ region of ^{1}H-NMR which arise from hydrogen-bonded imino protons have been the focus of attention. Some information has also been obtained from the C–H base protons and amino protons which lie in the 7–9 δ region and methyl and methylene protons in the 1–3 ppm region.

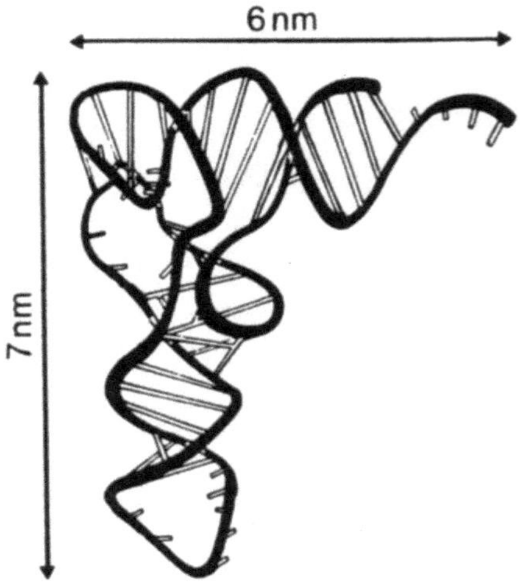

Fig. 3.10.2. Crystal structure of yeast t-RNA (based on Ref. [207])

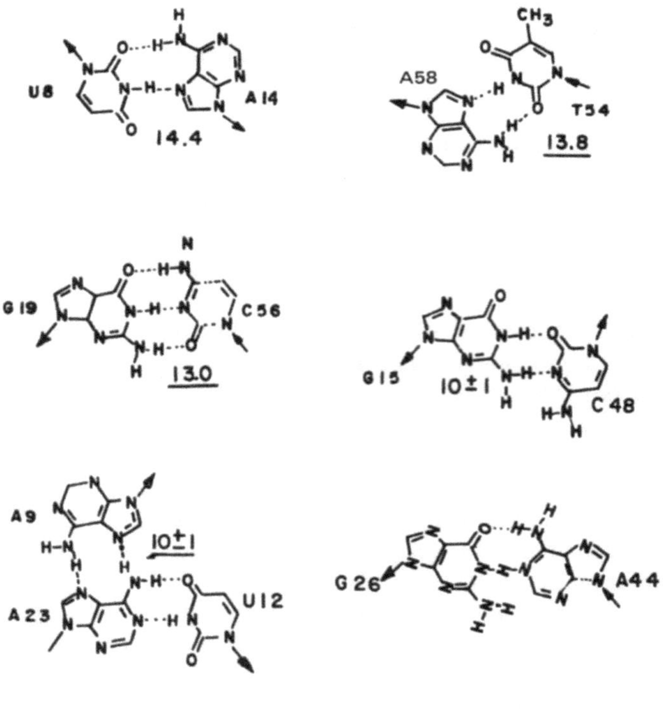

Fig. 3.10.3. Tertiary interactions in t-RNA detected by NMR spectroscopy (reprinted with the permission from Ref. [480]). The bases and their sequence positions are indicated

One straightforward application is to integrate the lowfield region and count the number of imino and thus the hydrogen-bonded base pairs involved in secondary and tertiary structures [472, 474]. In initial years of t-RNA work, some difficulties were experienced and the number of imino protons was considerably less than the present number of 26 ± 2. Since the clover-leaf model includes 20 base pairs, the remaining base pairs are involved in tertiary interactions. Fig. 3.10.3 outlines some of the tertiary structures which have been identified by proton NMR. In the absence of Mg ions, the tertiary structures melt at a lower temperature (around 45°) and the integrated intensities show a loss of four to five protons. Another method which has been used to differentiate between the secondary and tertiary structures utilizes paramagnetic ions. These ions preferentially bind to sites involved in tertiary structures [484].

One of the most interesting results which has emerged out of t-RNA work is that these molecules exhibit the same tertiary structure in solutions though the secondary structures may be slightly different to accommodate the extra bases. In other words, the chain folding of different t-RNA molecules in solutions is very similar.

Attempts have been made to correlate the solution structures of t-RNA with the four crystal structures referred above. In this approach, the crystal structures are used as models for the determination of the base stacking and the hydrogen-bonded-base pairs. The chemical shifts arising from two effects discussed above are combined and the low-field imino proton spectrum is simulated. The four structures predict quite different spectra from one another. Further, at this stage none of the predicted spectra matches exactly with the observed spectrum (see, for example, Fig. 3.10.4). Though it is recognized that none of the reported structures are perhaps very accurate, it is difficult to say from the NMR data alone whether the solution conformation differs from the crystallographic data, because of the difficulties in estimating different contributions to chemical shifts accurately and the limitation of the crystallographic models at the present degree of refinement.

In addition to the imino protons in the low field regions which provide information on the structure of hydrogen-bonded base pairs, use has been made in recent years of resonances from aromatic protons (in the region 6.5 to 8.5 ppm) and methyl groups of minor bases (1–4 ppm) and from nuclei such as ^{13}C and ^{31}P. With all the NMR probes thus available attempts have been made to study the structure-function relationship of t-RNA. Such problems are not amenable to X-ray crystallography.

t-RNA can exist in metastable denatured states which cannot be aminoacylated. NMR has been used to study the conformation of denatured t-RNA with the aim of recognizing structural changes responsible for the loss of activity and conformational mobility of t-RNA [474, 479, 484]. In the denatured conformers, the hydrogen-bonding arrangements of the hU and anticodon loop are altered.

Denaturation of t-RNA can be accomplished by raising the temperature. Thermal denaturation occurs due to breaking of hydrogen-bonded base pairs and proceeds in a sequential fashion [485, 486]. The unfolding of t-RNA is manifested in the NMR spectra as a broadening of the resonances in the low-field region and loss of intensity due to sequential "melting" of hydrogen-bonded base pairs [484, 487]. In the case of $\text{t-RNA}^{\text{Leu}}$ [484] about 9 hydrogen-bonded base pairs are lost between 35–52 °C. These can be attributed to the loss of the "minor stem" and non-clover leaf helix. Four more resonances disappear in the temperature range 52–62 °C because of the melting of TψC

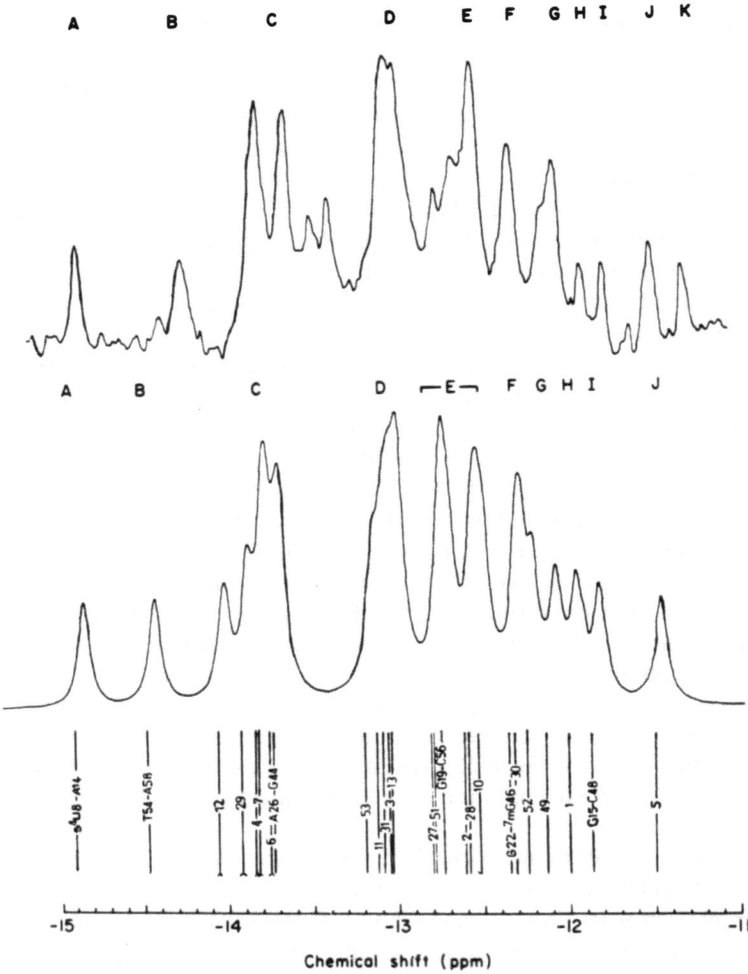

Fig. 3.10.4. Low-field spectrum of t-RNA along with the computer-simulated spectrum based on the crystallographic model (reprinted with permission from Ref. [476]). A, B, C, etc. refer to peak identifications

loop. The base pairing in the amino acid-acceptor end is quite stable and this structure is retained even beyond 62 °C.

NMR has been applied to study the dimerization of E. coli t-RNA^{Tyr} [453]. A comparison of the lowfield spectra of the monomer and dimer indicates a loss of six base pairs per molecule during dimer formation. A model for the hydrogen-bonding pattern in the dimer has been proposed based on NMR and other data.

The effect of modification of anticodon-loop on t-RNA conformation has revealed a large number of changes in the low-field region of ^{1}H-NMR [488]. The removal of the Y-base from the anticodon loop, results in significant effects on the conformation of the entire helical segments and the change is propogated into other parts of the molecule.

Photochemical cross-linking of S^4U_8 and C_{13} in E. coli t-RNA has been analyzed by NMR [489]. It has been observed that cross-linking causes only local conformational changes in t-RNA which explain the observation that photocross-linked molecules retain their biochemical activity.

Thus, NMR studies have added to our knowledge of a wide range of problems arising in the structure-function relationships of t-RNA.

3.11 Drug-Nucleic Acid Interactions

Efforts toward a rational basis for drug design have stimulated studies on the mechanism of drug-nucleic acid interactions. Antibiotics and antitumour reagents can combine with nucleic acids in a variety of ways. The phosphate groups are negatively charged at neutral pH and offer a convenient site for electrostatic interactions. Several proton donor and acceptor groups are available in the nucleotidyl fragments, the most important from the biological viewpoint being the sites on nucleic acid bases. Intercalation of drugs with base pairs in nucleic acid duplexes has been found to be a frequent mechanism of drug-nucleic acid interactions [490, 491]. These complexes are stabilised by stacking interactions between the aromatic moieties of the drug and nucleic acids. Finally, some drugs

Table 3.11.1. Studies on drug-nucleic acid interactions

Nucleus	System	Ref.
^1H	Oligonucleotides–actinomycin D	[459]
^1H	DNA-actinomycin D	[492]
^1H	Deoxytetranucleotide–actinomycin D	[490]
^1H	Deoxydinucleotide–actinomycin D	[493]
^1H	Deoxyhexanucleotide–actinomycin D	[450]
^1H	Deoxyoligonucleotides–actinomycin D	[451]
^1H	d(pGpC)–actinomycin D	[494, 512]
^1H, ^{31}P	Oligonucleotides–actinomycin D	[495]
^1H	Dinucleoside diphosphate–9-aminoacridine	[496]
^1H	Nucleoside phosphates–acridine dimer	[497]
^1H	Miracil D–oligonucleotides	[498]
^1H	Deoxytetranucleotide–proflavine	[490]
^1H	DNA, RNA fragments–propidium	[490, 499]
^1H	Deoxydinucleotides–ethidium bromide	[500]
^1H	Deoxytetranucleotide–ethidium bromide	[491]
^1H	Poly(U)–ethidium bromide	[501]
^1H	Dinucleoside monophosphate–ethidium bromide	[502]
^1H	Oligonucleotide–ethidium bromide: model describing interactions	[503]
^1H	Antimalarial compounds–DNA: model	[510]
^1H	Poly(dA–dT)–steroid diamine	[506]
^1H	Poly(A) poly(U)–miracil D	[511]
^1H	t-RNA–ethidium bromide	[513, 514]
^1H	Poly(dA–dT)–netropsin	[508]
^1H	Poly(dA–dT)–Bleomycin–A	[1092]

such as *cis*-dichlorodiammineplatinum(II) react through the formation of covalent linkages [454].

NMR studies on drug-nucleic acid interactions have been initiated rather recently. A large amount of work has been reported on drugs which undergo linkage through intercalation since the base protons and the protons on the drug can be utilized for monitoring the binding. A review on this topic has appeared recently [454]. Table 3.11.1 gives a compilation of the published work on drug-nucleic acid interactions.

The mode and stability of drug-nucleic acid complexes depends to a large extent on the complementary topologies of the drug and receptor molecules [504–508]. Of course, in many cases the topology itself is changed on complexation. Therefore, many drugs reveal a structure-dependent specificity in their binding behavior. Actinomycin, for example, binds to DNA but not to RNA or DNA-RNA hybrids [450–451, 490–491]. Further, it shows a specificity for the d(G-C) base pairs and intercalates at the purinyl-3′,5′-pyrimidine site. Ethidium bromide on the other hand also binds through intercalation but prefers a pyrimidinyl-3′,5′-purine site [491].

Binding of proflavin to poly (dA-dT) has been monitored by examining the chemical shift changes in the sugar and base protons of the nucleic acid and the protons on the proflavin rings. A model for the stacking of the proflavin ring with A-T base pairs has been developed from the observed ring-current shifts [490] coupled with the conformational calculation of the linked atom [509] (Fig. 3.11.1).

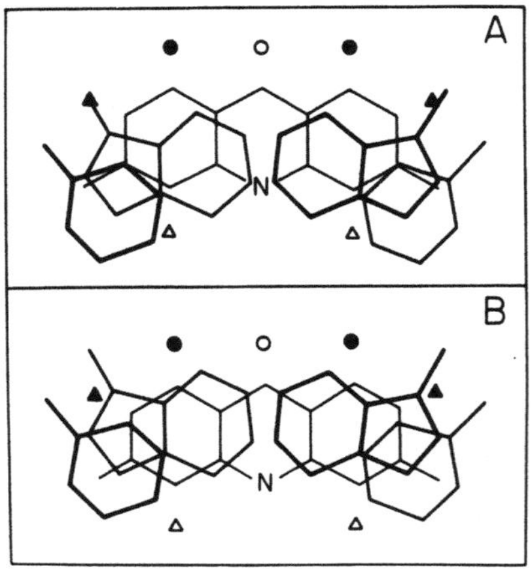

Fig. 3.11.1. Interaction model of proflavin at a pyrimidinyl-3′,5′-purine site in B-DNA based on linked-atom conformational calculations [509] (A) and optimized using the upfield shift of proflavin protons (B) based on ring-current contributions [490]. The symbols represent positions of proflavin protons in the two cases (reprinted with the permission from Ref. [490])

In the case of the poly (dA-dT) complex with steroiddiamine, the lines of the nucleotidyl protons become narrower on complextion. Thus, the nucleic acids show considerably higher segmental motions in these complexes [506].

Propidium diiodide interacts with DNA and RNA fragments by inducing a conformation transition in the former such that the furnose ring in the complex has a ^{3}E conformation [499].

Recently, studies have been carried out on the binding of Ethidium bromide to t-RNA [513–514]. The results have provided information on the specific drug-binding sites of t-RNA.

4 Amino Acids, Peptides and Proteins

4.1 Description of the Structure of Peptide Units

Proteins and peptides are responsible for a wide variety of biological functions. These molecules play a vital role in life processes ranging from biosynthesis to molecular transport. Enzymes are proteins which catalyze biochemical reactions. Certain polypeptides are involved in molecular recognition and information transfer. Hemeproteins are involved in transport of oxygen, and in biological oxidation reactions. Small linear and cyclic peptides are found to act as hormones, antibiotics, ionophores, antitoxins and are involved in a wide range of other biological functions. Several textbooks dealing with the structure and function of peptides and proteins [515–519] are available and may be consulted by readers who are not familiar with the field.

The repeating backbone unit in polypeptide chains is structurally simpler than the building block of nucleic acids. Polypeptides and proteins consist of a repeating sequence of peptide units –NH–C$^{\alpha}$HR–C′O– (Fig. 4.1.1). However, the number of possible side chains in proteins is much larger (Table 4.1.1). In addition to the 22 amino acids listed in Table 4.1.1 some other residues and modified amino acids occur in peptide hormones

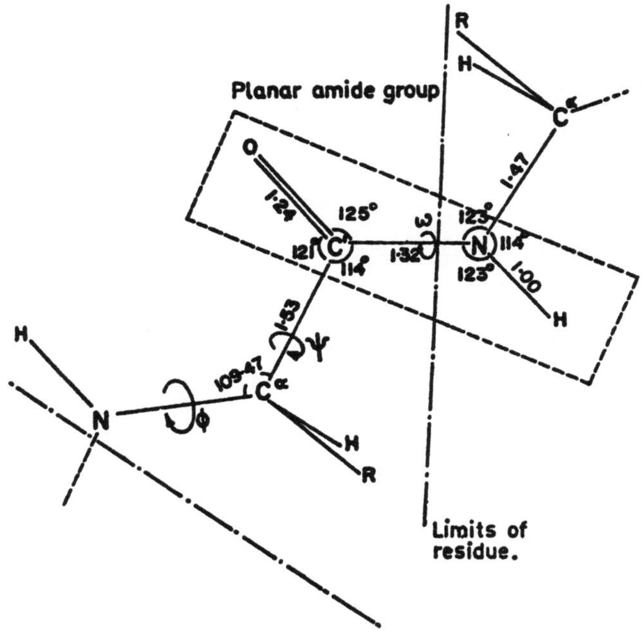

Fig. 4.1.1. Polypeptide chain showing the standard dimensions of the peptide unit and torsion angles which fix the conformational structure of each amino acid residue

Table 4.1.1. Side chains (R) present in common amino acid residues along with notations for side chain atoms

Amino acid	Abbreviation	Side chain (R)
Glycine	Gly	H
Alanine	Ala	CH_3
Valine	Val	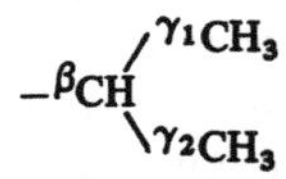
Leucine	Leu	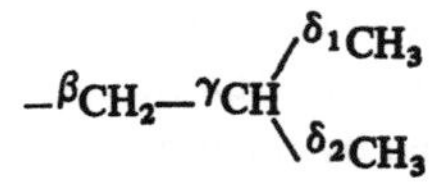
Isoleucine	Ile	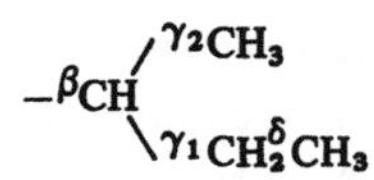
Serine	Ser	$-CH_2OH$
Threonine	Thr	$-CH(OH)CH_3$
Cysteine	Cys	$-CH_2SH$
Methionine	Met	$-{}^{\beta}CH_2-{}^{\gamma}CH_2-{}^{\delta}S-{}^{\epsilon}CH_3$
Aspartic acid	Asp	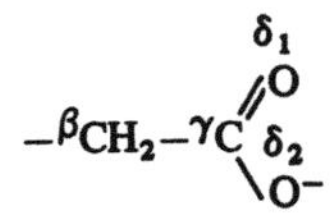
Asparagine	Asn	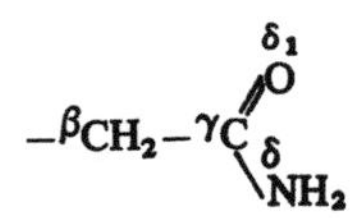
Glutamic acid	Glu	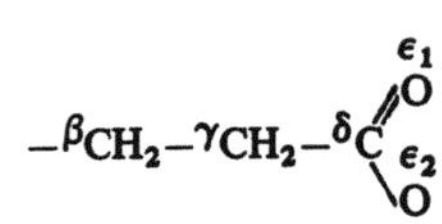
Glutamine	Gln	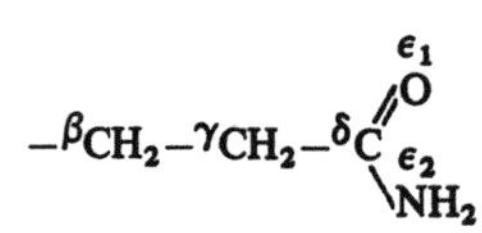
Lysine	Lys	$-{}^{\beta}CH_2-{}^{\gamma}CH_2-{}^{\delta}CH_2-{}^{\epsilon}CH_2-{}^{\zeta}NH_2$

Table 4.1.1 (continued)

Amino acid	Abbreviation	Side chain (R)
Arginine	Arg	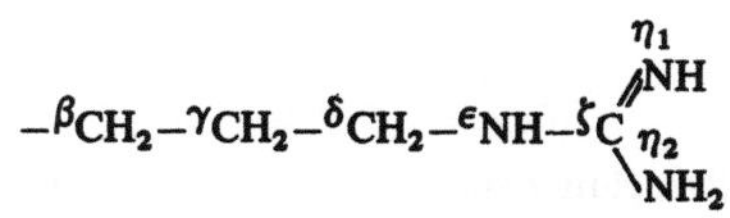
Ornithine	Orn	$-^{\alpha}CH_2-^{\beta}CH_2-^{\gamma}CH_2-^{\delta}NH_2$
Phenylalanine	Phe	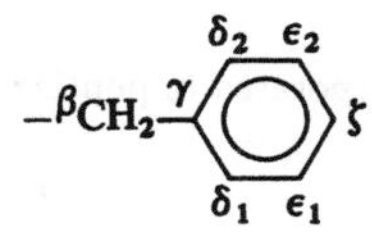
Tyrosine	Tyr	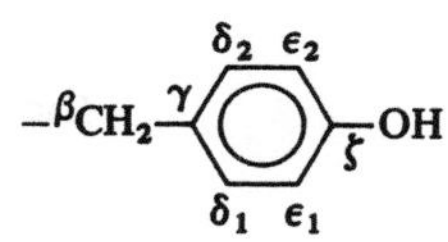
Tryptophan	Trp	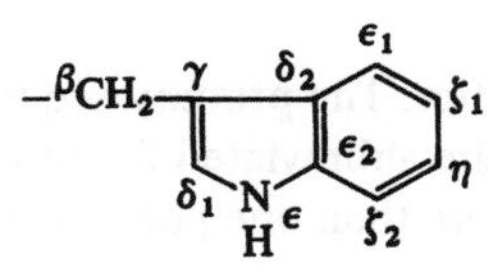
Histidine	His	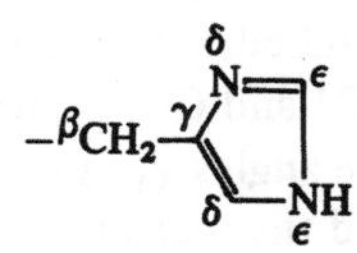
Proline	Pro	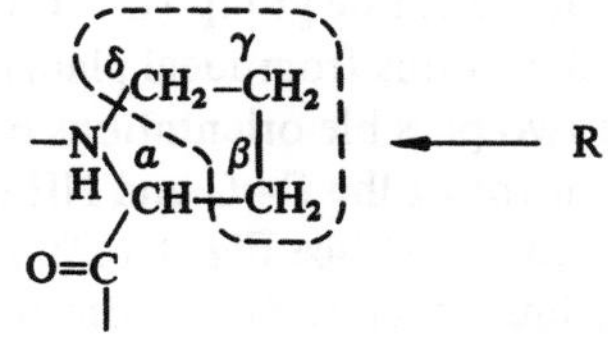
Hydroxyproline	Hyp	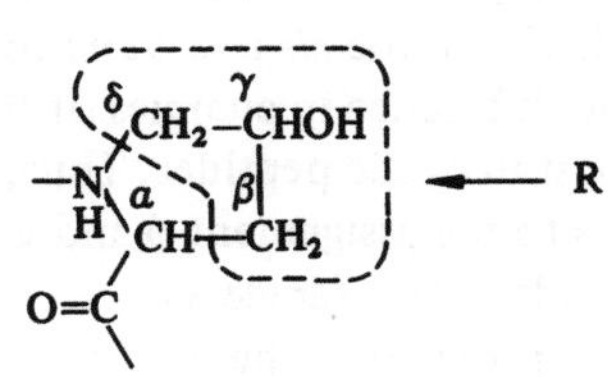

Table 4.1.1 (continued)

Amino acid	Abbreviation	Side chain (R)
Pyroglutamate	<Glu	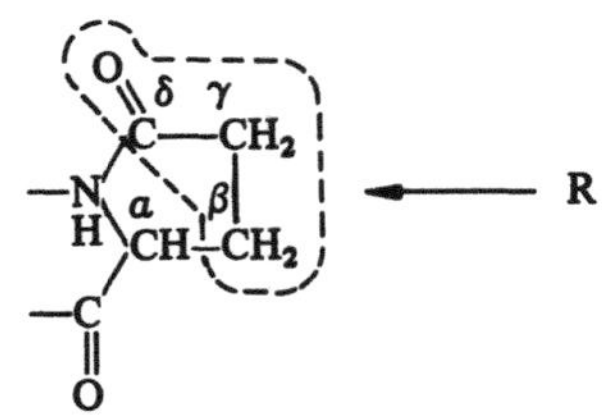
Modified units incorporated in peptide backbone:		
Sarcosine	Sar	$-N(CH_3)-CH_2-CO-$
α-Hydroxyisovaleric acid	Hy-Iv	$-O-CH-CO-$ $\vert$ $CH(CH_3)_2$
Lactic acid	Lac	$-O-CH-CO-$ $\vert$ CH_3

and antibiotics. The primary sequence of polypeptides and proteins is represented by specifying the abbreviated forms of amino acid residues. Larger proteins may be composed of more than one polypeptide chains. Hemoglobin, for example, consists of four polypeptide chains held together by electrostatic interactions.

The three-dimensional structure of a single polypeptide chain can be represented by specifying the dihedral angles ω_i, ϕ_i and ψ_i which denote torsion angles around the bonds C′–N, N–C^α and C^α–C′, respectively [520–522]. The side chain rotations are represented by the angles χ_1, χ_2 etc. There is a strong interaction between the NH and the C=O group and the π electrons of the carbonyl group are delocalized into the C′–N bond which consequently has a partial double bond character [520]. Thus, the atoms in the amide group (C^α–C′O–NH–C^α) form a rigid planar structure. Of course, minor deviations from ideal planarity may be observed, particularly in cyclic peptides. Of the two possible orientations of the planar amide group, the one involving a *trans* arrangement of the C=O and NH group is energetically favored in linear peptides [520, 523–525] (see Fig. 4.1.2). A *cis* arrangement may however occur with amino acid residues such as Pro, hydroxy-Pro, Sarcosine (Sar = *N*-methyl-Gly) and certain cyclic compounds such as diketopiperazines. Here, because of the special structure of the side chain, the *cis* and *trans* conformations have similar structural configuration and the difference in energies of the two conformers is small. The *cis* conformation may also exist in cyclic peptides. Thus, the conformational problem of peptides generally reduces to the assignment ϕ and ψ angles in the peptide backbone and the orientation of the side-chains (angle χ).

The problem of the notation and the convention followed in the measurements of torsion angles in peptides are in a much more settled state than is the case with nucleo-

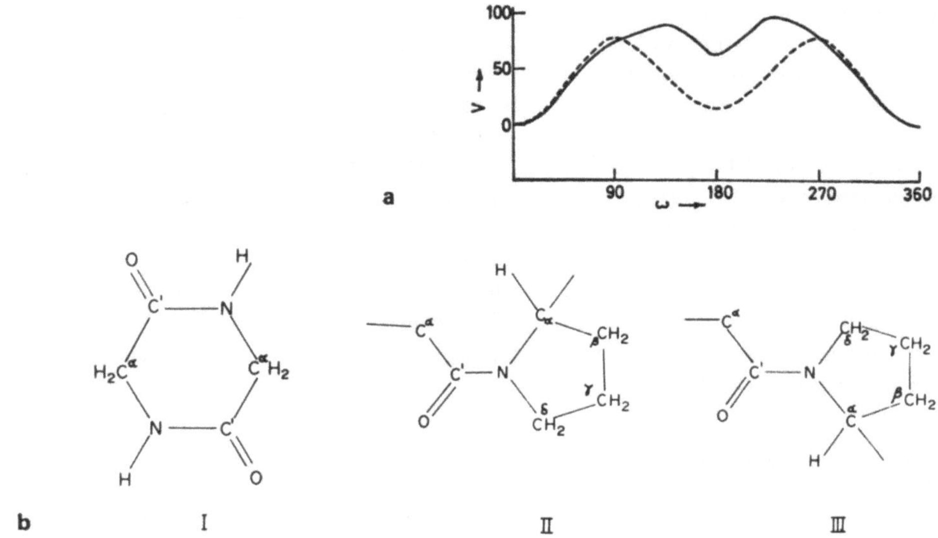

Fig. 4.1.2. (a) Energy difference V between *cis* and *trans* peptide units calculated by the EHT method, (b) 2,5-diketopiperazine (I) with *cis* peptide units and proline residues which can coexist both in *cis* (II) and *trans* (III) conformations because of special spatial structure (reprinted with the permission from Ref. [525])

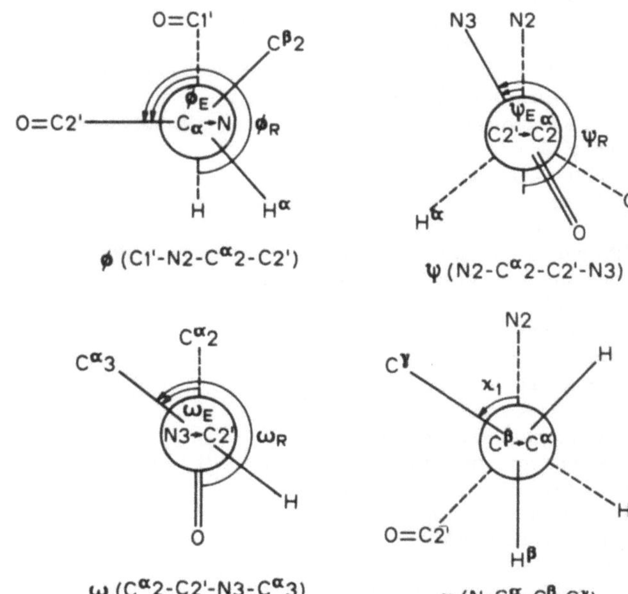

Fig. 4.1.3. Newman projection diagrams for the torsion angles ϕ, ψ, ω and χ. Subscript "E" denotes conventions proposed by Edsall et al. [522], which have been accepted by IUPAC and are now almost universally used. Subscript "R" denotes conventions followed by Ramachandran's group [520] in their earlier work

tides discussed in the previous chapter. The convention accepted by the International Union of Pure and Applied Chemistry (IUPAC, [522]) is shown in Fig. 4.1.3. According to this convention, torsion angles are measured with the *cis* conformation as 0° and in accordance with the general convention discussed in Chapter 1. However, the ranges of the torsion angles for the peptide backbone (angles ϕ, ψ and ω) are generally chosen as −180 to +180°. Thus, an angle of −60° in this convention corresponds to 300° in the convention discussed earlier and corresponds to a g^- state. The values of the side chain torsion angles (χ) are however expressed in the range 0 to 360°.

Prior to the acceptance of IUPAC recommendations, certain authors had used alternative conventions. The most notable of these involved measurement of ϕ, and ψ angles with the *trans* conformation as 0° [520]. These values differ by 180° from the values of the IUPAC conventions. The interrelationship between the two conventions is illustraded in Fig. 4.1.3.

Table 4.1.2. Reviews and general papers dealing with NMR studies on the conformation of oligopeptides, polypeptides and proteins

System and information	Ref.
Polypeptide conformation: ^{1}H- and ^{13}C-NMR	[526]
Cyclic peptides: use of lanthanide shifts	[527]
Oligopeptides: aromatic residues, variable distances	[528]
Polypeptides, cyclicpeptides: ^{1}H- and ^{13}C-NMR	[119]
Peptides in water solution: lanthanide complex	[529]
Peptide structure and conformation	[530]
Cyclic peptides: CD, potential energy complementary approaches	[531]
Cyclic peptides: X-ray and NMR studies	[532]
Peptides and proteins: molecular interactions, periodical reports	[533, 541, 542, 543]
Hormonal peptides	[534]
Polypeptide conformations	[535]
Peptide conformations with special reference to gramicidin S	[536]
Cyclic polypeptides: review	[537]
NMR, fluorescence methods for determining peptide conformations	[197]
Peptide conformations: detailed review, other physical methods	[538]
Cyclic peptides: conformation and reactivity	[539]
Membrane-active peptides: review	[540]
Protein conformations: ^{1}H- and ^{13}C-NMR	[544]
NMR in protein research: review	[545]
Large peptides, small proteins: review	[546]
Internal motions, globular proteins	[547]
Polypeptide conformations: combined use of NMR and potential energy calculations	[548, 549]
NMR methods: conformational analysis	[550]
Biological research: NMR	[551]
Proteins: ^{13}C-NMR	[26]
Amino acids, peptides, proteins: conformations	[125]
Peptides and polypeptides: film dialysis	[552]
Peptides and proteins	[553]
^{13}C-NMR: Hormone structure and function	[554]
Heme proteins	[555, 556]

Certain protein structures are stabilized by intramolecular hydrogen bonds involving NH and C=O groups. The usual convention used in the representation of such bonds is $i \rightarrow j$ where i is the amino acid residue of the donor (NH) group and j the amino acid residue containing the acceptor group.

The conformational structures of amino acids, peptides and polypeptides and small proteins have been extensively studied by NMR techniques. Several review articles have been written on various aspects of this subject. A comprehensive list is provided by Table 4.1.2. The impact of NMR on this area of research can be easily assessed from this table. In fact, in the realm of small peptides and proteins, NMR has outclassed other physical techniques available for studies of structure-functional relationships. One of the most important contributions of NMR has been the detection of conformational preferences and mobilities of these molecules in solutions. Larger proteins, however, still present unsurmountable difficulties to the NMR spectroscopist. The interpretation of results in these cases has to based predominantly on X-ray structures and the availability of synthetically modified analogs.

4.2 Theoretical Considerations. The ϕ, ψ Maps

As in the case of nucleotides, potential energy calculations have proved to be of great help in the interpretation of NMR data. Isoenergy maps for dipeptides have been obtained using a variety of semiempirical methods. (For reviews see [520] and [521].) The problem here is simpler since, in general, only two backbone trosion angles ϕ and ψ have to be considered. In the following, the main features of such calculations are summarized.

The isoenergy map for the simplest dipeptide, Gly-Gly is characterized by a relatively large area of conformational space of low energy (Fig. 4.2.1a). The only areas where the energy is large correspond to regions where ϕ or ψ has a value of 0°. Since the two hydrogen atoms attached to the α-carbon atoms are equivalent, a center of symmetry arises at (0°, 0°) and (180°, 180°). It is thus seen that the Gly residues possess an important structural property. They can easily fold in a variety of conformations and thus fit to the secondary structure of protein and peptide chains at positions where amino acid residues with bulkier side chains will be excluded. The situation may be compared with Gly-Ala where the replacement of one of the hydrogen atoms at C^α by a methyl group causes an asymmetry in the conformational map. The isoenergy maps for D- and L-amino acid residues illustrate mirror image relationships. A typical conformational map for the L-residues is shown in Fig. 4.2.1(b). The map for the D-residues can be obtained by inverting it with respect to the point (0°, 0°). It may be noticed that potential energy calculations on such systems show that the regions of conformational stability are limited. There are two broad low-energy regions which account for nearly all the observed conformations of Ala residues in polypeptides and proteins.

It is thus observed that the introduction of a bulkier side chain at C^α imparts "rigidity" to the peptide backbone structure by limiting the range of available conformational space. For example, Fraser et al. [557] synthesized copolymers of γ-ethyl L-glutamate (*e*) and glycine (*g*). It was observed that polymers of the type $(ee)_n$ exist predominantly in the α-helical form, $(ge)_n$ in the β-conformation and $(gg)_n$ tends to

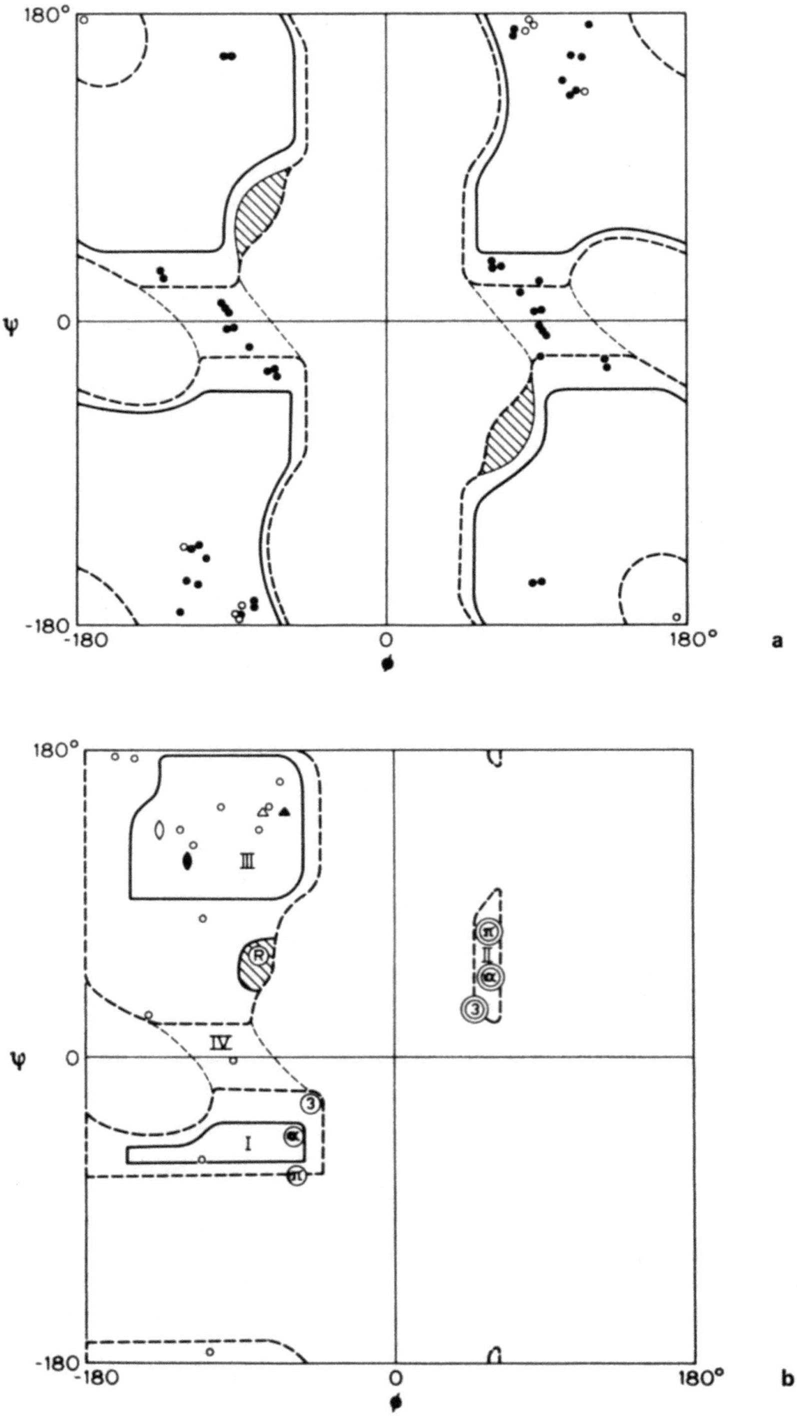

180°
ψ
0
-180
-180
0
180°
ϕ
a
180°
III
IV
I
II
ψ
0
-180
-180
0
180°
ϕ
b

assume a random coil structure. In the series $(eeg)_n$, $(eeeg)_n$, $(eeeeg)_n$ and $(eegeee)_n$, the helicity was found to increase with decreasing proportion of *g*. Further, while the characteristic ratio (C_∞) for polyalanine and other poly-L-peptides is around 9, polyglycine has a very low C_∞.

Another amino acid which plays a key role in the secondary structures of proteins and peptides is proline. Proline residues, as already stated, can exist both in the *cis* and *trans* conformation around the C′–N bond. Thus, although the conformational flexibility around the (ϕ, ψ) angles for Pro is limited this residue plays a vital structural role due to flexibility caused by the torsion angle ω. Proline residues predominantly determine the conformation of structural proteins such as collagen [558]. Both Gly and Pro residues are involved in the reversal of the direction of peptide chains in globular proteins [559, 560]. The structural role of Pro residues has been established for a large number of peptide hormones, peptide antibiotics and peptide antitoxins. For reasons mentioned above, both Gly and Pro residues have a higher probability of forming cyclic peptides.

The striking difference in the conformational behavior of the maps for Gly-Gly and Gly-Ala clearly demonstrates the importance of side chain interactions in the determination of the backbone conformations of polypeptide chains. An additional factor which one has to take into account in long-chain polymers is the fact that certain conformations can be stabilized by long-range intra- and interchain interactions. For example, the right-handed α-helix (α_R) is stabilized further by an intrachain hydrogen bond while the structure of collagen consists of three polypeptide chains held together by interchain hydrogen bonds. Polar groups in the side chains such as Ser, His, Lys, Thr, Asp, etc. can interact with the backbone both through electrostatic and hydrogen bonding interactions. Thus, the backbone conformation of amino acid residues in polypeptides or proteins depends on a number of factors and dipeptide maps such as in Fig. 4.2.1 can be used only as a rough guide line in the exploration of the three-dimensional folding of proteins. Limitations of dipeptide models to conformational studies of polypeptides and proteins have been reported [561, 562]. In spite of these complications, the observed three-dimensional arrangements of proteins and polypeptides are limited. Some of the important conformational structures which occur in proteins and polypeptides are discussed below (Fig. 4.2.2).

(a) *Right-handed α-helix* (α_R) corresponds to values of $-58°$ and $-46°$ for ϕ and ψ and is stabilized by an intrachain N–H···O=C hydrogen bond between every fifth and first amino acid residue in the polypeptide chain (5 → 1 hydrogen bond). This region corresponds to a low-energy region even in a dipeptide. Segments of such structures occur in a large number of globular proteins. There are 3.6 units per turn of α_R and the unit rise is 1.5 Å.

◀ **Fig. 4.2.1.** (a) Conformational map for Gly-Gly based on the contact criterion. The solid lines enclose the "fully" allowed conformation while "partially" allowed regions are shown by dotted lines. (b) Allowed regions (I–IV) for Gly-Ala. Note the asymmetry of the map and relatively restricted regions of energetically favored conformations. Some of the classical protein structures are shown in the figure.
Ⓐ = right-handed α-helix, ◎ = left-handed α-helix, Ⓡ = ribbon structure, ③ = 3_{10} helix, (π) = π helix, ♦ = β-structure, ▲ = collagen (reprinted with the permission from Ref. [520])

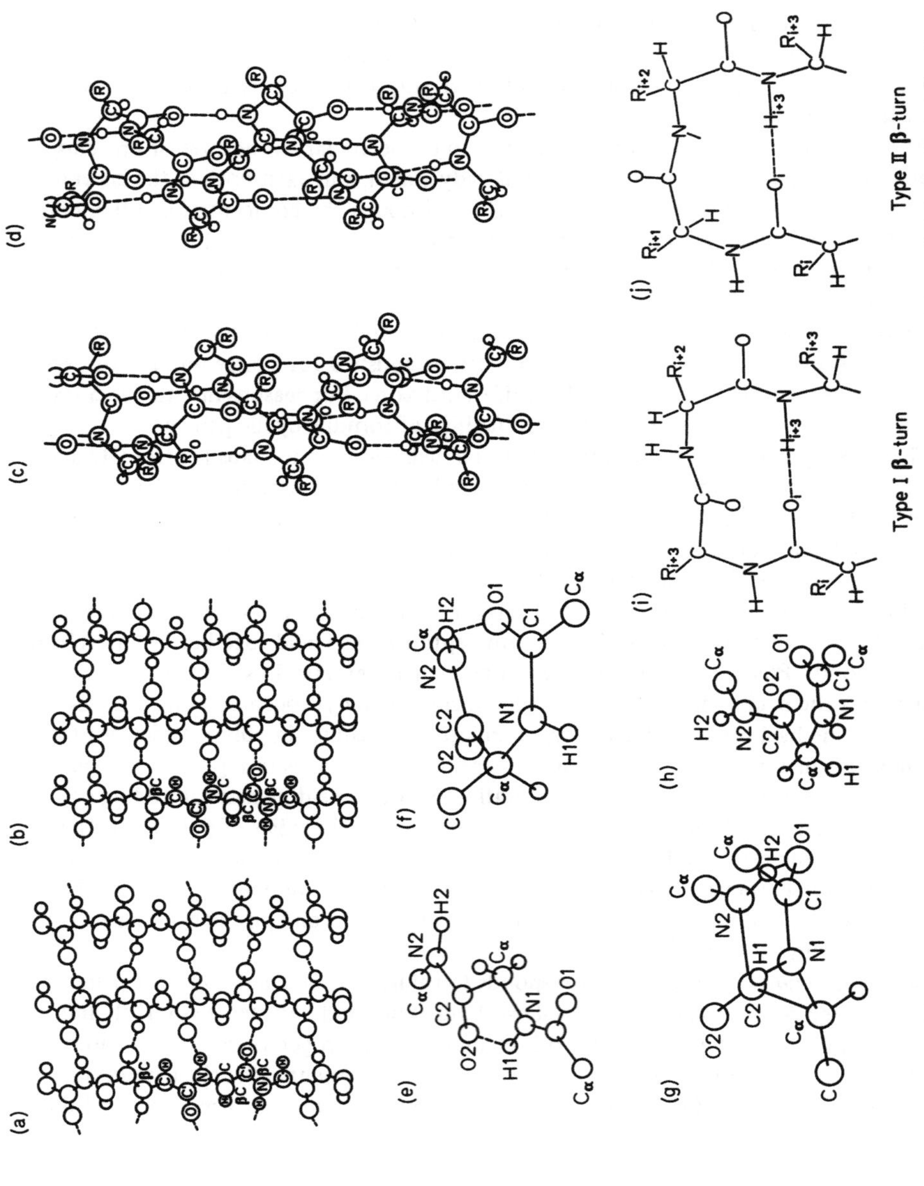

Fig. 4.2.2. Certain selected peptide conformations present in proteins and polypeptides. (a) parallel chain-pleated sheet, (b) antiparallel chain-pleated sheet, (c) left-handed α-helix, (d) right-handed α-helix, (e) C_5-structure, (f) $C7^{ax}$, (g) $C7^{eq}$, (h) M, (i) type I β-turn, (j) type II β-turn

(b) *Left-handed α-helix* (α_L) is the enantiomorphic conformation of α_R, stabilized by a hydrogen bonding pattern similar to α_R but is intrinsically less stable because of steric interactions when a bulky side chain is present. Of course, for -Gly, α_R and α_L are equivalent.

(c) β-Structures: These correspond to nearly extended structures and are stabilized further by intra- or interchain hydrogen bonds. The two adjacent chains forming such pleated sheet structures may be parallel or antiparallel. β-Structures may be formed in globular proteins through folding of the polypeptide chain on itself. In fibrous proteins, several chains may be held together by hydrogen bonding interactions. β-Structures occur as structural components of several proteins.

(d) *β-Turn:* When a polypeptide chain folds back on itself forming a β-sheet, then the conformation at the folding point involves a hairpin loop. Venkatachalam [563] has made an extensive investigation to explore possible ways in which a peptide chain may fold on itself. Of the various suggested β-loop structures, at least three, type I, type II and type II′, have been observed in crystal structures. All the three structures form a $1 \rightarrow 4$ hydrogen bond and utilize a tripeptide section in the turn but differ in the configuration of amino acid residues in the peptide sequence. Type I turn contains both the intervening peptide units (2 and 3) in the L configuration and the $\phi_2, \psi_2, \phi_3, \psi_3$ angles are –60, –30, –90 and 0°. Type II turns are formed when the amino acid residues have the sequence LD and values of torsion angles are –60, 120, 80 and 0° while the II′ turn exhibits a DL sequence with the torsion angles 60, –120, –80 and 0°. β-Turn conformations frequently occur in cyclic peptides.

(e) *3_{10} Helix:* This helix has ϕ, ψ angles of –49 and –26° and occurs in the conformational map close to the α_R. It contains 3 units per turn and the axial rise is 2 Å.

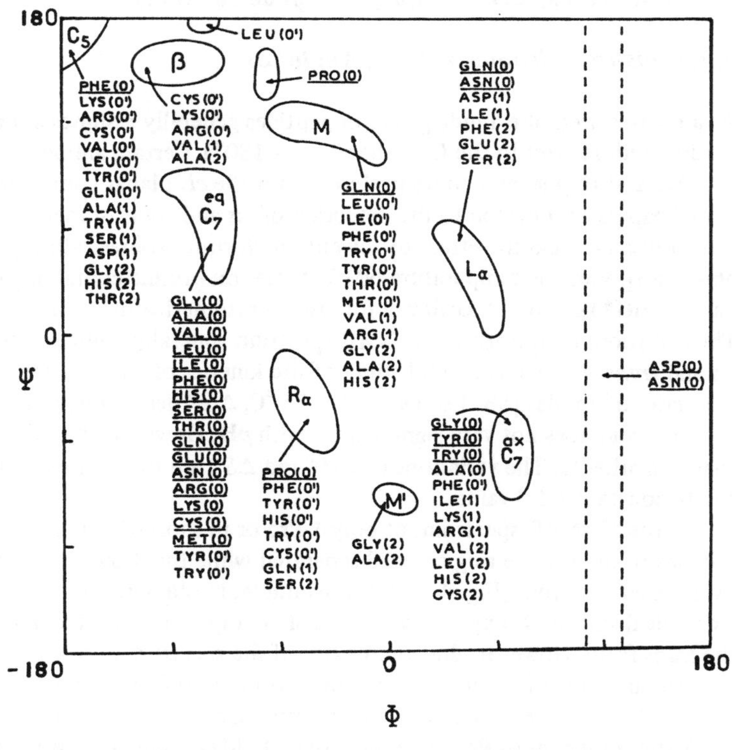

Fig. 4.2.3. Stable conformations of α-amino acid residues of proteins as predicted by the PCILO method. The most stable conformations for each residue is underlined. The numbers in brackets after the names of the residue indicate the energy (kcal/mol) relative to the global minimum for that residue (adapted from Ref. [561])

(f) *C–5 Conformation:* This conformation also contains ϕ, ψ angles of 180 and 180°) but unlike β-structures, the conformation is stabilized by a weak 1 → 1 intramolecular hydrogen bond at the dipeptide level. In the presence of β-carbon atoms, these conformations have higher energies than the conformation at C5.

(g) *$C7^{eq}$ and $C7^{ax}$:* These conformations are important for dipeptides as they are stabilized by 3 → 1 hydrogen bonds between adjacent peptide units. The ϕ and ψ angles are –80 and 40° for $C7^{eq}$ and 80 and –60° for $C7^{ax}$. The two conformations differ in the orientation of the β-CH_3 group. These conformations have been postulated in solutions of dipeptides in carbon tetrachloride and form P-type helices. However, they rarely occur in oligo- and polypeptides.

(h) *M conformation:* The ϕ and ψ angles are 0 and 60°. Its enantiomorph (M′) generally has a much higher energy.

The relative stabilities of the above conformations and hence their frequency of occurrence in peptides and proteins depend on a number of short- and long-range interactions. A PCILO study [561] at the dipeptide level has been performed on all amino acid residues. The results are summarized in Fig. 4.2.3.

4.3 NMR Techniques in the Study of Peptide Conformations

4.3.1 *cis-trans Isomerization (Angle ω)*

As already stated, the amide group in peptides generally assumes a *trans* planar conformation with respect to the C′–N bond (ω = 180°). Certain amino acid residues such as Pro, Sar and Hyp show a finite preference for the *cis* planar conformation. A large number of papers are devoted to the detection of *cis* and *trans* isomers, dynamics of their interconversion and the effect of solvent on *cis-trans* isomerization [136–141, 219, 564–586]. Variable temperature NMR studies on proline-containing peptides yield a barrier (ΔG^*) of 16–21 kcal/mol for *cis-trans* isomerization around the X-Pro bond. Thus, at room temperature, the NMR spectrum generally contains separate signals for *cis* and *trans* isomers. A careful study of the kinetics of the rotation of the peptide bonds in a series of amides [587] shows that at 25 °C, ΔG^* lies in a narrow range of 16.4–21.4 kcal/mol and does not vary significantly with *p*H. However, the values of ΔH^* and ΔS^* change markedly. The variations in ΔH^* and ΔS^* are thus compensated leading to a nearly constant ΔG^* value.

In the ^{1}H-NMR spectrum, of poly-L-Pro or in short-chain oligopolymers, the α protons of the proline ring are well separated while the β and γ protons give rise to an overlapped spectrum (Fig. 4.3.1). For example, in oligoprolines, H^α in the *cis* form has a chemical shift of 4.5 ppm relative to TSP, compared to a value of 4.8 ppm for H^α in the *trans* form. However, the complexity of the spectrum may prevent the detection of small amounts of conformers in either *cis* or *trans* forms. The ^{13}C chemical shifts are even better suited for studies on *cis-trans* isomerization. In the *cis* conformation, the δ and γ carbons of Pro occur at a lower field compared to those in the *trans* conformation (Fig. 4.3.2). The α and β carbons can likewise be distinguished and readily assigned to either the *cis* and *trans* conformation. An extensive study of proline-containing peptides [581] reveals that the γ carbon resonance of the *trans* conformer about

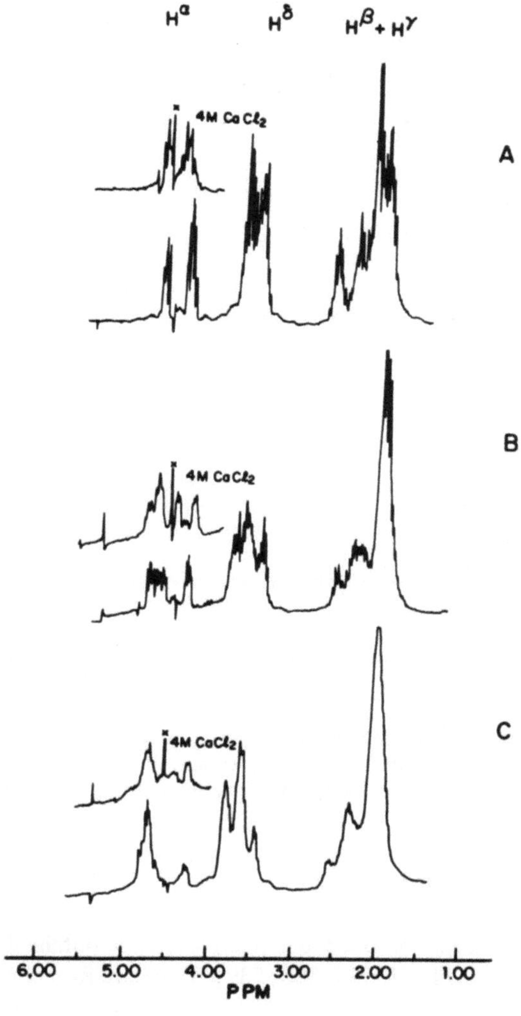

Fig. 4.3.1. 220 MHz ^{1}H-NMR of (A) $(Pro)_2$, (B) $(Pro)_4$ and (C) $(Pro)_6$ relative to TSP at 52 °C, $pH = 4$. The effect of salt (4M $CaCl_2$) on the H^α region is shown as inserts (reprinted with the permission from [141])

the X-Pro bond occurs in the range 25.1 ± 0.5 ppm while the resonance for the *cis* conformer conformer occurs at 23.4 ± 0.3 ppm. The pattern of α and β carbon resonances shows a wider spread of chemical shifts and is influenced by other factors, in addition to *cis-trans* isomerization. Nevertheless, these resonances are used as complementary evidences in the conformational studies of peptides containing Pro residues.

The difference in free energy (ΔG) between the *cis* and *trans* isomers of Pro is rather small (ΔG lies between –2.0 and 2.0 kcal/mol). The two forms thus coexist in most linear peptides at ambient temperatures. The population ratio of the two states depends on a number of factors such as solvent, *p*H, ionic concentration, nature of the substituent X attached to the Pro residue and the length of polypeptide chain. For example, the low molecular weight oligo$(Pro)_n$ with $n < 5$ contains a random distribution of *cis* and *trans* residues; the conformation is predominantly *trans* for systems with $n \geqslant 6$.

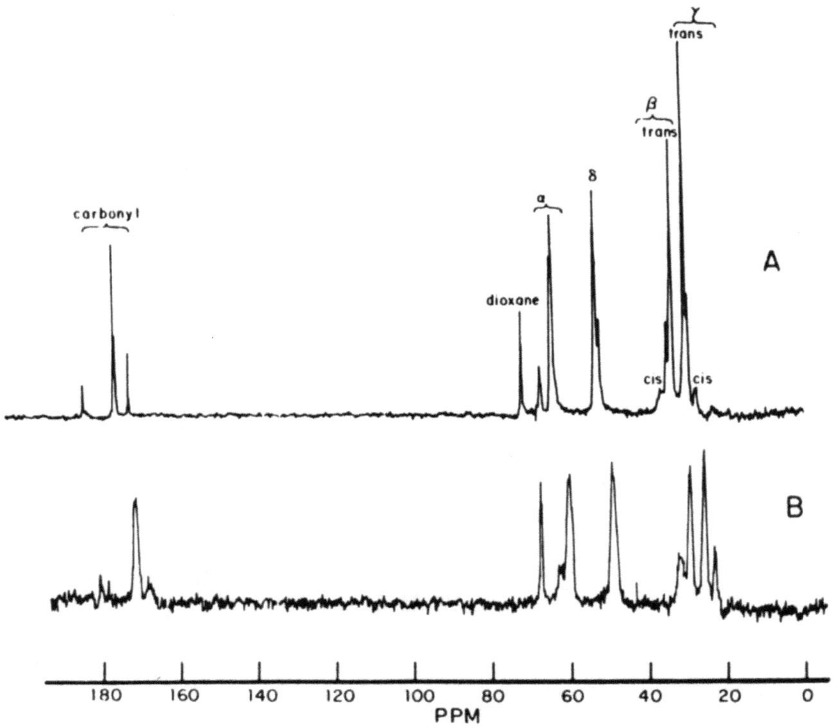

Fig. 4.3.2. ^{13}C-NMR spectra of $(Pro)_6$ in aqueous solution at 50 °C and *p*H 4. The bottom trace is in 4M $CaCl_2$ solution. Dioxane is employed as internal standard and its position (67.4 ppm from TMS) is used to fix the bottom scale (reprinted with the permission from Ref. [141])

4.3.2 φ and ψ Angles

Information on the ϕ and ψ angles can be obtained by use of vicinal coupling-constants [60, 70, 74, 77, 546, 588, 589]. There are four coupling constants (Fig. 4.3.3) which depend on the angle ϕ, while the value of ψ can only be estimated using $J(H{-}C^{\alpha}{-}C'{-}^{15}N)$. The dependence of the *vicinal* coupling constants on the angle ϕ is shown in Fig. 4.3.4. The angular dependence of two of the four coupling constants has been investigated both theoretically and experimentally. The hatched areas in Fig. 4.3.4 denote the limits of ^3J values based on the scatter of experimental observations. The values of the other two $^1H{-}^{13}C$ coupling constants are based on theoretical calculations [73]. For the angle ψ, only the three-bond $^1H{-}^{15}N$ coupling constant is available and so far very little experimental data are available to verify its applicability [82].

A five-bond proton-proton coupling ^5J has found application in limiting the range of backbone conformations of peptides [114]. Likewise, the use of $^{13}C{-}^{13}C$ coupling constants may prove useful in the study of peptide conformations [91]. Several other one- and two-bond coupling constants have advantageously been applied to conformational studies of peptides. The literature on these aspects has been reviewed recently [60].

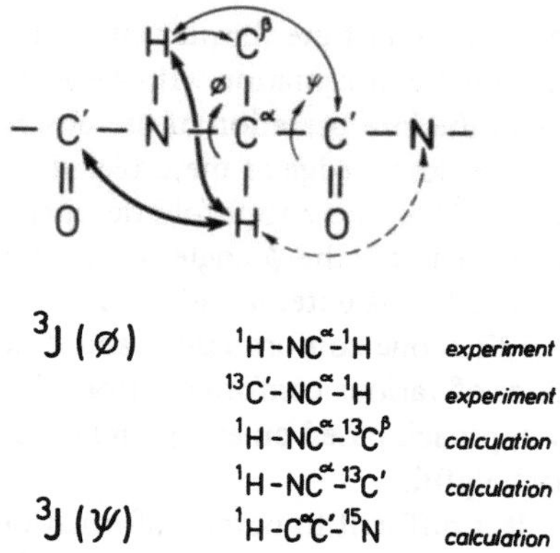

Fig. 4.3.3. Set of the three-bond coupling constants which are related to the peptide backbone angles ϕ and ψ (reprinted with the permission from Ref. [546])

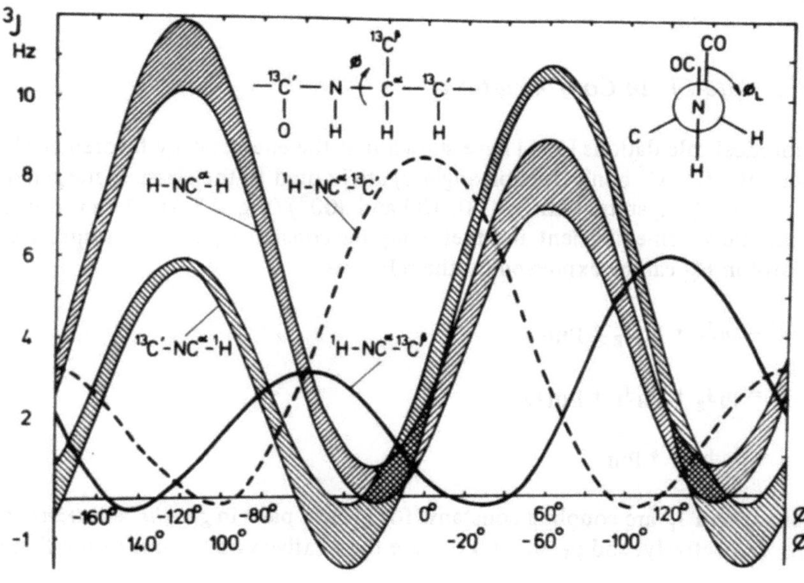

Fig. 4.3.4. Angular dependence of the three-bond coupling constants dependent on the torsion angle ϕ (reprinted with the permission from Ref. [546])

For low-molecular weight peptides, values of all the coupling constants can be obtained. Use of double resonance and INDOR techniques can be made wherever necessary to complete the assignment [590]. To translate this information into values of torsion angles, it is necessary to make a guess whether the molecule exists in a single rigid conformation or in more than one conformational states. The existence of intramolecular hydrogen bonds or the insensitivity of NMR parameters to changes in solvent and temperature are good evidences of conformational purity. Whenever the X-ray structure is known, one can use NMR coupling constants to determine whether such a structure predominates in solution as well. Further, when the coupling constant values are near the maximum of the Karplus curve (i.e. $^3J(H–N–C^{\alpha}–H) > 10$ Hz), this suggests a relatively rigid conformation with respect to the particular angle. Serious problems may arise in the interpretation of the observed data in cases where one has to start without any prior knowledge of the molecular conformation of the molecule. Firstly, a particular value of $^3J_{obs}$ has a four-fold degeneracy in terms of solutions for ϕ and ψ angles. Here, the problem for the ψ angle is even more serious since only one coupling constant is available for its determination. Secondly, the molecule may be flipping rapidly between more than one conformations and it may not be possible to calculate the relative populations of various conformers from the observed values $\langle J \rangle$. In such cases, a multidisciplinary approach based on energy minimization, CD and other spectral properties can be very helpful.

It is difficult to obtain information on J values in larger proteins and polypeptides because of intrinsically larger line widths and difficulties in spectral analysis and assignments.

4.3.3 Side-Chain Conformations

Theoretical calculations [591] have shown that the energetically favored conformations around the $C^{\alpha}–C^{\beta}$ bond (torsion angle χ) are limited to the classical staggered states g^+, t and g^- (i.e., states with $\chi = 60$, 180 and 300°) (Fig. 4.3.5). The protons H'_{β} and H''_{β} are generally non-equivalent and their coupling constants (J′ and J″ respectively) with the proton H_{α} can be expressed by the relations

$$J' = p_I J_t + p_{II} J_g + p_{III} J_g \qquad (4.3.1)$$

$$J'' = p_I J_g + p_{II} J_t + p_{III} J_g \qquad (4.3.2)$$

$$1 = p_I + p_{II} + p_{III} \qquad (4.3.3)$$

where J_g and J_t are coupling constants for proton pairs in *gauche* and *trans* conformations, respectively, and p_I, p_{II} and p_{III} are the relative populations of the three staggered

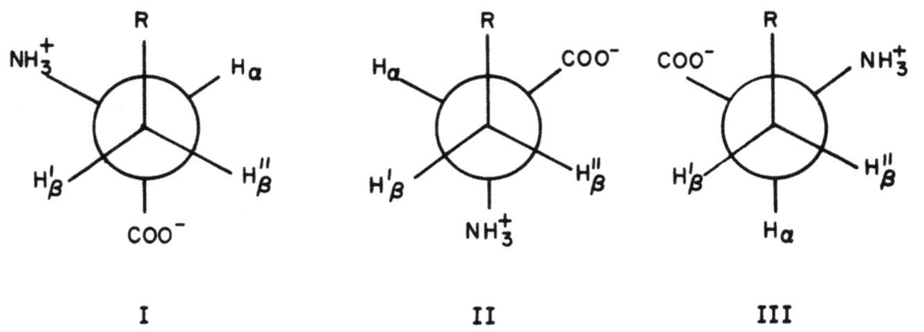

Fig. 4.3.5. Staggered conformations (I, II, III) of amino acids relative to the torsion angle χ

conformations. Typical values of J_g and J_t are 2.60 and 13.6 Hz (592). Kopple et al. [67] have proposed the following relation for the H–C$^{\alpha}$–C$^{\beta}$–H coupling constants:

$J = 9.4 \cos^2 \Theta - 1.4 \cos \Theta + 1.6$ which gives $J_t = 12.4$ Hz and $J_g = 3.25$ Hz. In a recent report [117], a critical analysis of both ^{1}H–^{1}H and ^{1}H–^{13}C coupling constants has been made. The following set of values have been recommended: For ^{1}H–^{1}H, $J_g = 2.4$, $J_t = 13.3$ Hz. For ^{1}H–^{13}C, $J_g = 1.2$ and $J_t = 10.0$ Hz.

One of the problems in the study of side-chain conformations involves the assignment of the two protons attached to the β-carbon atom. In recent years, this problem has been solved by the simultaneous use of three-bond ^{13}C′–C$^{\alpha}$–C$^{\beta}$–H coupling constants involving carbonyl carbon and β protons [102] and by stereoselective deuteration [593].

The second problem in the analysis of conformer population is the use of component coupling constants described above. As mentioned in Chapter 2, the electronegativity effects of substituents are stereospecific and it is doubtful if the same values of J_g and J_t can be used in conformers (Fig. 4.3.5). The estimates thus obtained are therefore only semiquantitative.

4.3.4 Detection of N–H···O=C Hydrogen Bonds

The above discussions show that even in favorable cases where angles ω, φ and χ can be uniquely fixed, the total conformation of even a small peptide remains undetermined.

As discussed in Sect. 4.2, several peptide conformations are stabilized by intra- or interchain N–H···O=C bonds. The temperature coefficient of the NH chemical shift can be used to detect the presence of such bonds (Sect. 2.2). The breaking of hydrogen bonds leads to an upfield shift and can be induced by increasing the temperature. Thus, for peptides forming hydrogen bonds with solvent, the chemical shift will depend significantly on the temperature. The temperature effect on intramolecularly hydrogen-bonded systems is relatively low.

The NH proton exchange rate can be monitored by observing the intensity of the NH peak in deuterated solvents. A slow rate of exchange indicates internal hydrogen bonds. Caution is needed when judging intramolecular hydrogen bonds by either of the above methods since erroneous results are sometimes observed [119].

Another method which has been suggested to detect sequestered NH protons uses low concentrations of a stable nitroxyl radical (e.g. 3-oxyl-2,2,4,4-tetramethyloxazolidine) in peptide solutions. The line broadening is less for sequestered protons than for those exposed to solvent [594]. This method has the advantage that it does not seriously disturb the distribution of peptide conformations.

Similar techniques can be used to detect the solvent exposed or sequestered carbonyl groups using ^{13}C-NMR. Solvent titrations show that the peptidyl carbonyl which is intramolecularly hydrogen-bonded exhibit less variation in the chemical shift on going from DMSO to a proton donor solvent [124].

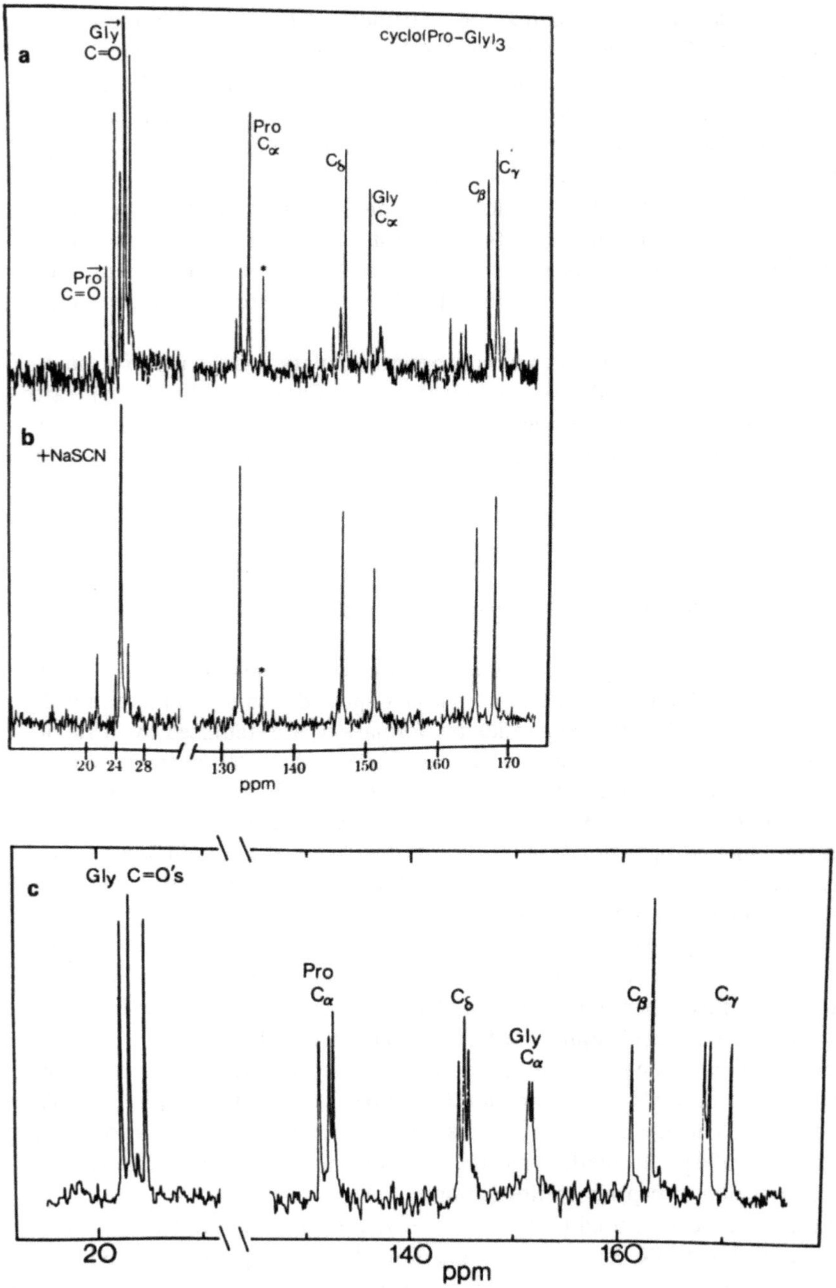

Fig. 4.3.6. Fourer Transform ^{13}C-NMR spectra of cyclo(Pro-Gly)$_3$. Chemical shifts are relative to CS_2, the Gly carbonyl carbon is enriched with ^{13}C. The spectra are (a) free peptide (0.04M) in acetonitrile-d_3/chloroform (1 : 1; V/V), (b) sodium thiocyanate added to (a), molar ratio salt : peptide = 1 : 4.4, (c) peptide in D_2O solution (reprinted with the permission from Ref. [144], copyright 1974, American Chemical Society)

4.3.5 Molecular Symmetry

NMR methods are ideally suited to the detection if the molecules fold symmetrically. This information can be readily obtained by examining the multiplicity of lines for a particular group. ^{13}C-NMR, due to the larger range of chemical shifts, gives the most unambiguous results. Symmetry considerations have provided invaluable help in building models of cyclic peptides.

We can illustrate the potentialities of ^{13}C-NMR in the detection of molecular symmetry by use of a typical cyclic peptide: cyclo-(Pro-Gly)$_3$ (Fig. 4.3.6). In solution, NMR data reveal that this peptide can exist in three molecular conformations. Two of these (S and S_1^*) have a C_3 symmetry while the third is asymmetric (A). In the conformations with S_3 symmetry, the three Pro and Gly units show magnetic equivalence. Furthermore, there is a low barrier to interconversion between S and S_1^*. Thus, in the presence of NaSCN, where the presence of A is small, the ^{13}C spectrum shows a single resonance line for each chemically inequivalent site. With increasing salt concentration, the molecule adopts an S_1^* conformation. This results in an increase of coupling constants J(HNC$^\alpha$H) for the Gly residue from 3.0 and 4.5 Hz (characteristic of S) to 5.0 and 5.5 Hz (S_1^*). The antisymmetric conformation is stabilized in solvents such as D_2O and gives separate signals for each Pro-Gly unit (Fig. 4.3.6c). In acetonitrile-$CDCl_3$, both symmetrical C_3 and A conformations are present. The S conformation exhibits all the three peptide bonds in *trans* conformation while one of the bonds possesses a *cis* conformation in A. Thus, separate signals for A and S forms are detected (Fig. 4.3.6a). Finally, the usual NMR criteria reveal that the symmetrical structure S is stabilized by a N–H···O=C hydrogen bond [144].

4.3.6 Conformational Mobility

The flip from one rotational state to another modulates the dipolar interactions. The ^{13}C spin-lattice relaxation time which is dominated by ^{13}C–^{1}H dipolar interactions thus acts as a source of information on the molecular flexibility in the amino acid side chains in polypeptides. (For a review see [530].) It may be pointed out that such motions may occur even when the backbone is relatively rigid. A theory to extract detailed information on motions from T_1 measurements has been developed [595]. Extensive use of ^{13}C spin-lattice relaxation times (T_1) has been made to investigate the conformational flexibilities of peptide residues. For peptide hormones and oligopeptides, measurements of T_1 can be carried out routinely on commercial NMR spectrometers. However, for more complex peptides serious sensitivity problems will arise.

When the dipole-dipole interaction with directly attached ^{1}H provides the sole relaxation mechanism and the overall motion of the peptide group is isotropic and rapid with respect to the ^{13}C-NMR time scale, one can use the following equation

$$\frac{1}{NT_1} = \left\langle \frac{1}{r^6} \right\rangle \hbar^2 \gamma_C^2 \gamma_H^2 \tau_{eff} \qquad (4.3.4)$$

where N is the number of protons attached to the carbon atom, r the C–H distance and τ_{eff} the effective correlation time for reorientation of the C–H vector and γ_C and γ_H are the gyromagnetic ratios for ^{13}C and ^1H, respectively. NOE measurements and comparison of ^{13}C line widths are useful tests to check whether T_1 is dominated by a dipolar relaxation mechanism [595] or not. When the molecular motion is anisotropic, τ_{eff} has to be decomposed into components describing the motion about each principal axis. Anisotropic tumbling can be recognized by examining the peptide α-carbon T_1. If variations in T_1 can be correlated with the preferred modes of rotation of the molecule (as inferred from the molecular geometry), then anisotropic motions should be considered in the interpretation of T_1 data.

It can be seen from Eq. (4.3.4) that NT_1 values are proportional to the mobility of the carbon atom. Thus, values of NT_1 can be used to detect the overall molecular tumbling in solutions, the internal molecular flexibility and segmental motions in peptides. It is possible to separate τ_{eff} into components arising from overall molecular tumbling (τ_{mol}) and internal motions (τ_i) at different nuclear sites in the molecule using certain simplifying assumptions. For example, one can assume that the lowest NT_1 value for a specific peptide reflects τ_{mol} [596]. For a rigid isotropically tumbling sphere of radius a, tumbling in a medium of viscosity η, and delete

$$\tau_{mol} = 4\pi a^3 \eta / 3kT \tag{4.3.5}$$

If anisotropic tumbling can be ruled out, then T_1 variations can be correlated to the additional frequency components at the nuclear site due to internal motions in the peptide.

A general theory for internal liberation and rotation has been developed [597]. For CH_3 groups in proteins and polypeptides, the internal rotation is slower than liberational motions.

The pyrrolidine ring of proline displays considerable conformational mobility in peptides and peptide hormones. This is manifested in the carbon –13 spin lattice relaxation times [598]. The motions responsible for the ring mobility arise from the ring puckering vibrations.

4.3.7 ^1H and ^{13}C Chemical Shifts

In a protein or peptide which does not have a preferred solution conformation, resonances from the same type of amino acid at different locations in the polypeptide chain are usually indistinguishable. For a polypeptide exhibiting a well defined structure, each amino acid residue has a unique environment and the resonances are shifted by a few ppm from the shifts corresponding to random coils. Such secondary shifts can be used for the detection of the local environment of a particular amino acid residue. Especially, amino acids close to an aromatic ring show pronounced secondary shifts and the theory to translate this information into geometrical parameters is relatively firmly founded. The magnetic anisotropy of the carbonyl group should, in principle, also provide useful information but the results so far have not been very encouraging. Titration of an acid-base functional group produces changes in the chemical shifts of groups in its immediate neighborhood. Other methods of a similar nature involve binding of a paramagnetic ion or of ligands. A technique which has shown promising results is the use of lanthanide shift reagents when the protein has a specific binding site for the lanthanide ions [167]. Use of lanthanide shift reagents has been made both in the case of small and in large polypeptides [527, 529]. In heme proteins [555, 556], iron acts as natural paramagnetic probe in structural studies.

The ^{13}C chemical shifts depend on a number of factors and it is difficult to quantitatively analyze the results of secondary shifts even when the factors causing such shifts (ring current, hydrogen bonding, configuration etc.) are known. However, such effects can be analyzed using semiempirical rules. D- and L-amino acid residues in linear peptides also influence carbon-13 NMR parameters [599].

4.4 Conformations of Amino Acids

Tabulations of the spectra, coupling constants and chemical shifts of L-amino acids were reported several years before the period covered in this report (for recent reviews see [2–11, 125, 530]). One of the motivations for such studies has been to understand the NMR of building blocks of proteins so that the resulting information can be applied to larger systems. The earliest spectra were reported already in 1957 (600) though meaningful conformational information emerged only after 1960.

Several attempts [593, 601–606] have been made to assess the relative conformer populations around the C^{α}–C^{β} bond (Fig. 4.3.3) using *vicinal* proton coupling constants. The relative populations depend on the state of ionization of the amino acid although all the three conformers show significant populations at all *p*H, values. The relative population of II is however less than ca. 25% and the equilibrium is thus dominated by I and III [593]. The relative residence times of a number of amino acids related to wool structure have been determined [602–605]. For histidine-containing peptides, the rotamer distribution is influenced by a *p*H-dependent intramolecular hydrogen bond between the imidazolyl and carboxylate group [607].

^{13}C-NMR data on amino acids have accumulated in recent years [608–610]. Like the proton spectral features, the ^{13}C parameters are also sensitive to *p*H [606, 611]. The chemical shifts are sensitive to conformation [612] while T_1 values have proved useful for studying intramolecular motions in the side chains. The conformational dependence of ^{13}C-NMR parameters in amino acids has been reviewed by Deslauriers and Smith [530].

The ring puckering of proline, both as free amino acid and incorporated into various peptides, has received much attention [613–620]. The theoretical treatment of ring puckering is similar to that discussed for the five-membered furanose ring in Chap. 3 [621, 622] and the conformation can be described in terms of two parameters. The most common conformational states of the molecule correspond to C^{α}–C^{β} half chairs. Two energy minima are found in the conformational space corresponding approximately to C^{γ}-*endo* and C^{γ}-*exo* conformations with a barrier of about 3 kcal/mol. The proton *vicinal* coupling constants show a fast *endo* $\rightleftarrows$ *exo* interconversion between the two states of the molecule discussed above [615–618]. Studies on carbon-13 spin-lattice relaxation times (Fig. 4.4.1) also indicate that proline exhibits a considerable mobility

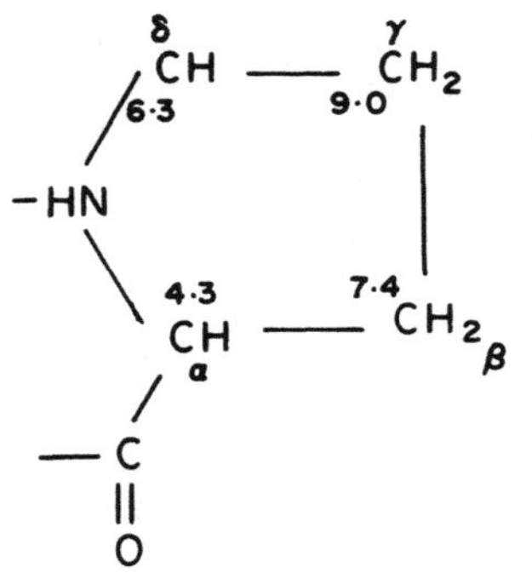

Fig. 4.4.1. Values of NT_1 (s) of proline *p*H 4 (data from Ref. [623])

which is reflected in shorter NT_1 values for α carbon atoms compared to β, γ and δ carbon atoms. The results can be interpreted in terms of rapid flips between approximate C^{β}–C^{γ} half chairs with C^{γ} exhibiting a somewhat greater angular displacement than C^{β} [614]. A typical range of motion for the C^{γ}–H vectors is 50–70°. Some general trends in the proline ring conformations have been deduced from studies of a number of proline-containing peptides [623]. The conformational flexibility of the ring is decreased in oligopeptides. Substitution at the nitrogen atom restricts the mobility of the δ-carbon.

To summarize, amino acids exist in solutions as equilibrium mixtures of several conformers. The rate of interconversion between such states is fast on the NMR time scale. The intramolecular motions are manifested in the ^{13}C spin-lattice relaxation times.

4.5 Linear Peptides

Small peptides play an important role in biology. Their function as hormones has been recognized for several years now. Synthetic peptides have found applications in medicinal chemistry. Certain peptides selectively transport ions across membranes. Microbial peptides containing not only the usual L-amino acids but also D-residues and some unusual amino acid residues and hydroxy acids have attracted considerable interest because of their antibiotic and toxic properties. In last twenty years, a great deal of efforts has been devoted to the synthesis of new physiological active peptides and to understand their mode of action at a molecular level. Biologically active peptides cover a wide range from dipeptide derivatives to polypeptides containing more than fifty amino acid residues.

NMR has been widely used to investigate the three-dimensional structure of small peptides in solution. However, even in such cases, the number of low-energy conformations is quite large. The conformational flexibility poses serious problems in the interpretation of experimental results and of the correlation of the peptide structures with their biological function. The conformation obtained by the NMR method is an average of a manifold of energetically favored structures which may coexist and rapidly interconvert in solution. The equilibrium between such states of the peptide depends on conditions such as solvent, ionic strength, concentration, temperature, and the presence of other species. Thus, meaningful structure-activity correlations can be derived only when either the peptide conformation is so rigid that it can be assumed that it will not be significantly modified by the receptor or if an unequivocal criterion can be established between the mean conformation and the conformation in solution. The problem is further complicated because very little is known about hormone or drug receptors. Thus, attempts at establishing structure-activity relationships have met with only modest success. Nevertheless, NMR studies have very fruitfully disclosed conformational features of peptides in solution and pointed out pitfalls involved in correlating crystal conformations with biological activity. Recent NMR results of linear peptides are compiled in Table 4.5.1.

Table 4.5.1. NMR studies on linear peptides

Nucleus	System and information	Ref.
	(a) Dipeptides and peptide analogs	
^{13}C, ^{15}N	Ala-Pro, ^{13}C, ^{15}N labelling: J	[624]
^{1}H	Dipeptides, aromatic: diastereomers	[625]
^{1}H	IR, models, potential energy calculations	[626]
^{1}H	Coiled conformations	[627]
^{13}C	Protected peptides: lanthanide shifts, sites, spectral assignment	[628]
^{1}H	Pd(II) complexes: CD, electronic spectra shi	[629]
^{1}H	Shift reagent studies	[630]
^{1}H	Models, IR, potential energy calculations	[631]
^{13}C	Prodipeptides: *cis-trans* isomerism	[137]
^{1}H	Gly-Sar, acetyl-Sar: rotation rates for peptide bonds	[587]
^{1}H	GLY-L-Phe, L-Phe-Gly: Pd(II) complexes	[632]
^{1}H, ^{13}C	Dipeptides containing Thr, Ser, Asp, His	[633]
^{1}H	Nitrobenzyl-Asp: model peptides, rotamer populations	[634]
^{1}H	*S*-Benzyl-cysteinyl, Cys-Asp, Val-Cys-Gly	[338]
^{1}H	Flavin-histidine peptides: fluorescence	[635]
^{1}H	Gly-dipeptides: structure-property relationship	[636]
^{1}H	Gly-Sar: magnetization transfer *cis-trans* isomerisation	[165]
^{13}H	L-Pro-L-4Hyp: isomerization rates	[574]
^{1}H	Pd(II) complex, Gly-L-Tyr	[637]
^{1}H	*N*-Me-L-Ala: *cis-trans* isomerization	[638]
^{13}C	Diproline: *cis-trans* isomerization	[136]
^{1}H	*cis-trans* Isomerization: use of double resonance	[567]
^{1}H	Peptide models: intramolecular interactions, IR	[639]
^{1}H	Acyl-Pro, acylthio-Pro: *cis-trans* isomerization	[572]
^{1}H, ^{13}C	Peptides containing Sar, Pro: *cis-trans* isomerization, *p*H effect	[565]
^{1}H	*N*-Substituted acetamide, substituent effects	[640]
^{1}H	L-Cys peptides: disulfide stereochemistry	[641]
^{1}H	Depsipeptides: β-turns	[642]
^{13}C	Enriched amino acids and peptides: effect of *p*H on *cis-trans* isomerizaton	[606]
^{1}H	L-Phe-Gly: specific isotope labelling	[643]
	(b) Tripeptides	
^{1}H, ^{19}F	Gly-fluoro prolyl-Trp	[644]
^{1}H	Di- and tri-peptides, wool: conformer populations	[645]

Table 4.5.1 (continued)

Nucleus	System and information	Ref.
^{1}H	Conformer populations; Asp, Cys	[601]
^{13}C	X-Pro-Pro: *cis-trans* isomerization	[138]
^{1}H	Leu-Gly-Phe: methylene chemical shift non-equivalence	[646, 134]
^{1}H	*p*H Dependence of conformation, γ-turn	[647]
^{1}H	Acetyl-Gly-Pro-methylamide, CD	[569]
^{1}H, ^{13}C	Thyrotropin-releasing factor TRF)	[648–653]
^{1}H, ^{13}C	Melanocyte-stimulating Harmone (MSH)	[654–656]
	(c) Tetrapeptides	
^{1}H	Tetrapeptides related to α-chymotrypsin	[657]
^{1}H	Proline containing peptides, β-turn	[658]
^{1}H	Frangulaine; tyrosine tetrapeptide	[659]
^{1}H	Tentoxin	[660]
^{1}H	Enkephalin related fragments	[662]
^{13}C	Gly-^{13}C-enriched peptide: preparation	[663]
^{1}H	Tropoelastin-repeating peptide: energy calculations	[664]
^{1}H, ^{13}C	Tetragastrin, Trp-Met-Asp-Phe-NH_2: T_1	[665]
^{1}H	Titration shifts of amide protons	[142]
^{13}C	Tetra- and pentapeptides: *cis-trans* isomerization	[566]
^{1}H	Pro-containing tetrapeptides: β-turns	[666]
^{1}H	Di-, tri- and tetrapeptide fragments of somatotropin release inhibiting factor	[667]
^{1}H	Elastin tetrapeptide	[668]
^{1}H, ^{13}C	Elastin repeating peptides	[669, 670]
	(d) Pentapetide	
^{1}H	Pentagastrin	[671]
^{1}H, ^{13}C	Elastin pentapetide	[672, 673]
^{13}C	Gly-Gly-X, Gly-Gly, X = Ser, Thr, Asp, Glu, Asn, Gln	[674]
^{1}H, ^{13}C	H-Gly-$(Pro)_4$-OH, CD, protein-saliva repeating peptide	[675]
^{1}H	Met-enkephalin, opiate-like peptide	[676]
^{1}H	Enkephalins, zwitterionic, cationic forms	[677]
^{13}C	Met-enkephalin: T_1	[678]
^{1}H, ^{13}C	(Leu^5)-enkephalin: β-bend, X-ray, energy calculations	[679]
^{1}H	(Leu^5)-enkephalin: conformers	[680]
^{1}H, ^{13}C	Enkephalin, derivatives	[681–683]
^{1}H	Tropoelastin pentapeptide, repeating pentapeptide	[109, 684]
	(e) Hexapeptides	
^{1}H	H-Gly-Trp-$(Gly)_n$-Trp-Gly –) H (n = 0.2)	[685]
^{1}H	Oxidized L-Cys-L-Lys-L-Ala-Gly-Gly-L-Cys	[686]
^{1}H	Tropoelastin repeating peptide: energy calculations	[687]
^{1}H	Elastin hexapeptide, Ca-binding: CD	[669]
^{1}H, ^{13}C	Elastin hexapeptide	[688, 689]
	(f) Larger peptides and peptide hormones	
^{1}H	Methionine heptapeptide	[690]
^{1}H	Linear oligopeptides containing Tyr, spatial arrangement of the aromatic moiety	[691]

Table 4.5.1 (continued)

Nucleus	System and information	Ref.
^{1}H, ^{13}C	Bradikinin, Arg-Pro-Pro-Gly-Phe-Ser-Pro-Phe-Arg: *sic-trans* isomerization	[692]
^{13}C	Human encephalitogenic nonapeptide: synthesis	[693]
^{1}H	Physalemine, *C*-terminal peptide: CD	[694]
^{1}H	Corticotropin: *cis-trans* isomerization	[586]
	Angiotensin	
^{1}H	Angiotensin II: solvent saturation	[189]
^{13}C	Angiotensin-proline analog	[695]
^{13}C	Flexibility in angiotensin II: T_1	[696]
^{1}H	Angiotensin: hydrogen exchange	[697]
^{13}C	X-Pro bond in angiotensin and other peptides	[650]
^{13}C	Angiotensin structure	[698]
^{1}H	Angiotensin II hydrogen bond: proton exchange	[121]
^{1}H, ^{13}C	Angiotensin conformation	[699, 700]
	Luteinizing Hormone – Releasing Factor	
^{1}H	LH-releasing factor analogs	[701]
^{1}H	LHRH and component peptides	[702]
^{1}H	LHRH: intramolecular charge transfer complexes	[703]
^{13}C	LHRH, flexibility: T_1	[704]
^{1}H	conformation in solutions	[534]
	Insulin	
^{1}H	Despentapeptide (B26–30) of insulin: aggregation	[705]
^{1}H	Des (B chain *C*-terminal pentapeptide)	[706]
^{1}H	Linewidths, molecular mobility	[719]
	Parathyroid hormone	
^{1}H	Human parathyroid hormone, synthetic fragment 1–34	[707]
	Alamethicin	
^{1}H	Zuzukacillin: sequence, conformation, relatino to alamethicin	[708]
^{13}C	Polymer-bound peptide conformation, alamethicin	[709]
^{1}H	Peptides containing α-aminoisobutyric acid	[710]

4.5.1 Dipeptides

Dipeptides are simplest models to test the conformational flexibility of peptides in the (ϕ, ψ)-space and to check theoretical predictions on the preferred conformations of such molecules. Attempts have therefore been made to determine the conformation of dipeptides as dilute solutions in CCl_4 using NMR in conjunction with IR, dipole moments and other physical techniques [626, 627, 631, 639, 711]. Under these conditions, the behavior of the molecules should mimic theoretical calculations where solvent interactions are ignored. It has been concluded that under these conditions the molecule prefers two conformations:

(a) an extended conformation with a weak interaction between N1–H1 and C2–O2 (C_5 structure) and

(b) a folded seven-membered ring conformation (C_7) with a strong intramolecular hydrogen bond between N2–H2 and C1–O1. However, the three-bond $J(HNC^{\alpha}H)$ can also be interpreted in terms of a random conformation of dipeptides [712, 713].

An interesting paper on the conformation of L-PheGly makes use of the specific ^2H and ^{15}N isotope substitution of the Gly residue to obtain the mean conformation of this molecule [643]. The proposed structure involves hydrogen bonds between the terminal $\overset{\oplus}{N}H_3$ and $COO^{\ominus}$ groups.

The coupling of *N*-acyl-α-amino acids with α-hydroxymethyl amides results in depsipeptide molecules containing two chiral centers and one ester group inserted between two amide *functions*. Such linkages occur in certain biologically active peptides such as valinomycin. NMR, IR, X-ray and theoretical investigations have shown that these molecules fold due to the 4 → 1 hydrogen bond to yield conformations similar to those of the β turns in peptides [642].

A large number of papers on dipeptides are devoted to *cis-trans* isomerization in Pro- or Sar-containing peptides [136, 137, 165, 565, 567, 572, 574, 587, 606, 638]. NMR has also been used to study the stereochemistry of disulfide bonds [641] and structure of lanthanide and transition-metal ion complexes [628–630, 637].

4.5.2 Tripeptides

The thyrotropin-releasing factor (TRF) which controls the secretion of the thyrotropic hormone from the anterior pituitary glands belongs to this class. TRF has the sequence Glu-His-Pro-NH_2 (Fig. 4.5.1). In order to study the structural requirements for its activity, many biochemical and physical studies have been undertaken. CPF calculations have predicted a number of low-energy conformations for TRF [714]. These results show that there is no interaction between the cyclic moieties and that the pyrrolidine ring of pyroglutamate is not involved in any intramolecular interaction. The His-Pro peptide bond has a *trans* conformation. Several models with or without intramolecular hydrogen bonds have been postulated. Thus, these studies predict a flexible conformation for TRF.

Both ^1H- and ^{13}C-NMR have been applied to elucidate the solution conformation of TRF [648–653]. In aqueous solutions, the His-Pro peptide bond predominantly assumes a *trans* conformation. However, a small population of *cis* isomer is present. The

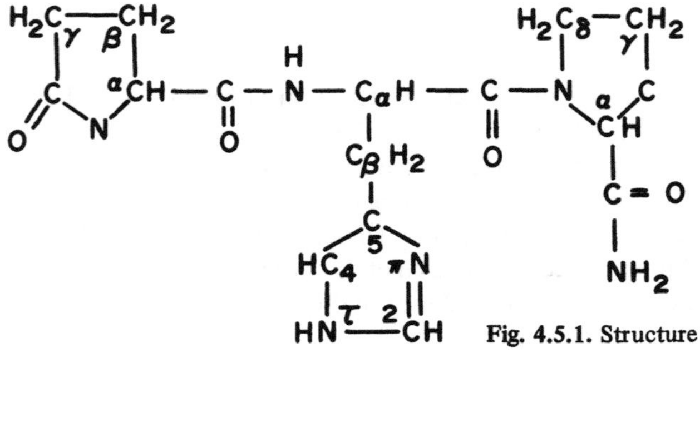

Fig. 4.5.1. Structure of the thyrotropin release factor

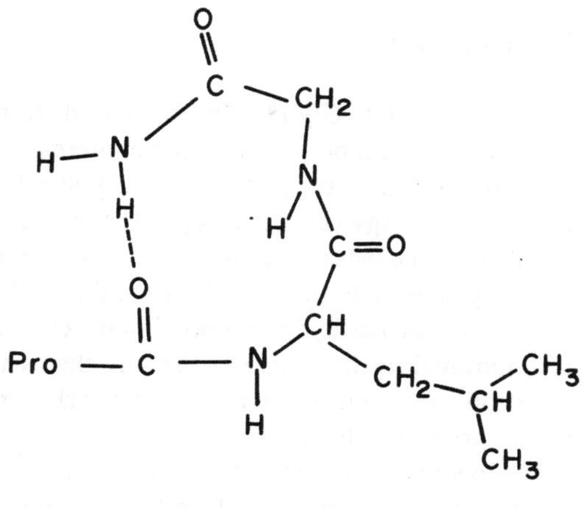

Fig. 4.5.2. β-Turn structure of melanocyte – stimulating hormone-release inhibiting factor

${}^{13}C$ chemical shifts reveal that the N^{τ}–H tautomer is mainly populated in basic solutions [652]. The activity of TRF derivatives has been correlated with pK_a of His. However, the ${}^{13}C$ spin-lattice relaxation times do not exhibit a large change in relative mobilities at different pH values. Thus, it is inferred that the proposed hydrogen bond between N^{π} of the imidazole ring and the His peptide nitrogen, if present, does not immobilize His side chains. It has been suggested that the N^{τ}–H tautomer of a neutral His residue is present in the active form of TRF [652].

The tripeptide Pro-Leu-Gly-NH_2 has been proposed to act as an inhibiting factor in the release of the melanocyte-stimulating hormone from pituitary. Its conformational structure has been studied by ${}^{1}H$- and ${}^{13}C$-NMR [122, 654–656], by CPF calculations [715] and by X-ray crystallography [716]. These studies indicate a ten-membered β-turn structure involving a hydrogen bond between the *trans* carboxamide proton and the prolyl carbonyl carbon (Fig. 4.5.2) as one of the preferred conformations. A conformational transition is observed when D_2O is replaced by DMSO. ${}^{13}C$ relaxation time measurements indicate a very flexible structure for MSH–R–IF [656]. Thus, more than one conformations may coexist in solution.

4.5.3 Tetrapeptides

If sharp bends in the peptide backbone such as β-turns result from interactions between adjacent peptide units, then it should be possible to demonstrate such conformers in small molecules. Derivatives of tetrapeptides which are likely to form β-turns have been examined by ${}^{1}H$-NMR and saturation transfer experiments [594, 666]. Sequences such as Gly-L-Pro-D-Val-Gly, Val-L-Pro-Gly-Gly, and Gly-L-Pro-L-Asn-Gly fold in β turns in methanol solutions. However, the sequence Gly-D-Ala-L-Val-Gly does not form such structures.

${}^{1}H$- and ${}^{13}C$-NMR data of common amino acid residues (X) in linear tetrapeptides H-Gly-Gly-X-L-Ala-OH have been obtained [66, 661]. The influence of X on the neighboring units Gly 2 and Ala 4 has been estimated. These data are suitable for the determination of "random-coil" parameters in conformational studies of polypeptide chains in D_2O and H_2O solutions.

4.5.4 Pentapeptides

Enkephalin (Tyr-Gly-Gly-Phe-Phe-Leu) binds to the same receptor sites as rigid opiates. Therefore, there has been considerable interest in the solution conformation of this endogenous analgesic peptide [676–683]. Studies using X-ray crystallography [717] and potential energy calculations [718] have also been performed. These studies indicate that enkephalin exists in a well-defined conformation both in crystals and in solutions. In dimethyl sulfoxide solution (Leu^5)-enkephalin forms a β-bend structure stabilized by an intramolecular hydrogen bond between Leu^5 NH and CO of Gly^2. Considerable conformational rigidity exists in the side chains and a portion of the molecule has an orientation analogous to the rigid framework of morphine which may account for its analgesic properties [679].

The low molecular weight proline-rich proteins are metabolic precursors in the biosynthesis of glycoproteins. Investigations on human parotid saliva reveals that the repeating unit of this protein is Gly-Pro_4. In their zwitterionic form, all the Pro residues adopt a *trans* conformation [675]. When the pD decreases, residues 2 and 5 remain in the *trans* form whereas residues 3 and 4 reveal the presence of *cis-trans* isomerism.

4.5.5 Hexapeptides

Tropoelastin, the precursor protein of the biological elastin fiber contains three repeating sequences: Val-Pro-Gly-Gly, Val-Pro-Gly-Val-Gly and Ala-Pro-Gly-Val-Gly-Val. In order to understand the function and pathology of biological elastic fibers, ^{1}H-NMR studies have been carried out on these repeating sequences [664, 672, 687]. In conjunction with potential energy calculations, molecular models for such repeating units have been postulated. Several sets of intramolecular hydrogen bonds have been detected in organic solvents leading to the formation of β- and γ-turns.

4.5.6 Larger Peptides. Peptide Hormones

Corticotropin is a peptide hormone which intervenes in the stimulation and regulation of steroid genesis. It contains 39 amino acid residues. This hormone adopts a partially ordered structure in solution [586] and undergoes a *cis-trans* isomerization at Pro-24. The proton signals of Tyr-23 and Val-22 have been used to monitor changes at Pro-24.

Bradikinin is a nonapeptide whose conformational structure has been studied by ^{1}H- and ^{13}C-NMR [692]. The three X-Pro are in a *trans* conformation. Rapid internal motion is indicated by the ^{13}C spin-lattice relaxation times for all peptide side chains. The results point to a disordered structure in solution.

4.5.6.1 Angiotensin

Angiotensin II is a natural pressor agent and its analog angiotensinamide II has been prepared synthetically. It is a linear octapeptide with the primary sequence Asn-Arg-Val-

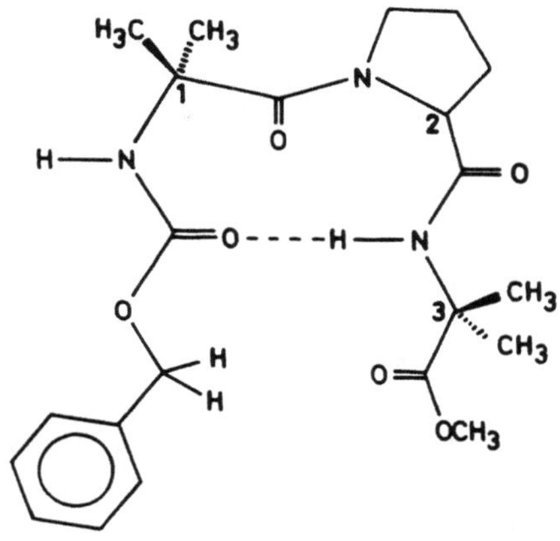

Fig. 4.5.3. Type III β-turn conformation of protected tripeptide Z-Aib-Pro-Ala-OCH_3 (reprinted with the permission from Ref. [710], copyright 1979, American Chemical Society)

Tyr-X-His-Pro-Phe where X = Val when the source is bovine and Ileu for hog and horse. A large number of models has been suggested for the conformational structure of angiotensin II [699]. Extensive ^1H- and ^{13}C-NMR studies [121, 189, 650, 695–700] have failed to provide a suitable model for the structure involved in biological activity. The peptide aggregates in concentrated aqueous solution or in DMSO. The molecule also displayes a tendency to fold on itself leading to β-turns [699]. The Pro residue adopts a *trans* conformation. Titration of the acid-base functional groups also results in conformational transitions [700]. Thus, the conformational equilibrium involves several states and is strongly influenced by environmental conditions. Considerable flexibility in the backbone is indicated by the ^{13}C spin-lattice relaxation times [696], with the exception of *C*-terminal Phe and the *N*-terminal Asp.

4.5.6.2 Leuteinizing Hormone-Releasing Hormone (LRF)

LRF acts upon the anterior pituitary causing secretion of follicle-stimulating and luteinizing hormones. The primary sequence for this peptide is Glu-His-Trp-Ser-Tyr-Gly-Leu-Arg-Pro-Gly-NH_2 and its solution conformation has been studied by ^1H- and ^{13}C-NMR [534, 701–704]. The temperature dependence of NH proton resonances in aqueous and DMSO solutions does not reveal any strong intramolecular hydrogen bonds [702]. The conformations in the two solutions are very similar and do not indicate any stacking of the indole and phenol rings. The Pro residue is predominantly in a *trans* conformation. The ^{13}C spectrum resembles that calculated from the shifts of constituent amino acids and the values of J(HNCH) range between 6.2 to 7.4 Hz. These evidences indicate a random-coil conformation for LRF.

4.5.6.3 Peptides Containing α-Aminobutyric Acid (A_{ib})

α,α-Dialkylamino acids introduce considerable steric hindrance to the conformation of linear peptides. Such amino acids occur in relatively large proportions in antibiotic alamethicin and the related peptide suzukacillin. NMR studies on peptides related to alamethicin have been carried out in recent years [708–710]. It has been observed that these molecules adopt well-defined conformations in solution. The protected tetrapeptide C_6H_5-CH_2-O-Aib-Pro-Aib-Ala-OMe for example, shows two consecutive type III β-bend structures.

4.5.6.4 Human Parathyroid Hormone (PTH)

PTH consists of one polypeptide chain containing 84 amino acid residues. Its major role is in calcium homeostatis in bones and kidneys. A biologically active fragment PTH (1-34) has been synthesized. Its NMR spectrum shows that the molecule exists in a flexible extended form but a short segment forms a rigid spatial structure [707] involving residues 20–24.

4.5.6.5 Insulin

Again, studies on this relatively large hormone has been restricted to fragments [705, 706, 719]. The line widths of polypeptides are considerably larger indicating a slow tumbling rate [719].

4.6 Cyclic Peptides

One of the most fruitful applications of NMR spectroscopy to biological molecules has been in the area of molecular conformations of cyclic peptides. The extent of such applications can be gauged from Table 4.6.1 which outlines recent studies on the subject. For earlier work, the reader is referred to other reviews in this area [125, 328, 527, 531, 532, 534, 536, 537, 539–541, 554].

Cyclic peptides have several important biological functions. They act as hormones, antibiotics, antitoxins and ion carriers. The constraints introduced by cyclization considerably reduce the large number of conformational structures possible for a linear peptide containing the same number of amino acid residues. Moieties such as proline which occur in a large number of cyclic peptides further reduce the conformational freedom of the molecule.

4.6.1 Dipeptides

Cyclic dipeptides or diketopiperazines (DKP) have been investigated by X-ray crystallography [863] and theoretical calculations [864]. The overall shape of the six-membered DKP ring can be best described by conformations shown in Fig. 4.6.1 [633]. Both the peptide bonds are in the *cis* configuration. The overall ring conformation is decided by the interaction between the DKP ring and the side chains. In peptides containing an aromatic amino acid residue (X, where X = L-Phe, L-Tyr, etc.), the overall shape of the molecule will be such that a maximum overlap between the aromatic and DKP rings

Table 4.6.1. NMR studies on cyclic peptides

Nucleus	System and information	Ref.
	(a) Cyclic dipeptides	
^1H	Cyclo(Gly-X), X = aromatic group	[720–722]
^1H, ^{13}C	Cyclo(Pro-Leu); D- and L-amino acids	[723]
^1H	Cyclo(Gly-Sar)	[724]
^1H, ^{13}C	Pro-containing cyclic peptides: lanthanide shift reagents	[725]
^1H	Aromatic dipeptides	[625]
^1H	Diketopiperazine, sarcosine containing dipeptides	[726]
^1H	Cyclo(sar-*N*-tosylanthraniloyl): rotation barrier	[727]
^1H	Cyclo(Thr-His), griselimycin	[728]
^1H	Cyclic dipeptides, tryptophan: excited state	[729]
^{13}C	2.5-Diketopiperazine, dipeptides	[730]
^1H	Containing His and tryptophan residues	[731]
^1H	Cyclo(L-Thr-L-Thr), cyclo (L-Thr-L-His)	[732]
^1H	Thio analogs, piperazinethione	[733]
^1H	Diketopiperazine containing Pro	[734]
^1H	Cyclo(L-Pro-L-Pro), cyclo(L-Pro-D-Pro): lanthanide shifts	[735]
^1H, ^{13}C	Cyclo(L-cystine): UV, CD, ORD	[736]
^1H	Cyclo(L-His-L-Ser), cyclo(L-His-L-Asp), cyclo(L-His-L-His)	[737]
^1H	Pro, pipecolic acid and/or 2-azetidine-carboxylic acid	[738, 739]
^1H	Simple cyclic peptides containing Gly, Sar, Ala	[740]
^1H	Cyclic peptides of sarcosine	[741]
^1H	Diketopiperazine, proline, hydroxyproline	[620]
^{13}C	Pro-containing cyclic dipeptides	[742]
^1H	*cis*-2,5-Diketopiperazine, Phe: deuterium labelling	[743]
^1H, ^{13}C	Frangulin: solvent-induced conformation	[744]
^1H, ^{13}C	Cyclic dipeptides containing Thr, Ser, Asp and His	[633]
^1H	Cyclic dipeptide, proline, dipeptides containing Sar	[745]
^{13}C	Cyclo(D-Leu-L-Leu), cyclo (L-Leu, L-Leu)	[746]
^1H	*N*-Monosubstituted cyclic dipeptides, strain	[747]
	(b) Cyclic tripeptides	
^1H	Cyclo(L-Aze)$_2$, cyclo(L-Aze)$_3$, cyclo(L-Pro)$_3$	[748]
^1H	Cyclo(Pro)$_3$	[749]
^{13}C	Cyclo(Pro)$_3$	[116, 595]
	(c) Cyclic Tetrapeptides	
^1H, ^{13}C	Cyclo(L-Pro-Gly)$_2$, slow *cis-trans* isomerization	[139]
^1H	Containing Gly, Ala, and Sar	[750]
^{13}C	Tentoxin, cyclic tetrapeptide: CD	[751]
^1H, ^{13}C	Tentoxin, isolation and bioassay: CD	[752]
^1H, ^{13}C	Cyclo(Gly)$_4$	[753]
^1H	Cyclo(Sar)$_4$	[754]
^1H	Serratamolide	[755, 756]
^1H	Diasterioisomeric dihydro-AM-toxin I, tetradepsipeptide analog	[757]
	(d) Cyclic pentapeptides	
^{13}C	Cyclo(Gly-Ala-Gly-Gly-Pro): two conformations	[758]
^1H	Cyclo[Phe$_n$Gly)$_{5-n}$]	[198]

Table 4.6.1 (continued)

Nucleus	System and information	Ref.
^{15}N	Cyclicpentapeptide	[759]
^{1}H	Cyclo(Gly-Pro-Gly-D-Ala-Pro)β and γ-turns	[199]
^{1}H	Cyclo(Phe$_3$-Gly$_2$), deuterated cyclopentapeptide	[760]
^{1}H	Cyclic pentapeptide: β and γ-turns	[761, 762]
^{1}H	Cyclic pentapeptides: hydrogen bond	[763]
^{1}H	Cyclic pentapeptides, sarcosine, alanine, glycine	[764]
^{1}H, ^{13}C	Cyclo(Pro-Phe-Gly-Phe-Gly)	[765]
^{1}H	Cyclic pentapeptides, review	[766]
^{1}H	Cyclo(Sar)$_5$: crystal structure	[767]
^{1}H	Cyclic pentapeptides	[768]
^{1}H	Actinomycin model: actinocylbissar-cosylvaline	[769]
^{1}H	Actinomycin-related peptides	[770]
^{1}H	Actinomycin pentapeptide lactone	[771]
^{13}C	Actinomycin pentapeptide lactone rings	[772]
	(e) Cyclic hexapeptides	
^{1}H	Cyclo(Ser-Pro-Gly-Ser-Pro-Gly)	[773]
^{1}H	Cyclo(Sar-Sar-Gly)$_2$, cyclo(Sar)$_6$, cyclo(Sar-Gly-Gly)$_2$	[774]
^{1}H	Cyclo(Pro-Ser-Gly-Pro-Ser-Gly)	[775]
^{1}H	Cyclo-tri(D-hexahydromandelyl-L-Val)	[776]
^{1}H, ^{13}C	Cyclo(Pro-Gly)$_3$	[777]
^{1}H	Containing *N*-methylamino and α-hydroxy residues: IR, CD	[778]
^{1}H	Containing L-Val, L-Leu, L-norvaline, Gly residues: IR, CD steric factors	[779]
^{1}H	Cyclo(Pro-Sar-Gly)$_2$: solvent-induced shifts, interaction with benzene	[191]
^{1}H	Evolidine	[780]
^{1}H	Containing *N*-methylamino acids: steric, CD	[781]
^{1}H, ^{13}C	Cyclo(Pro-Gly)$_3$, CD	[144]
^{1}H	Containing side chains at 2, 3, 5 and 6	[782]
^{1}H	Cyclo(D-Phe-L-Pro)$_3$	[783]
^{1}H	Cyclo(L-Ala-L-Ala-L-Ala-L-Ala-L-Ala-D-Ala)	[784]
^{13}C	Cyclo(Gly-L-Pro-Gly-Gly-I-L-Pro-Gly): assignments	[785]
^{1}H, ^{13}C	Proline peptide: β-structure	[786]
^{1}H	Cyclo(Gly-L-Pro-Gly-Gly-L-Pro-Gly)	[787]
^{1}H	Cyclic hexapeptides: β-structure, CO, substituent effects	[788, 789, 792, 794]
^{1}H, ^{13}C	Cyclo(Pro-Gly)$_3$	[790]
^{1}H	Cyclo(Gly-Gly-D-Ala-D-Ala-Gly-Gly)	[791]
^{1}H	Cyclo-tris(L-CO-benzyl-tyrosyl-Gly)	[793]
^{1}H	Cyclo(L-Val-L-Pro-Gly-L-Val-L-Pro-Gly)	[795]
^{13}C	Cyclo(X-Pro-Y)$_2$	[596]
^{13}C	Cyclo(L-Pro-Gly)$_3$	[796]
^{1}H	Monamycins: cyclohexadepsipeptides, antibiotic	[797]
^{1}H	Ferrichrome peptides: Structure and strain contributors	[798]
^{13}C, ^{15}N	Ferrichrome, peptide hydrogen bonding, ^{13}C shifts	[799]
^{13}C	Ferrichrome conformation dependence of ^{13}C, alumichrome	[800]
^{13}C	Alumichrome: assignments of ^{13}C signals	[801]
^{1}H, ^{13}C	Ferrichrome: charge relay hydrogen bond, solvent effect	[802]
^{1}H, ^{13}C	Metal-free ferricrocin, ferrichrysin	[803]
^{1}H	Tropoelastin, cyclic analog of repeating hexapeptide	[804]

Table 4.6.1 (continued)

Nucleus	System and information	Ref.
	(f) Larger cyclic peptides: peptide hormones and antibiotics	
	General peptides	
^1H	Cyclo(Gly-L-Leu-Gly)$_3$	[805]
^1H	Cyclic peptide-Leu, Tyr fragments: interaction with pepsin and chymotripson, kinetics	[806]
^1H	Cyclic deca-L-peptide, antamanide	[807]
^1H	Medium-size depsipeptides	[808]
^1H, ^{13}C	Cyclo(Pro-Gly-Gly-Pro)$_2$: CD	[809]
^{13}C	Cyclolinopeptide A, conduction IR, K^+ ion complex hydrogenbond	[810, 811]
^1H	Cyclo(Pro-Gly)$_4$: CD, metal and amino acid complexes	[812]
^1H	Cyclo(depsipeptides), destruxin B insectidal cyclodepsipeptides	[813]
^1H	Cation-specific bicyclic peptide	[814]
	Neurohypophyseal hormones oxytocin and vasopressin	
^1H	Pencillamine oxytocin, peptide hormones	[815]
^1H	C-terminal tripeptides: CD, vasopressin analog	[816]
^1H, ^{13}C	Hormone pituitary, review	[817]
^1H	Oxytocin analog	[818]
^{13}C	Oxytocin, vasopressin, orginine vasotocin, neurohypophysical hormones: review	[819]
	Neurohypophyseal hormones: structure model	[820]
^1H	Effect of trifluoroethanol-oxytocin, vasopressin	[821]
^{13}C	Oxytocin, related peptides and analogs	[822]
^{13}C	Oxytocin, vasopressin, bacitracin, assignment	[823]
^1H	Lysine vasopressin	[824]
^1H	Oxytocin, vasopressin, peptide hormones	[825]
^1H	(1-Penicillamine, 2-leucine) oxytocin, synthesis	[826]
^{15}N	Oxytocin, Pro-Leu-Glycinamide	[827]
^{13}C	Oxytocin, arginine vasopressin, isotocin, mesotocin, glumitocin, aspartocin, other analogs	[828]
^{13}C	Oxytocin interaction with bovine neurophysins I and II	[829]
^{13}C	Oxytocin binding with bovine neurophysins I: ^{13}C enrichment	[830]
^{13}C	Oxytocin: neurophysin interaction	[831]
^1H	Oxytocin fragments	[832]
^1H	Oxytocin, (penicillamine) oxytocin: conformational flexibility and function	[833]
^1H	Tocinamide, retro-D analog: CD	[834]
^1H	Tocinamide: CD	[835]
^1H	Oxytocin: solvent saturation transfer	[193]
	Gramicidin S and analogs	
^1H	NMR-solvent saturation and titration, H-bonded amide protons	[189, 836]
^1H	Difference: NMR, INDOR	[590]
	(review)	[536]
^1H	*N*-Methylleucine-gramicidin S	[837]
^1H	N, N^1-Diphthalogyl-gramicidin S	[838]
^1H	Dimethylleucine gramicidin S: resonance assignments	[839]
^1H	Internal rotations and solution conformation	[840]

Table 4.6.1 (continued)

Nucleus	System and information	Ref.
^1H	*N*-Methylleucine gramicidin S	[841]
^1H	Biosynthesis of deuterated analogs: difference scalar decoupling for spectral analysis	[842]
^1H	Gramicidin S nitroxide: line broadening	[174]
^{13}C	Internal rotation in amino acid side chains	[843]
^{13}C	Hydantoinic scale: gramicidin S conformation	[844]
^{13}C	^{13}C-enrichment: spectral assignments	[845]
^1H, ^{13}C	Gramicidin: conformational mobility	[846]
^1H	Native gramicidin S: hydrogen bond	[847]
^1H	Native gramicidin S analogs: CD	[848]
^{19}F, ^{13}C	Gramicidin A: channel in phospholipid residues	[849]
^{15}N	Spectral assignments: relation to structure	[850, 851]
	Valinomycin	
^1H	Analog, cyclo(D-Val-L-Pro-L-Val-D-Pro)$_3$	[852]
^1H	Lanthanide shift reagents, backbone torsion angles	[853]
^1H	Valinomycin analogs: configurational alteration, ORD	[854]
^1H	Valinomycin, analogs: IR, ORD	[855]
^1H	Solvent-dependent conformations, CD	[856]
^1H	Valinomycin: homoallylic coupling	[857]
^1H	Sandwich complexation and membrane processes	[858]
^1H	Coupling constants, ϕ, ψ angles	[879, 880]
	Antibiotics	
^1H	Telomycin: hydrogen bonds, β-turn	[859]
^1H	Stendomycin: β-turn, α-helical segments	[860]
^1H, ^{13}C	Viomycin: intramolecular hydrogen bonds	[861]
^1H	Polymyxin B: folded structure	[862]
^1H	Ilamycin B$_1$: β-turn, peptide antibiotics	[35]

occurs [863] subject of course to steric constraints. Thus, peptides such as cyclo(Gly-X) adopt a flag pole-boat conformation. The introduction of an L-amino acid residue (Y) in place of Gly prevents flag pole orientation due to steric interactions between the two side chains. In such a case, a planar conformation can still maintain the interaction between DKP and the aromatic ring while the side-chain interference is removed. In the absence of aromatic side chains, several conformations have similar energies and the molecular structure is dictated by side-chain interactions and medium effects. Both planar and bowsprit conformations may exist in such cases. The peptides adopt a distorted conformation when side chains such as Ser or Thr are present since these residues form intramolecular hydrogen bonds through the OH group [633] in the side chain. NMR studies by Kopple and coworkers [720–722] have shown that the protons of the second residue of cyclic dipeptides containing aromatic moieties are shifted upfield, in conformity with the above rules. This, together with the values of $^3J(NH{-}C^{\alpha}H)$ coupling constants has facilitated the assignment of a "flag pole" conformation to cyclic peptides of the general formula cyclo(Gly-X). On the other hand, molecules such as cyclo(Y-X),

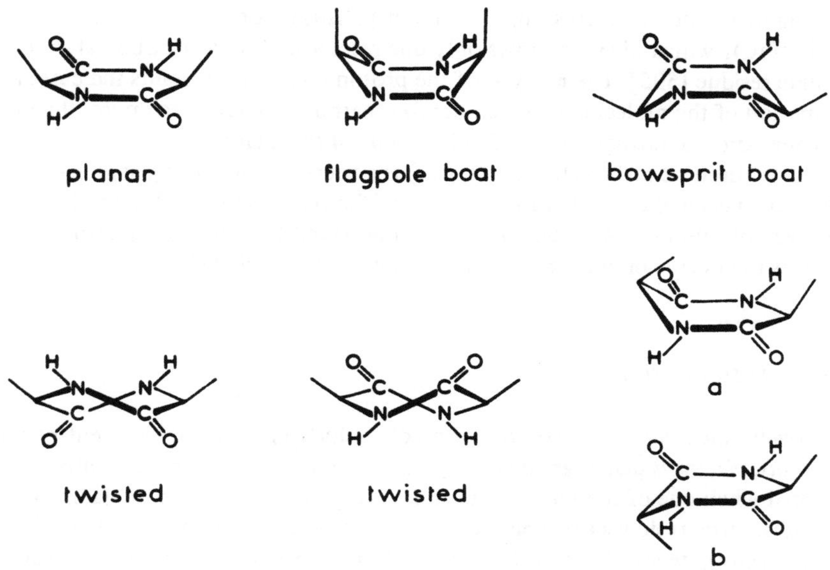

Fig. 4.6.1. Possible conformations of the six-membered ring of 2,5-diketopeperazine. In addition to the planar and boat forms, distorted forms may exist due to the folding of side chains in L (a) and D (b) residues (reprinted with the permission from Ref. [633])

where Y = L-Ala, L-Val, etc., prefer a planar conformation. In proline-containing cyclic dipeptides, both boat and planar conformations of the DKP ring coexist [742]. With polar side chains such as Thr, Ser, and Asp, short-range intramolecular interactions stabilize the distorted forms [633].

Carbon-13 spin-lattice relaxation times have been used to monitor the overall molecular motion as well as internal motions in 2,5-diketopiperazines [730]. The α-carbons of Gly residues in cyclic dipeptides containing such residues show enhanced intramolecular mobility when compared to the α-carbon of the second residue. The T_1 values of cyclo(L-Tyr-Gly) are influenced by the hydrogen bonding of the hydroxy group to the solvent [730]. To elucidate the effect of D- and L-amino acid residues, the T_1 values of cyclo(L-Pro-L-Leu) and cyclo(L-Pro-D-Leu) have been measured [723]. The Pro residue reveals a greater mobility in the former peptide. Moreover, the conformational structure is influenced when the configuration of one of the amino acid residues is changed from L to D [723, 725, 735].

4.6.2 Tripeptides

Studies on cyclic tripeptides have been limited to amino acid residues related to proline. Stereochemical considerations demonstrate that all peptide units must adopt a *cis* conformation [865] in such systems. The ^{1}H-NMR spectra of cyclo(Pro)$_3$ [749] reveal that the three proline rings are symmetrically arranged around the axis through the ring [116] leading to a structure possessing C_3 symmetry. This is confirmed by the ^{13}C-NMR of cyclo-(Pro)$_3$ which likewise shows only one resonance for each carbon atom in the proline residue [595]. The analysis of the proton coupling constants indicates a rigid conformation of the molecule in which the four carbon atoms of each pyrolidine ring are coplanar and the nitrogen atoms displaced out of this plane.

L-Azetidine-2-carboxylic acid (L-Aze) is a four-membered ring analog of proline. The biological and conformational effects of the replacement of Pro residues by Aze has been of interest [748, 866]. It has been observed that the introduction of L-Aze reduces the degree of conformational freedom of the angle Φ [748].

4.6.3 Tetrapeptides

The twelve-membered ring conformation of cyclo(Sar)$_4$ has a unique centrosymmetric *cis, trans, cis, trans* planar amide configuration [750, 754]. A similar conformation is adopted when one of the sarcosine residues is replaced by glycine or alanine or when two adiametric residues are replaced by glycine. On the other hand, ^{1}H and ^{13}C spectra of cyclo(Gly)$_4$ reveal that in this peptide all the four peptide units are equivalent [753]. The molecule has a S_4 symmetrical structure with the symmetry axis perpendicular to the cyclopeptide plane. The peptide groups differ from the standard *cis* or *trans* orientation and a rapid equilibrium between the two states is ruled out by NMR. It has been proposed that Gly residues in this case assume a non-planar *transoid* conformation. Evidence of a slow rotation around the C—C′ bond in cyclo-(L-Pro-Gly)$_2$ has been obtained [139].

Tentoxin is a phytotoxic cyclic tetrapeptide which causes a selective chlorosis when it is applied to the germination of certain plant seedlings. It exhibits the structure cyclo-(L-MeAla-L-LeuMe-Phy[(Z)Δ]-Gly). NMR studies reveal that its conformation in chloroform solution is related to that of cyclo(Sar)$_4$ except that the Gly-MeAla end of the ring undergoes a "ring flip" [751, 752].

The conformation for the naturally occurring antibiotic serratamolide has been determined using NMR and IR [755]. The ring possesses a two-fold axis of symmetry, as indicated by the NMR spectra. The NH protons are exchanged relatively rapidly with the medium and exhibit a large temperature-dependent chemical shift which conforms to a model devoid of intra-molecular N—H···O=C bonds.

4.6.4 Pentapeptides

Cyclic pentapeptides have been used as models for β and γ turns of peptide structures. Two peptides cyclo(Gly-L-Pro-Gly-D-Ala-L-Pro) and cyclo(Gly-Pro-Gly-D-Ala-Pro) have been studied by ^{1}H- and ^{13}C-NMR [761]. These compounds contain *all-trans* peptide bonds and two of the five N—H protons do not participate in intramolecular interactions. A model containing one β and one γ turn has been suggested for these peptides.

For cyclic (Gly-L-Ala-Gly-Gly-L-Pro), two conformations (*M* and *m*) have been detected in DMSO [758]. The species *M* has two and *m* has one N—H···O=C hydrogen

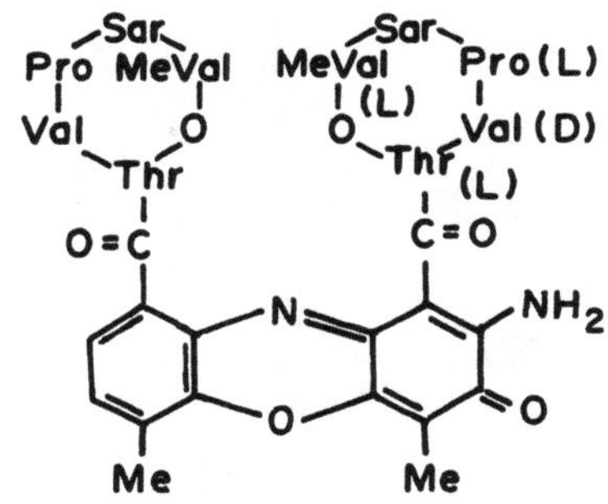

Fig. 4.6.2. Structure of actinomycin D (reprinted with the permission from Ref. [125])

bonds. The proline residue displays a *trans-trans'* conformation in *M* and a *cis-trans'* structure in *m*.

Unlike other cyclic pentapeptides, cyclo(Sar)$_5$ exists in solution in only one dominant conformation [767]..The *N*-methyl and methylene groups show chemical non-equivalence indicating the existence of an asymmetrical structure. Based on crystal structure data a *cis*, *cis*, *cis*, *trans*, *trans*-conformation has been assigned to the peptide groups.

Actinomycin D (Fig. 4.6.2) inhibits DNA-dependent RNA synthesis. It has two pentapeptide lactone rings. Attempts have been made to determine the solution conformation of this important antibiotic and its analogs [772, 769–772]. In crystals [867], the molecule assumes *cis* conformation for the sarcosine and proline residues and the structure is stabilized by hydrogen bonds involving D-Val. Some controversy exists as to whether the crystal structure also exists in solution. A hydrogen bond between the D-Val NH proton and the Sar carbonyl group is however confirmed by the slow exchange rate and temperature coefficient of the chemical shift of the NH proton [771].

4.6.5 Hexapeptides

Cyclic hexapeptides form a 18-membered ring system. Several molecules of this class of peptides are found in nature. The synthesis of cyclic hexapeptides and depsipeptides is relatively simple. Certain cyclodepsipeptides act as selective ion carriers. Schweizer and Ludescher [868] have proposed an antiparallel β-sheet structure with C_2 symmetry (Fig. 4.6.3) as a general backbone conformation for such peptides [868]. The hydrogen bonds enclose two ten-membered rings. Such a β-sheet structure has been observed in crystal structures [869, 870]. Minimum energy calculation based on CPF calculations [871] provide further support to the model proposed above. NMR studies have been carried out on a large number of cyclic hexapeptides and the structure has been assigned on the basis of line multiplicities, ^3J(HNCH) and bonded- and solvent-exposed NH proton resonances. A large number of cyclic hexapeptides show spectral features which are in general agreement with the β-structure [773, 775, 783, 785, 786, 788, 789, 792, 795]. For example, it has been found that two of the six NH groups are involved in intramolecular hydrogen bonds while the other four are exposed to solvent. In cyclopeptides containing L-Ala and Gly residues, two groups of NH resonances in the ratio

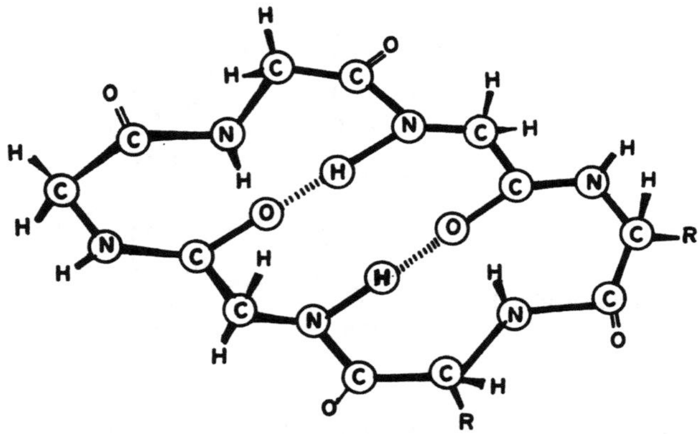

Fig. 4.6.3. Proposed antiparallel β-sheet structure for hexapeptides (reprinted with the permission from Ref. [125])

4:2 have been observed. This suggests a fast conformational isomerization resulting in a hydrogen bond rearrangement in the ring [789]. It may be pointed out that alternate low-energy conformations with one or no intramolecular hydrogen bond can be generated for cyclic hexapeptides several of which have fairly low energies and may coexist in dynamic equilibrium with the β-loop structure discussed earlier [784, 791]. Such alternate conformations may dominate the equilibrium when the formation of the β-loop structure is prevented due to lack of hydrogen-bonding protons. For example, the presence of three L-Pro residues in alternating positions in cyclo(Pro-Gly)$_3$ precludes the formation of an antiparallel β-type structure. The conformation of this molecule is flexible and strongly solvent-dependent [144, 777]. In CD_2Cl_2 solution, cyclo(Pro-Gly)$_3$ exists in a conformation displaying an average C_3 symmetry. An asymmetric conformation occurs in DMSO and in polar solvents [144]. Cyclo(Sar-Sar-Gly)$_2$ and cyclo(Sar)$_6$ adopt a number of conformations in solution some of which exhibit C_2 symmetry while others are asymmetric [774].

Binding with amino acids can bring about conformational changes in cyclic hexapeptides [796]. Similarly, cation binding can induce conformational variations in such systems [774, 790, 799, 800]. An interesting example is the solution conformation of ferrichromes (ferricyclohexapeptides) and related compounds which chelate the metal with high affinity. The Fe(III) ions in these systems can be replaced by Al(III) or Ga(III) resulting in diamagnetic complexes which can be studied conveniently by NMR [798–803]. The presence of metal ions holds these molecules in a tight rigid conformation. On the other hand, on release of the metal ion, the conformation becomes flexible and is dictated by the sequence of amino acids and solvent. In DMSO, the dominant conformer has a β-loop conformation similar to the one found in other cyclohexapeptides. In D_2O solution, the overall structure is a time average of several possible conformations. Very large ^{13}C shifts occur when the peptide binds to metal ions. Shift changes as large as ~5 ppm are observed in the C=O region which arise from a compensation of shifts resulting from the protonation of the carbonyl group and hydrogen bonding of the covalently linked NH group [800]. Recently, complete assignment of carbon signals of alumichrome has been made [801].

The carbon-13 spin-lattice relaxation times of cyclic hexapeptides having the sequence X-L-Pro-Y show an interesting correlation with the conformation of proline residues. Cyclic peptides where all the peptide groups have *trans* conformation lack internal motions in the peptide backbone. However, a cyclic peptide containing two *cis* X-Pro bonds exhibits considerable internal motions [596].

4.6.6 Larger Cyclic Peptides, Peptide Hormones and Antibiotics

4.6.6.1 Oxytocin and Vasopressin

Oxytocin and vasopressin (Fig. 4.6.4) are neurohypophyseal hormones which have received wide attention. Studies have been carried out using ^{1}H-, ^{13}C- and ^{15}N-NMR [193, 815–835, 872, 873]. A part of these nonapeptides (containing *C*-terminal end) forms a linear peptide chain. The cyclic moiety consist of six peptide units forming a 20-membered cyclic ring through disulfide bridges. The crystal structure of a derivative of the *C*-terminal tetrapeptide of oxytocin (*S*-benzyl-Cys-Pro-Leu-Gly-NH_2) has been determined; it includes a 1 → 4 β-loop [874]. Potential energy calculations [873] show several possible conformations of the cyclic moiety of oxytocin. A conformation involving a β 1 → 4 turn in the cyclic moiety and a similar loop in the terminal tripeptide was proposed on the basis of NMR studies [875] in DMSO solutions. However, the consensus among recent investigators is that both oxytocin and vasopressin consist of a blend of several low-energy conformers with a fast rate of interconversion (see for example [815, 817, 818, 825, 833]). It is interesting to note that (1-penicillamine)-oxytocin, which exhibits a more rigid conformation than oxytocin, acts as an antagonist while oxytocin itself is an agonist. Thus, the biological activity of oxytocin bears some relation to the conformational flexibility of this molecule [833].

Considerable efforts have been made in the assignment of the ^{13}C-NMR spectra of the two hormones and the problem is now more or less solved (see for example [119, 822, 823, 827, 872]). The approaches used in such studies include (a) NMR measurements of derivatives of the two peptides, (b) use of specific ^{13}C and ^{2}H labelling (c) studies of *p*H-induced shifts and (d) determination of ring closure effects. It has been possible to measure and assign spin lattice relaxation times (T_1) to individual carbon-13 nuclei in the two hormones and compare the relative flexibilities [158, 876]. The NT_1 values for C^{α} in the cyclic peptide moiety are essentially equal. The NT_1 values of the acyclic tripeptides increase with the distance from the cyclic hexapeptide ring (Fig. 4.6.4), showing increasing mobility in the acyclic peptides with the distance from the cyclic portion. The T_1 values for Lys β to ε carbons are characteristic of segmental motion in this residue which is observed in amino acids and Lys residues in short peptides.

An interesting application of ^{15}N-NMR to small biological molecules is the spectrum of oxytocin and its *C*-terminal tripeptide (Fig. 4.6.5). The ^{15}N chemical shifts are compatible with the exposure of most of the peptide linkages to solvent. The correlation times for the motion of the tocin ring derived from ^{15}N-NMR provide further support to the hypothesis that the tocin ring is relatively rigid [827].

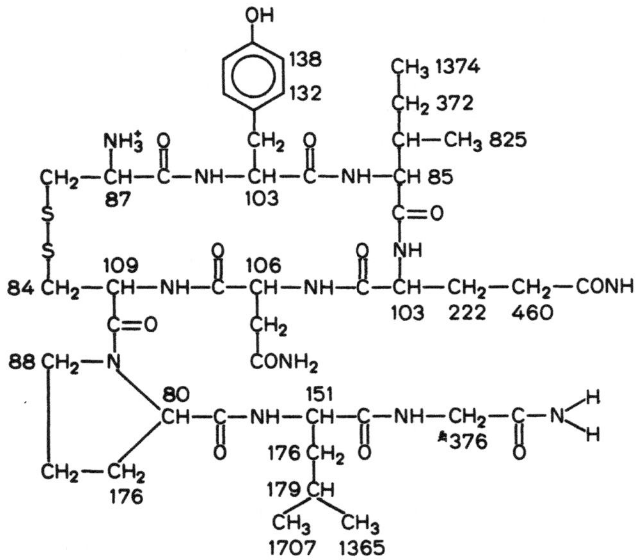

Oxytocin

a

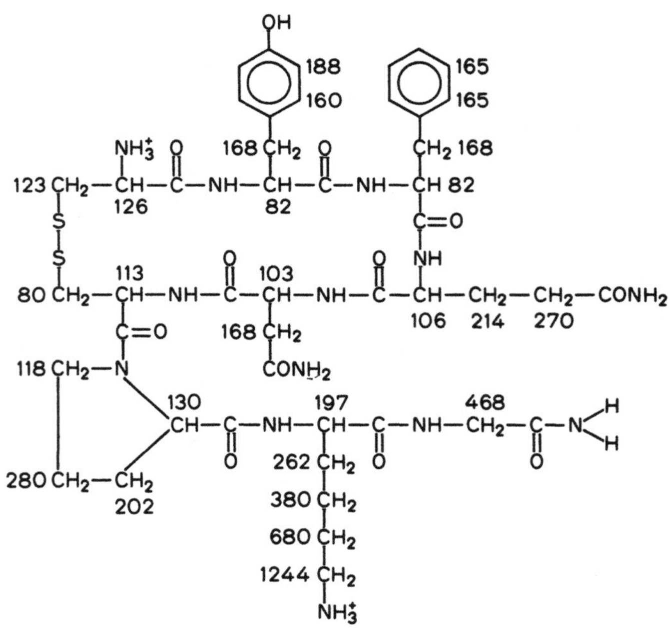

Lysine vasopressin

b

Fig. 4.6.4. Primary structure of (a) oxytocin and (b) vasopressin and ^{13}C spin-lattice relaxation times (reprinted with the permission from Ref. [876], copyright 1974, American Chemical Society)

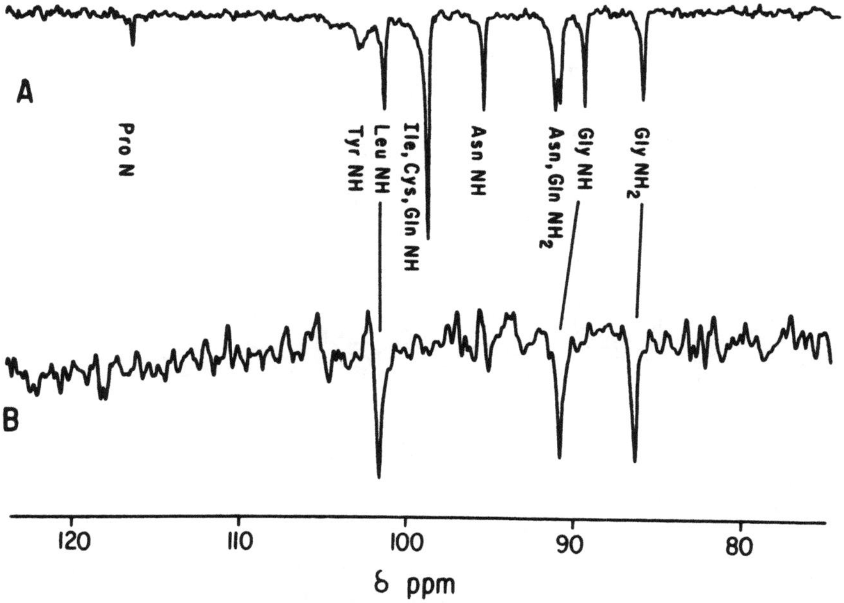

Fig. 4.6.5. ^{15}N-NMR spectra at 18.14 MHz of (A) oxytocin (100 mg/ml, H_2O, *p*H 4, 22,000 transients), (B) prolylleucylglycinamide (50 mg/ml, H_2O, *p*H 4.0, 2,000 accumulations) (reprinted with the permission from Ref. [827], copyright 1979, American Chemical Society)

Recently, studies of the interaction of oxytocin with neurophysins have been carried out using ^{13}C-NMR. Specifically labelled peptide hormones have been synthesized for this purpose [831]. Large shifts in ^{13}C-NMR occur as a result of binding and the dissociation rate depends on *p*H.

4.6.6.2 Gramicidin S

Gramicidin S is a decapeptide and possesses antibacterial activity [877]. It can be obtained fairly easily from natural sources. Its primary structure (Fig. 4.6.6) was established relatively early. Several peptides related to gramicidin S, e.g. tyrocidines, linear gramicidins, gramicidin J, and gratisin have also attracted attention. A distinct difference in biological activity of gramicidin S and its linear decapeptide emphasized the importance of the cyclic structure. A number of models of the conformational structure of gramicidin S have been proposed on the basis of X-ray data [878]. The one which has finally stood the tests based on other physical techniques is due to Hodgkin and Oughton [879]. This structure has a β-pleated sheet structure.

Extensive ^{1}H- and ^{13}C-NMR investigations have been carried out to establish the solution conformation and molecular mobility in gramicidin S and its analogs [174, 189, 536, 590, 836–851]. These studies generally favor a β-pleated sheet structure as one of the preferred conformational structures in solution. The structure displaying C_2 sym-

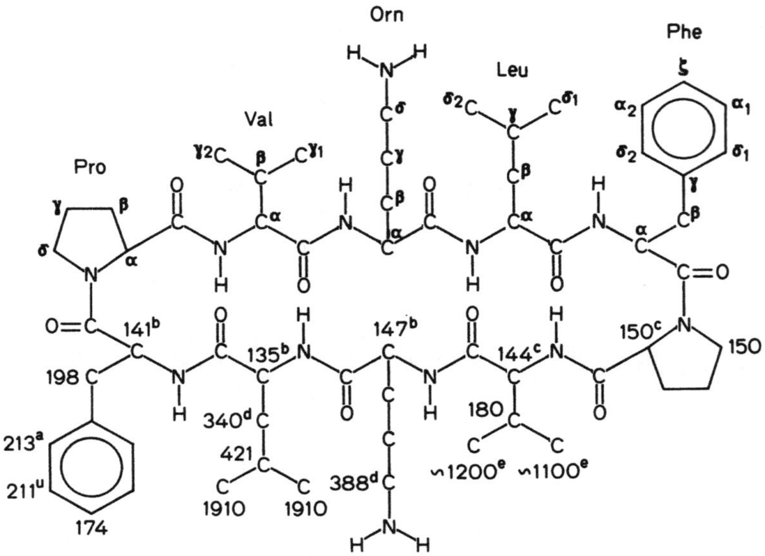

Fig. 4.6.6. Primary structure of gramicidin S along with NT_1 values (reprinted with the permission from Ref. [843], copyright 1973, American Chemical Society)

metry is stabilized by four intramolecular N–H···O=C bonds involving Val and Leu residues. The position of the β-turns is fixed by the fragments Pro-D-Phe. The conformation remains almost unchanged in solvents of different polarities. The five pairs of residues yielded only five carboxy, five C^{α} and five C^{β} carbon lines indicating a C_2 symmetry of the time-averaged conformation. Similarly, ^{15}N-NMR of gramicidin S is unusually simple consisting of only six lines.

Recently, the conformational preferences of gramicidin S in methanol-d_4 has been studied in great detail [840, 842]. Rotamer populations in each side chain and tertiary conformations of the whole molecule support the conclusion drawn from carbon-13 T_1 measurements that the molecule exhibits internal motions. Several tertiary structures coexist in solution.

4.6.6.3 Valinomycin

Valinomycin is an ionophore which selectively transports $K^{\oplus}$ across natural and artificial membranes. It has a 36-membered depsipeptide ring formed from twelve amino acid residues (Fig. 4.6.7). The conformation of this molecule has been extensively studied in solution both in its free form and as a complex with metal ions such as $K^{\oplus}$, $Rb^{\oplus}$ and $Cs^{\oplus}$ [852–858, 879]. Almost all the conformation – dependent coupling constants (e.g. H–NC$^{\alpha}$–H, ^{13}C′–NC$^{\alpha}$–H, H–NC$^{\alpha}$–^{13}C$^{\beta}$) have been measured and values of ϕ, ψ and χ angles for each amino acid residue have been estimated. Based on NMR findings, a symmetrical (nearly C_3 axial) rigid "bracelet"-type structure has been proposed [546, 879, 880]. The structure is stabilized by means of β-turns with all six NH groups

a

b

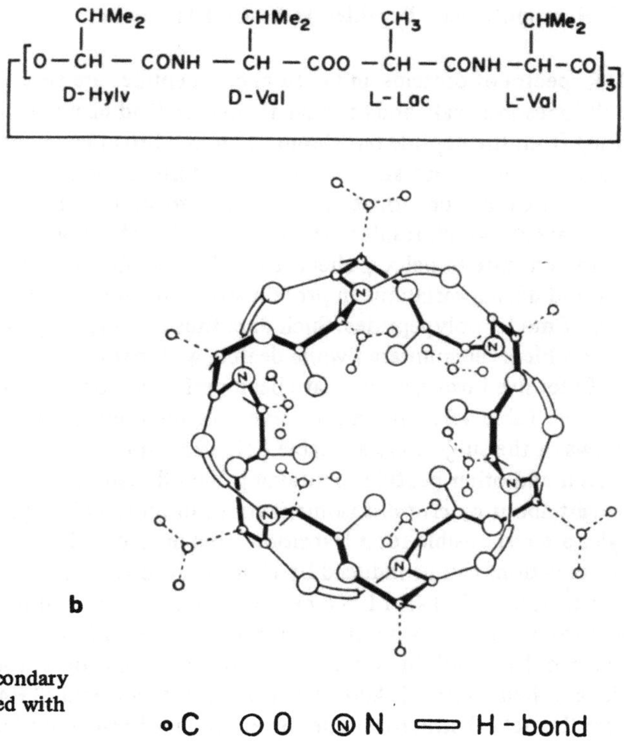

Fig. 4.6.7. (a) Primary and (b) secondary structure of valinomycin (reprinted with the permission from Ref. [546])

hydrogen-bonded to the carbonyl groups through 1 → 4 intramolecular bonds (Fig. 4.6.7). A molecular cavity lined with six ester carbonyl groups is formed in this structure. The cavity can hold non-hydrated $K^{\oplus}$, $Rb^{\oplus}$ or $Cs^{\oplus}$ but is too large for $Na^{\oplus}$. The solution conformation derived from NMR data for the valinomycin-$K^{\oplus}$ complex is in good agreement with a crystal structure reported recently [881]. The ions in the cavity are held together by ion-dipole interactions and the carbonyl groups are screened from the solvent by intramolecular hydrogen bonds.

According to these proposals the conformation of the complex and of the free ionophore in non-polar media are very similar. The main difference is that the ester oxygen atoms are turned toward the exterior in the absence of metal ions. The hydrogen bonding pattern is also similar except that one of the β-turns is less stable. This weaker hydrogen bond migrates at a high rate on the NMR time scale. In solvents of medium polarity only three hydrogen bonds are retained. In polar solvents, all intramolecular hydrogen bonds are disrupted. The proposed structure of valinomycin can explain the $K^{\oplus}/Na^{\oplus}$ selectivity and a number of other biological properties of these important class of peptides.

4.7 Homopolymeric Peptides. Helix-Coil Transition

NMR spectra of proteins and natural polypeptides are very complex and it is often difficult to resolve peaks and to assign them to individual nuclei. In particular, the resonances arising from the peptide backbone occur as a broad envelope of lines in the aliphatic region. To gain information on the characteristics of such resonances and their relation to the backbone conformation, it is useful to study homopolymeric peptides. The shifts of resonances which result when a polypeptide chain passes from a disordered (coil) to an ordered state (α-helix, β-sheet, etc.) give valuable information on the spectral changes expected during variations in protein structure and on the stability of ordered structures of model polypeptides. Such data thus provide the basis for studies on more complex problems encountered while dealing with natural proteins and polypeptides.

Extensive investigations have been performed on the coil-helix transition in polypeptides. Table 4.7.1. provides a review of the literature published in recent years. Earlier reviews of the subject are available [27, 125, 944–947]. In homo- or copolymers of Pro, Sar and other peptides containing *N*-methylamino acids such a transition is usually brought about by *cis-trans* isomerization. In other polypeptides, rotations around ϕ, ψ angles are responsible for a transition from an ordered to disordered state. The helix-coil transitions can be induced by changes in solvent, temperature, ionic strength or *p*H. For example, poly-L-Tyr has a random-coil structure in dimethyl sulfoxide [912]. In the presence of water or trimethyl phosphate, the equilibrium is shifted toward the helical state. Similarly, poly(α-amino acids) in $CDCl_3$ solution at room temperature generally exist as α-helices [946]. Addition of trifluoroacetic acid (TFA) to such solutions breaks down the helical structure. The solvent-induced transition from a coil to helical structure is accompanied by an upfield shift for the α-CH proton while the N-H peak moves downfield by an amount which is solvent-dependent [895, 946, 948]. In some systems, the two resonances move as single peaks with variations in the line width. However, in most cases, two peaks for the resonance of the α-carbon proton are observed and have been assigned to the resonances of the helix and random-coil structure (usually referred as the "double peak" phenomenon).

The ^{13}C-NMR spectrum reveals a similar behavior. Figure 4.7.1 shows the ^{13}C-NMR of poly(L-Glutamic acid) as a function of *p*H. At *p*H 4.25 and 4.50, peak doubling is seen for both the α and β carbons as well as for the carbonyl carbon on the backbone (C^o_{bb}). The *p*H region covered here encompasses the transition from helix to coil. While the separation of the α proton chemical shifts is of the order of 0.2–0.5 ppm that of the α carbon chemical shifts is 2–3 ppm [896]. The changes in helicity as measured by NMR methods correlate closely with ORD measurements.

The observation of double peaks in the helix-coil transition region for polypeptides has been the subject of extensive discussions [891, 906, 915–918, 940–942, 947, 949–952]. The lifetime of the helix-coil transition as measured by temperature and electric jumps [953, 954], dielectric dispersion [955, 956], ultrasonics [957, 958], electric field pulse methods [959] and by theoretical methods [915, 960] lies in the range 10^{-6}–10^{-7} s. On the other hand, the detection of separate NMR signals for helix and coil states implies that the lifetime of each states should be greater than 10^{-3} s. Several attempts have been made to resolve this discrepancy [947] which has been attributed to polydispersity in molecular weight [915–918], contribution of oligomer contents

Table 4.7.1. Homopolymeric peptides: helix-coil Transition

Nucleus	System and information	Ref.
^{1}H	*cis/trans*-Polysarcosine, Monte Carlo calculation	[882]
^{1}H	Alternating and random poly(benzyl D, L-glutamate)	[883]
^{1}H	Polypeptides, polydepsipeptides, poly(aminobutyric acid)	[884]
^{13}H	Copolymer benzyl asparatate – benzyl glutamate	[885]
^{1}H	Synthetic poly(α-amino acids)	[886]
^{1}H	Homooligopeptides: UV, IR	[887]
^{1}H	Alanine oligopeptides	[888]
^{1}H	Oligopeptides: conformational equilibrium and stability	[889]
^{13}C	Polypeptides: molecular mobility of Pro	[890]
^{1}H	Polypeptides: lifetimes in helix-coil transition region	[891]
^{13}C	Poly Pro: chain conformation, isomerization	[564]
SH	Synthetic polypeptide: NMR, review	[892]
^{13}C	Poly-L-methionine, helix-coil transition	[893]
^{1}H	Random DL copolypeptides containing Gly, Asp	[894]
^{35}Cl	Line width in cysteine-glutamine copolymer	[47]
^{1}H	Polypeptide: helix-coil transition, non-protonating solvents	[895]
^{13}C	Poly(γ-benzyl L-glutamate): rotational T_1, NOE	[896]
^{13}C	Polyglutamate: ^{13}C shift accompanying helix formation	[897]
^{1}H, ^{13}C	Poly-L-Pro conformation: *cis-trans* isomerization	[140]
^{13}C	^{13}C-peak doubling in transition region	[898–901]
^{13}C	Poly-L-Lys, in helix-coil transition	[902]
^{1}H, ^{13}C	Poly(γ-benzyl L-glutamate), intermediate state in transition	[903]
^{1}H	Poly(benzyl aspartate): helix-coil transition	[904]
^{1}H	Poly-L-Ala: helix-coil transition	[905]
^{1}H	Kinetic description, helix-coil transition Ising model	[906, 907]
^{1}H	Poly(hydroxy-L-Proline): ring conformation	[908]
^{1}H	Random poly(benzyl L-Glutamate-benzyl-L-Asp)	[909]
^{1}H	Side-chain interactions poly(amino acids), aromatic residues	[910]
^{1}H	Synthetic copolypeptide, Glu, Tyr	[911]
^{1}H	Poly-L-Tyrosine, IR, solvent interaction	[912]
^{1}H	Poly-Ala: solvent effect, ORD	[913]
^{1}H	Copoly[(benzyl-L-Glu)$_n$(benzyl-L-Asp)$_m$]: optical spectra	[914]
^{1}H	Polypeptides: polydispersity effects in helix-coil transitions	[915–918]
^{1}H	Poly-Ala: α-proton shift, Zeeman effect	[919]
^{1}H	Poly(γ-benzyl-L-glutamate): relaxation time	[916]
^{1}H	Poly-L-Lys: CD	[920]
^{1}H	Poly(2-hydroxypropylglutamine): pressure effect	[921]
^{13}C	Oligo proline: *cis-trans* isomerization	[142]
^{13}C	Poly-L-Lys, poly-L-Arg, poly-L-ornithine salt induced transition	[922]
^{1}H	Solid poly-L-Ala, poly-L-Leu, poly(L-val-Gly)	[923]
^{1}H	Poly oxyethylene – bound oligopeptide conformations	[924]
^{1}H	Poly-Ala oligomers	[925]
^{1}H, ^{13}C	Sarcosine oligopeptides: *cis-trans* isomerization	[570]
^{13}C	Poly(γ-hydroxy-L-Pro): *cis* residue	[571]
^{1}H, ^{13}C	Poly-Glu: ionic strength	[926]
^{1}H, ^{13}C	Poly(L-azetidine-2-carboxylic acid)	[927]
^{13}C	Poly(γ-benzyl-DL-Glu) in alpha and double helical conformations	[928]
^{1}H	Poly-L-His: molecular motions in solid state	[929]
^{1}H	Poly-Ala: solvent effect	[930]

Table 4.7.1 (continued)

Nucleus	System and information	Ref.
^{1}H	Oligo-L-Pro, synthesis	[573]
^{1}H	Norvaline oligopeptide: molecular association	[931]
^{13}C	Poly(γ-benzyl-L-Glu): rotational diffusion	[932]
^{15}N	Poly-Sar: related polypeptides	[36]
^{13}C	Poly-L-Glu: helix-coil transition	[933]
^{1}H, ^{13}C	Poly(γ-benzyl-DL-Glu): double-stranded conformational	[934]
^{1}H, ^{13}C	Poly(β-benzyl-L-Asp), helix-coil transition	[935]
^{1}H	Poly(ethylene imine): spin echo	[936]
^{1}H, ^{13}C	H(L-Ala)$_n$-L-Pro-OH, *cis-trans* isomerization	[575]
^{1}H, ^{13}C	X-Pro bond: NMR probe	[937]
^{1}H, ^{13}C	Oligo-L-Pro: *cis-trans* isomerization	[576]
^{1}H	Cooligopeptides with alternating L-isoleucine, D-alloisoleucine residues	[939]
^{13}C	Poly(D-glutamic acid): T_1	[938]
^{13}C	(Pro-Pro-Gly)$_{10}$: helix-coil transition	[577]
^{1}H	Polypeptides: slow nucleation theory	[891, 906, 940–942]
^{1}H	Poly(Pro-Pro-Gly): triple helix	[943]

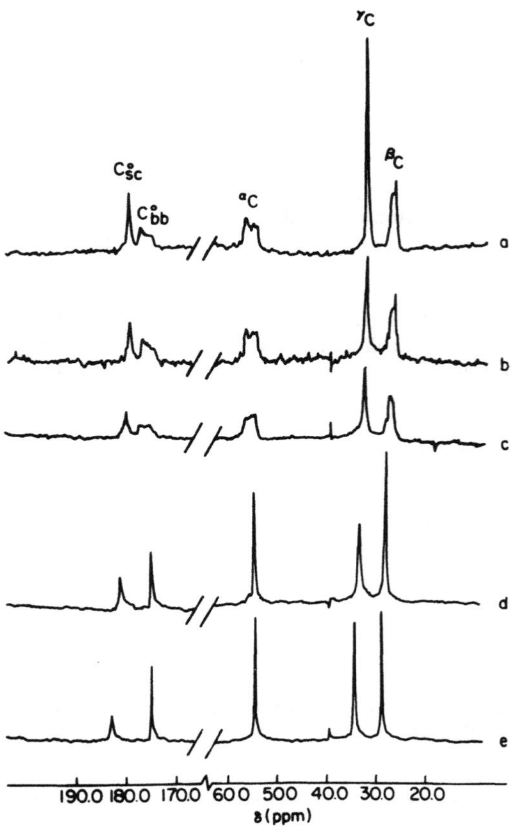

Fig. 4.7.1. 67.9 MHz ^{13}C-NMR of poly(L-glutamic acid), molecular weight 8,100 in D_2O, 0.2 M NaCl. The concentration in (a) is 13%. All others are about 2.5%. The pH values and number of scans are (a) 4.25, 42,066; (b) 4.25, 9,000; (c) 4.50, 8,358; (d) 5.15, 8,200; (e) 7.15, 2,650 (reprinted with the permission from Ref. [933]). C^{o}_{sc} and C^{o}_{bb} are the carbonyl carbons in the side chain and backbone respectively.

[905, 949], slow nucleation of helix formation [891, 906, 940–942], and solvation effects [951, 952].

In some of the earlier explanations, it was proposed that the two peaks correspond to the two states of solvation of the molecules in the coil form, the signals from helical states being too broad to be detected [951]. In contrast to this explanation, it is observed that the sum of the areas of the two peaks remains constant throughout the transition region and even when the molecules exhibit a highly ordered structure [961]. The double-peak behavior occurs even in dimethyl sulfoxide solvents thus ruling out the theory based on the protonation of the polypeptide molecule [952].

The rate of helix formation can also be responsible for peak doubling in the helix-coil transition region [891, 906, 940–942]. Ferretti et al. could simulate the line shape of NMR observed in the transition region using lifetimes of $\sim 10^{-2}$ s estimated from the total line shape analysis. However, this number may not be reliable since the analysis did not take into account the effect of molecular weight polydispersity [916]. According to this theory one of the two lines is due to the coil state which varies in area but not in position and the other one is attributed to signals arising from partially ordered segments. One of the drawbacks of this theory is that it cannot predict single shifting resonance lines detected for some polymers although the simulated double peaks are in excellent agreement with experimental observations.

The most satisfactory explanation is based on the concept that synthetic polypeptide samples are polydisperse, helices having different chain lengths display various stabilities and give rise to the NMR spectrum in which the lines with different chemical shifts are superimposed according to their helical content. The doubling of peaks can be interpreted as a natural consequence of this model and occurs where the helix-coil transition is highly cooperative. The helix content in such systems strongly depends on the molecular weight of the polypeptide chain, particularly of shorter chain lengths. Thus, in the helix-coil transition region, some molecules will be fully helical while others fully random, depending on their molecular weight. The peak doubling will thus occur even when the interconversion rate is fast on the NMR time scale [917]. However, for a polypeptide that displays a "weak cooperativity", and exhibits a Lorentzian molecular weight distribution, only a single peak is expected [916].

From conformational studies of alanine oligopeptides [905, 949] Goodman et al. have shown that the peak doubling in polydisperse polypeptides occurs because of the presence of significant amounts of oligomers. For example, the downfield peak in the α-CH region can be assigned to oligomers which are too short to form helices in mixed solvents such as $CDCl_3$-TFA. The upfield peak is due to a combination of helix and coil states of polypeptides. The two classes of peptides solvate differently because of their different chain length and exhibit different chemical shifts.

Since polydispersity of the polypeptide chain is a very important factor for the observation of peak doubling, attempts have been made to study well characterized molecular weight fragments. However, these attempts have led to contradictory results [916, 947, 950]. Possibly more than one factor discussed above is important and further work with carefully fractionated polypeptide samples is needed to solve the problem.

Another useful method in the study of helix-coil transitions is the use of carbon-13 spin-lattice relaxation times. Such studies give useful information on molecular motions in the two states [890, 896, 898, 902, 903]. In the case of poly(Lys) [902], the backbone carbonyl relaxation time T_1 in the coil state is around 1.15 s. The value drops rapidly at high *p*H where the system exists in a helical structure. The behavior of ^{13}C-NMR parameters (δ, T_1, T_2 and NOE) of poly(γ-benzyl-L-glutamate) [898, 903], has been studied extensively. The value of T_1 has been found to decrease by a factor of two in the transition from coil to helical states while the line widths increase by a factor of four. The values of T_1 steadily increase as one proceeds from the backbone toward the side chains. Thus, both in helix and coil states the overall reorientation of the molecule is slower than the segmental motions. In the helical state, the ^{13}C spin-lattice relaxation time of the α-carbon nuclei in deuterodimethylformamide is shown to be governed by the axial rotational diffusion coefficient [932]. The values of T_1 increase linearly with the molecular weight of the chain, due to rigidity of the chain and the molecular weight dependence of the axial diffusion coefficient of the helix. In systems where a double peak behavior is observed in the helix-coil transition, the ^{13}C lines arising from the signals of the helical and coil states have different T_1 values due to the greater segmental motion in the coil [896]. For example, in the case of poly(γ-benzyl-L-glutamate), the nuclear Overhauser enhancement of the α-C atoms in the coil is twice that of the helix. Thus, the intensity of the C^α line assigned to the helical structure is attenuated relative to that of the random coil.

The copolymers of D- and L-amino acids and polymers of aspartate esters have been studied by several workers [883, 885, 894, 904, 909, 928, 934, 935, 939]. Such polymers reveal a preference for the left-handed α-helix and the studies provide information on the dependence of chemical shifts on the handedness of the helix and the relative stabilities of left- and right-handed helices.

In polymers of L-proline and *N*-methylated amino acids, the helix-coil transition occurs because of rotation around the amide bond (angle ω). In contrast to transitions caused by ϕ, ψ torsion angles, the barrier to internal rotation around the C′—N bond is relatively large and the transitions proceed more slowly on the NMR time scale. Poly(L-Pro)(I) has a regular helical structure with *all-cis* peptide bonds. When dissolved in water, it isomerizes to *all-trans* form II. In concentrated aqueous salt solutions, the polymer chains consist of random sequences of *cis* and *trans* residues. Separate resonances for the two conformations are detected both in ^{1}H- and ^{13}C-NMR [140, 141, 564, 573, 576, 927]. In addition to the *cis-trans* isomerization, the proline ring itself displays conformational flexibility [613, 908]. Studies have also been carried out in mixed copolymers such as oligo(Pro-Pro-Gly), poly(Pro-Gly), poly(Gly-Gly-Pro-Gly) [890, 577]. Thus, studies of the molecular conformations and dynamics of proline-containing peptides present a challenge to the NMR spectroscopist with a range of time scales.

4.8 Characterization of Protein Structures by NMR

One of the triumphs of the NMR technique is its successful applications to macromolecules such as proteins in solutions. The number of NMR parameters which can be utilized for studying the structure and function of such molecules is quite large and almost all the atoms in proteins are amenable to NMR studies. Further, structural studies can be made under conditions which are closely related to physiological environment. The determination of the complete three-dimensional structure of large molecules is still beyond the present state of development of NMR techniques. This task can only be handled by X-ray crystallographers. NMR, however, provides complementary structural information on proteins whose crystal structures are known. Important information can also be obtained on proteins which are not amenable to X-ray studies for technical reasons. The various types of information on protein structure and function which can be obtained by NMR are summarized in the following.

4.8.1 Spatial Arrangement of Atoms in the Molecules

The spectrum of a randomly coiled protein can be simulated from the corrected shifts of the amino acid residues. The secondary shifts caused by the ordering of the three-dimensional structure can in principle be applied to the determination of the spatial environment of each nuclear site in the protein. In practice, however, with the present state of art, the geometrical information is limited to environmental effects arising from aromatic rings, paramagnetic ions, *p*H titration of acid-base groups and hydrogen bonding. Further, conformational changes induced by *p*H, substrate binding or chemical modifications can be detected by NMR. Tracer groups such as ^{2}H- and ^{19}F-labelled amino acids can be incorporated into proteins which can monitor local conformational features.

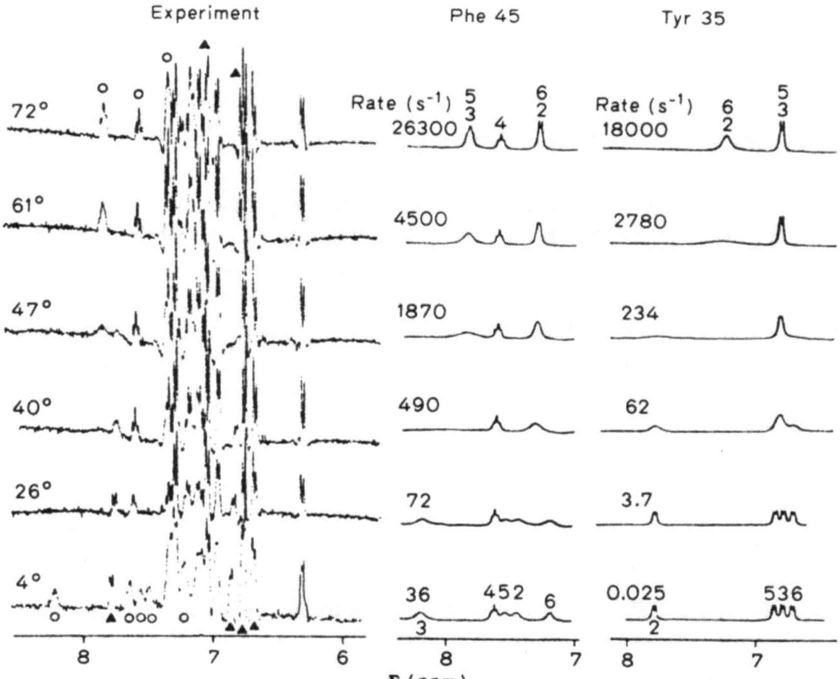

Fig. 4.8.1. Temperature dependence of the ^{1}H-NMR spectrum of BPTI. Left-hand side shows experimental spectrum – Phe 45 (○) Tyr 35 (△). Simulated spectra are shown on the right-hand side. In the high-temperature spectra, the mobility in the amino acid side chains averages the secondary shifts of *o, m* and *p* protons (reprinted with the permission from Ref. [547])

4.8.2 Protein Mobility

Classical descriptions provide two limiting models of protein structure. One is the highly ordered conformation based on average atomic positions provided by X-ray crystallographic results. The second is the flexible and unstructured "random-coil" from denatured proteins. In recent years, evidences of the flexibility in globular proteins have come from many sources [962, 963]. These results yielded new insights into the mechanism of enzyme actions. One of the most exciting results of NMR studies is that it has provided a quantitative description of internal motions [547, 964–970] in proteins and polypeptides. Evidence of internal rotations can be obtained when part of the molecule flips between two conformational states with different secondary shifts. A typical case is the resonances of aromatic side chains such as in Tyr and Phe. If the residues are rigidly fixed in the protein, then each proton in the aromatic residue will show a different chemical shift leading to a ABCDE spectrum in Phe (ABCE for Tyr). However, the chemical shifts can be averaged due to 180° flips around the C^{α}–C^{β} bond. A typical situation in basic pancreatic trypsin inhibitor (BPTI) illustrates this point (Fig. 4.8.1). Dynamic mobilities of Tyr and Phe rings have been detected in several proteins now [41, 547, 964, 966, 967, 971].

We have earlier discussed the use of T_1 in the detection of segmental motions in small peptides. Such motions are also present in proteins though these are somewhat localized.

4.8.3 Enzyme – Substrate Binding

Since the time scale of observation in NMR is relatively large, fast enzyme kinetics cannot be followed by this method. Thus, NMR investigations are limited mostly to binding of inhibitors, activators or slowly reacting substrates. The proton transfer or substrate exchange rates greater than 1 s can be conveniently measured by following spectral changes as a result of binding, i.e. broadening of lines during proton transfer or substrate binding, or disappearance of lines of a particular isotope during isotope exchange. The saturation transfer experiments can also be employed to study transfer rate between two binding sites.

4.8.4 Thermodynamic Parameters

One of the most useful applications of NMR techniques involves determinations of thermodynamic parameters such as ligand binding constants, *p*K values of individual titrable groups in proteins, redox potentials, and the influence of local environmental factors on these parameters. This subject is beyond the scope of the present review but details can be obtained from other review articles on the subject [26, 27, 544, 545, 553].

Table 4.8.1. Protein structures determined by NMR

Nucleus	System and information	Ref.
	(a) Lysozyme	
^{13}C	Methine aromatic carbon atoms	[31]
^{1}H	Inhibitor-induced conformational changes	[965]
^{1}H	Lanthanide shifts	[167]
^{1}H	Denaturation, arginine residues	[972]
^{1}H	Mechanism, cell-wall tetrasaccharides, conformation intermediates	[973]
^{1}H	Lys-residue dissociation constant	[974]
^{1}H	Denatured proteins; deviation from random-coil behavior	[975]
^{19}F	Conformational changes, characterization	[43, 976]
^{1}H, ^{19}F	Protein-solvent interfaces	[45]
^{1}H	Water ordering on protein surface	[977]
^{1}H, ^{13}C	Solution conformations, review	[971]
^{1}H	Solution conformations, lanthanide shifts	[978]
^{1}H	Trp-indole NH-exposed and burried proton	]979]
^{1}H	*p*H-Induced conformational changes	[980]
	(b) Basic pancreatic trypsin inhibitor (BPTI)	
^{1}H	Molecular dynamics, proteins and peptides	[964]
^{1}H	Conformations and dynamics	[966, 970, 981]
^{1}H	Internal motions	[547]
^{1}H	Lys residues (360 MHz)	[982]
^{1}H	Lanthanide binding	[983]
^{13}C	^{13}C-enriched *S*-methylmethionyl residue	[984]
^{1}H	Effect of chemical modifications on conformation	[985]
^{1}H	Reduction of disulfide bond: effect on conformation	[986]
^{1}H, ^{13}C	Structure determined by high-resolution NMR	[987]
	(c) Dihydrofolate reductase	
^{1}H, ^{19}F	Inhibitor and coenzyme binding, ^{19}F- and ^{2}H labelling	[988–992]
	(d) Antibody combining site	
^{1}H	NMR and ESR investigation *p*H titrations	[993–996]
	(e) Myelin basic protein (MBP)	
^{1}H	Solution behavior: CD	[997]
^{1}H	Anion binding, trypan blue, sodium dodecyl sulfate	[998]
^{1}H	Protein and constituent peptides	[999]
^{13}C	Conformation and assignment	[1000]
^{1}H	Conformation in solution	[1001]
^{13}C	Enriched, liposome incorporation: T_1	[1002]
^{1}H, ^{13}C	*p*H, concentration, intermolecular aggregation	[1003]
^{1}H	Aggregation, medium effects	[1004]
	(f) Calcium-binding proteins	
^{1}H	Troponins, paralbumin, review	[1005]
^{13}C	Par valbumin, phospholipase	[1006]
^{1}H	Tropomyosin, troponin	[1007]

Table 4.8.1 (continued)

Nucleus	System and information	Ref.
^{1}H	Paralbumin	[1008, 1009]
^{1}H	Intestinal calcium-binding protein	[1010]
	(g) Elastin and tropoelastin	
^{13}C	Aortica-α-elastin, coacervate, fibrous state	[1011]
^{1}H, ^{13}C	Elastin, polyhexapeptide	[1012]
^{1}H, ^{13}C	Synthetic polypentapeptide	[672]
^{1}H, ^{13}C	Elastin polypeptide	[124]
	(h) Ribonuclease	
^{1}H	Structure and function: review	[1013]
^{1}H, ^{13}C	*S*-Peptide, *S*-protein, titration	[1014]
^{13}H, ^{19}F	Analogs, labelled enzyme, conformation model structure	[42, 1015]
^{19}F	Conformational changes	[43]
^{13}C	Peptide-protein interactions: ^{13}C labelled peptide	[1016]
	(i) Heme proteins	
^{1}H	Ferric cyanomyoglobin: hindered rotation of methyl group	[1017]
^{1}H	Met hemoglobin, met myoglobin, latex bond: T	[1018]
^{1}H	Ferric myoglobin and hemoglobin derivatives: spin equilibrium	[1019]
^{1}H, ^{13}C	Hemes, hemoproteins, review	[555, 556, 1020]
^{1}H	Ligand-protein interaction in hemeproteins	[1021]
^{1}H	Hemoglobin: half-ligated state, allosteric interactions	[1022, 1023]
^{81}B	Methemoglobin conformation	[48]
^{1}H	Hemoglobin: preferential ligand binding	[1024]
^{19}F	Hemoglobin: conformational changes	[43]
^{1}H	Hemoglobin: Bohr effect	[1025]
^{13}C	Hemoprotein interaction with CO, labelled isonitriles	[1026]
^{13}C	CO binding to heme proteins	[1027]
^{1}H	Horse heart ferricytochrome, *p*H	[1028]
^{15}N	Ferric low-spin cyanide (^{15}N) complexes of hemoproteins	[1029]
^{15}N	Ligand binding to hemoprotein-labelled myoglobin cyanide complexes, cytochrome c, *p*H dependence, structure-function relationship	[37, 1030–1033]
^{13}C	Sperm whale myglobin, ^{13}C enrichment: T_1	[1034]
^{1}H	Heme disorder in monomeric insect hemoglobins	[1093]
	(j) Collagen	
^{1}H	Collagen CNBr fragment: helix-coil transition	[1035]
^{13}C	Cross polarization studies: molecular mobility	[53]
	(k) α-Chymotrypsin	
^{1}H, ^{19}F	Solvent interfaces	[45]
^{19}F	Ser-195 labelling: active sites	[1036, 1037]
^{19}F	Binding of inhibitor	[1038]

Table 4.8.1 (continued)

Nucleus	System and information	Ref.
	Other proteins	
^{35}Cl	Horse-liver alcohol dehydrogenase, halide	[46]
H	NAD-linked enzymes: hydride transfer specificity	[1039]
^{1}H	Staphylococuss protease: rigidity	[1040]
^{1}H	Staphylococcus nuclease: deuteration	[23]
^{1}H	Neurotoxin II from COBRA	[1041]
^{1}H	Egg-yolk flavoprotein, wide-line NMR	[1042]
^{1}H	Phosphoglycerate kinase-active site: lanthanide shifts	[1043]
^{13}C	Carbonic anhydrase: internal motions	[1044]
^{1}H	Serum albumin: mobility	[1045]
^{19}F	Bovine serum albumin	[1046]
^{1}H, ^{19}F	Acyl-carrier protein from E. coli	[44]
^{1}H	Fibrinogen-like peptide, NOE	[1047]
^{1}H, ^{13}C	Fibrinogen-like peptides	[1048]
^{1}H	Ribosomal proteins, E. coli	[1049]
^{1}H	Histidine-binding proteins: interaction sites	[1050]
^{1}H, ^{13}C	*S*-Methylglucagon	[1051]
^{1}H	Isoinhibitor K from Helix pomatio	[1052]
^{1}H	Ribosomal protein S1	[194]
^{1}H	Ribosomal protein S4	[1053]
^{1}H	Phosphotransferase: tyrosine environment	[1054]
^{1}H	Bovine-heart protein kinase: active sites	[1055]
^{1}H	Wool proteins	[1056]
^{1}H	Silk fibroin: wide-line NMR	[1057]
^{19}F	Fluorotyrosine alkaline phosphatase: influence of metal ions	[40]
^{19}F	Fluorotyrosine alkaline phosphatase: mobility of Tyr residues	[41]
^{1}H	Protein dynamics: spin echo	[968, 969]
^{13}C	Tobacco mosaic virus and its protein: mobility	[1058]
^{1}H	Ribosomal protein S16	[1094]

4.8.5 Conformation of Specific Proteins

Table 4.8.1 reviews the recent literature dealing with protein structures as studied by NMR. Some of the more widely studied proteins are discussed in the following sections.

4.8.5.1 Lysozyme

Lysozyme consists of a single polypeptide chain of 129 amino acid residues with four disulfide bridges. Its X-ray structure has been determined to a resolution of 2.5 Å [1059]. This enzyme has been studied extensively by NMR methods [31, 43, 45, 167, 965, 971–980, 1060]. The results have been reviewed in a recent article by Dobson [971]. Lysozyme contains one histidine residue (His 15). Its proximity to Ile 88 and Val 92 found in crystal structure is confirmed by *p*H titration of His and monitoring of the shift changes in other spectral regions [978]. The enzyme contains six Trp residues whose indole NH protons have been resolved in the NMR spectrum and assigned. The rates of exchange of different NH protons are in good agreement with the exposure of these protons to solvent as deduced from crystal structures [979]. The proximity of the methyl groups of Val, Met, Leu and Ile and other assigned residues to the axes of aromatic rings is demonstrated by the presence of secondary shifts on these protons [1060]. Lysozyme has a strong binding site for lanthanides in the active site region, between Asp52 and Glu35 [167, 978]. A pronounced correlation exists between the observed relative distances of various nuclei from the bound Gd(III) ion used as a relaxation reagent in lysozyme and those calculated from X-ray structures. A similar behavior is observed when Pr(III) is used as a shift reagent. Thus, there is a close agreement between the crystal and solution conformations of lysozyme (Fig. 4.8.2). Further, it has been possible to compare the three-dimensional structures of human and hen lysozyme which differ by about 40% in the primary sequence. However, the active site region of the enzymes from two sources depicts a remarkable similarity.

Subtle changes in the lysozyme conformation have been noticed with changes in *p*H [980] and on binding with inhibitors such as *N*-acctylglucosamine (NAG) [965]. Upon linkage of $(NAG)_3$ to the protein at low temperatures, separate resonances from native and inhibitor bound protein can be detected. The rate constant for such a binding is $1.3 \times 10^2\ s^{-1}$ at 45 °C. The *p*H-induced conformational change is fast on the NMR time scale. In both cases, an active site region is involved in conformational changes.

4.8.5.2 Basic Pancreatic Trypsin Inhibitor (BPTI)

BPTI has a molecular weight of 6,500, consists of one polypeptide chain of 58 amino acid residues, 3 disulfide bonds and 8 aromatic residues. It is an ideal model compound for the study of protein folding in globular proteins. The crystal structure has been solved to 1.5 Å resolution [1061]. A large number of NMR investigations has been carried out to compare the conformation in crystals with that in solutions. Furthermore, medium-dependent structural changes, molecular dynamics and protein folding mechanisms [547, 964, 966, 970, 981–987] have been extensively studied. The 360 MHz ^1H spectrum is characterized by resolved resonances from C^α protons (0–6 ppm), aromatic side chains (6 to 7.8 ppm) and resonances of hydrogen-bonded amide protons (7–11 ppm) from the α helix and β-sheet in BPTI. Most of the amide protons are slowly exchanged in D_2O solutions. The ^1H and ^{13}C spectra arising from the twenty methyl resonances in BPTI (6 Ala, 1 Val, 2 Leu, 2 Ile, 3 Thr and 1 Met amino acid residues) are well resolved and could be assigned. One of the most striking result derived from the comparison of the NMR spectral data with X-ray coordinates has been that the average backbone molecular conformations in the crystal and in solution are remarkably similar. Some differences are, however, noticed on the surface. Further, the conformation is preserved in a group of modified BPTI. At the same time, the solution conformation displays a large degree of mobility and can be best described as an ensemble of "open" and "closed" structures which mediate the proton exchange process. Such molecular fluctuations consist primarily of intramolecular translational and rotational motions which leave the hydrophobic domains intact [964].

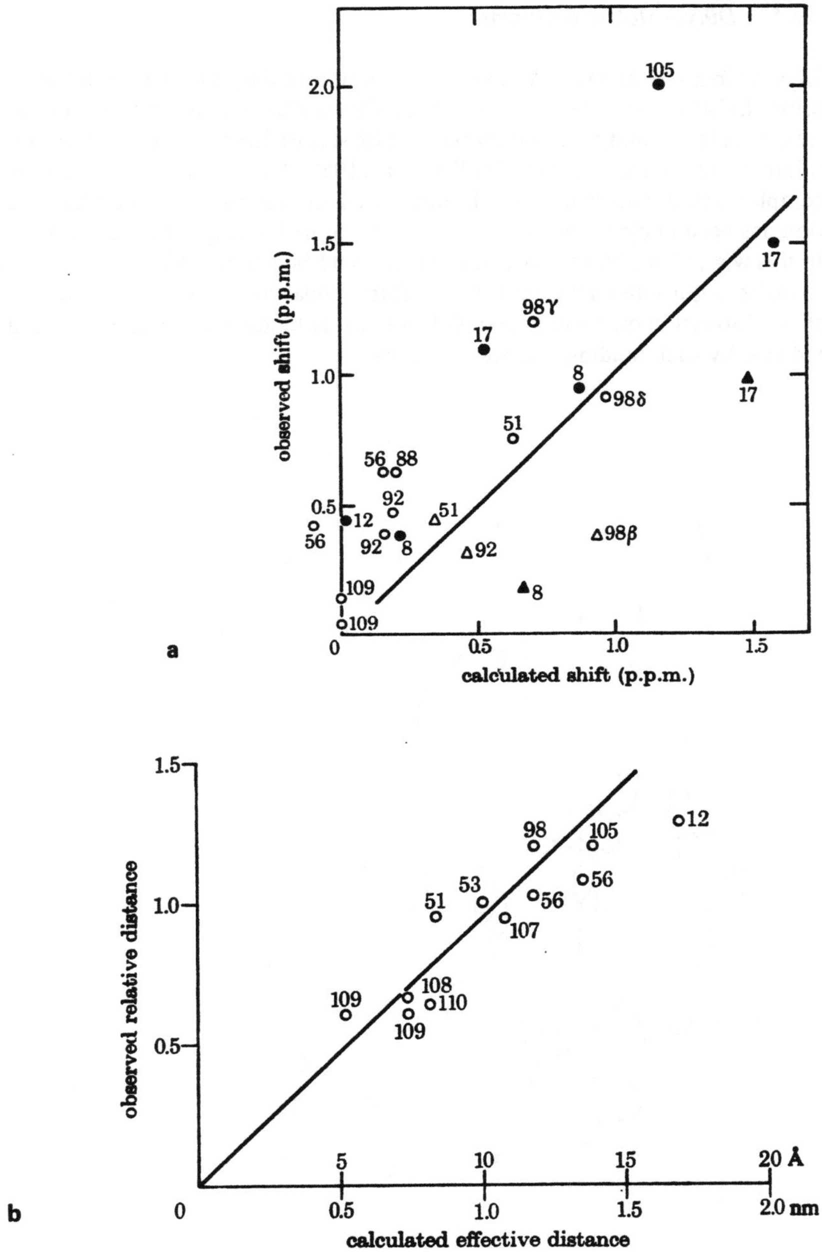

Fig. 4.8.2. Similarity between the crystal and solution conformation of lysozyme as deduced from NMR spectra: (a) using ring-current shift calculations; calculated shifts are those predicted from the crystal structure, ▲ refers to coupled CH resonances and ● refers to methyl resonances. △ and ○ refer to assignments by different methods. (b) using Gd(III) relaxation data; distances relative to Thy 51 = 1.0 (reprinted with the permission from Ref. [978])

4.8.5.3 Dihydrofolate Reductase

Dihydrofolate reductase catalyzes the reduction of dihydrofolate to tetrahydrofolate using NADPH as the coenzyme and thus plays a vital role in the biosynthesis of several amino acids, thymidylate and purines. NMR studies have been carried out on dihydrofolate reductase from L. casei (MW 18,000) [988–992]. The ^{1}H-NMR spectrum is rather complex and therefore use of ^{19}F substituted aromatic amino acids and selective deuteration has been made to examine the protein and its binding with coenzyme and inhibitors. In this way, about 20 amino acid residues could be studied. While this is not sufficient to make major inferences from the conformational structure of the enzyme itself, valuable information on coenzyme-inhibitor-enzyme binding and conformational changes induced by such binding has been obtained.

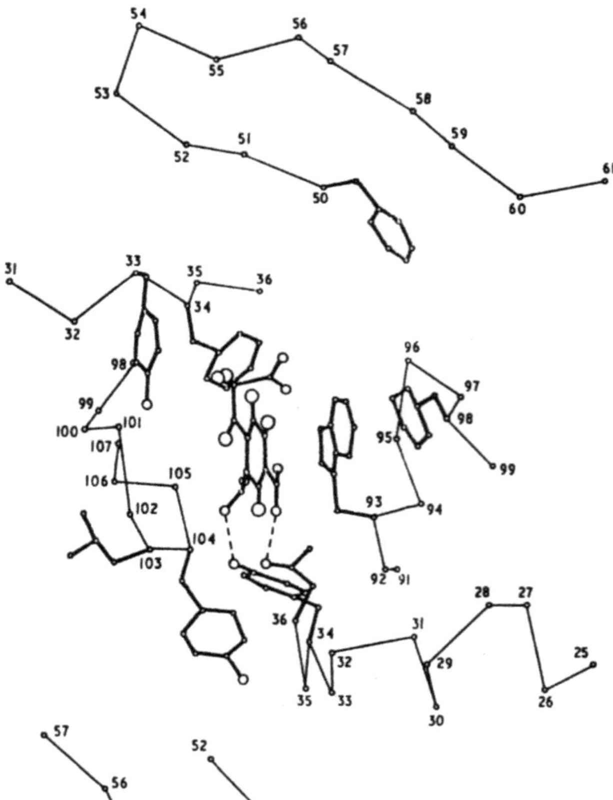

Fig. 4.8.3. Model of the combining site of protein 315 together with a small antigen Dnp glycine hapten bound parallel to Trp^{93} L (reprinted with permission from Ref. [993])

4.8.5.4 Antibody Combining Site

The immune response can specifically recognize and destroy any foreign substance (antigen) present in the body. The recognition is carried out by antibodies. Antibodies are relatively large molecules containing two heavy (MW 50,000) and two light (MW 25,000) chains. Each antibody contains two combining sites for antigen. All antibodies have a common structural feature by which they combine specifically with the antigen. NMR and spin-label studies have been carried out to investigate the structures of antibody combining sites [993–996].

Figure 4.8.3 shows a model of the combining site of the Fv fragment of protein 315 introduced on the basis of NMR chemical shifts, nuclear Overhauser effect, *p*H titrations, positioning of aromatic residues and lanthanide ions, and hapten binding studies. The orientation of the hapten is such that hydrogen bonds with Asn^{36} L and Tyr^{34} L are possible. The specificity of binding to haptens is effected by stacking interactions with Trp^{93} L, hydrogen bonding with Tyr^{34} L and Asn^{36} L, and hydrophobic interactions. Several other haptens studied combine in essentially the same way.

4.8.5.5 Myelin Basic Protein (MBP)

The highly ordered myelin sheets around axons of peripheral and central nerves contain three important proteins. In recent years, attention has been focussed on one of them, MBP, because it can act as a powerful antigen. Several NMR studies using ^1H and ^{13}C resonance have been carried out to understand the conformational structure of this protein [997–1004]. The picture which emerges from these studies is that the MBP structure is rather loose and shows considerable segmental motions. However, the presence of chain folding leading to a well-defined shape cannot be precluded.

4.8.5.6 Calcium-Binding Proteins

NMR studies on several calcium-binding proteins have been carried out [1005–1010]. A relatively flexible conformation is generally indicated. Tropomyosin, however, acquires a coiled α-helical structure.

4.8.5.7 Elastin and Tropoelastin

Fibrous elastin is the unstable core of the elastic fiber derived from tropoelastin. Tropoelastin is shown to contain a repeating tetrapeptide (Val-Pro-Gly-Gly), a repeating pentapeptide (Val-Pro-Gly-Val-Gly) and a repeating hexapeptide (Ala-Pro-Gly-Val-Gly-Val). Extensive ^1H- and ^{13}C-NMR studies have been carried out both on elastin as well as on the repeating component peptides [109, 124, 668–670, 672, 673, 684, 688, 689, 804, 1011, 1012]. On the basis of these results it is concluded that the β-turn is the most stable secondary feature of elastin molecule. Additional conformational characteristics develop around this unit. The conformation of the polymer can be a variety of β-spirals.

4.8.5.8 Ribonuclease

Ribonuclease containing 124 amino acid residues has been extensively studied by NMR spectroscopy for several years. The enzyme can be cleaved between the residues 20 and 21 giving S peptide and S protein; this cleavage has been of help in the assignment of ^{1}H- and ^{13}C-NMR. However, there are still serious difficulties in the assignment of the spectral lines [42, 1013–1015].

4.8.5.9 Hemoproteins and Hemoenzymes

Hemoproteins are a class of proteins which contain a ferroporphyrin (protoheme) as a prosthetic group in combination with a protein chain (globin). Many of these proteins are involved in respiratory functions. Hemoglobin (Hb) and myoglobin (Mb) are involved in oxygen transport and storage. Blood of polychaele worms contain chlorocruorin (Chl) which differs from hemoglobin in the structure of porphyrin. A related class of compounds are cytochromes which function as electron carriers in biological oxidation-reduction reactions.

In the heme group, iron is bound to the four nitrogen atoms of the protoporphyrin ring [1062, 1063]. These four nitrogens are at a distance of about 2.01 Å from the center and show very little radial expansion upon ligand binding (Fig. 4.8.4). For bulkier ions, the coordination is affected by the displacement of the ion normal to the plane containing the four nitrogen atoms. Iron can exist in two oxidation states Fe(II) and Fe(III), with a d^6 and d^5 electronic configuration, respectively. Furthermore, each of these

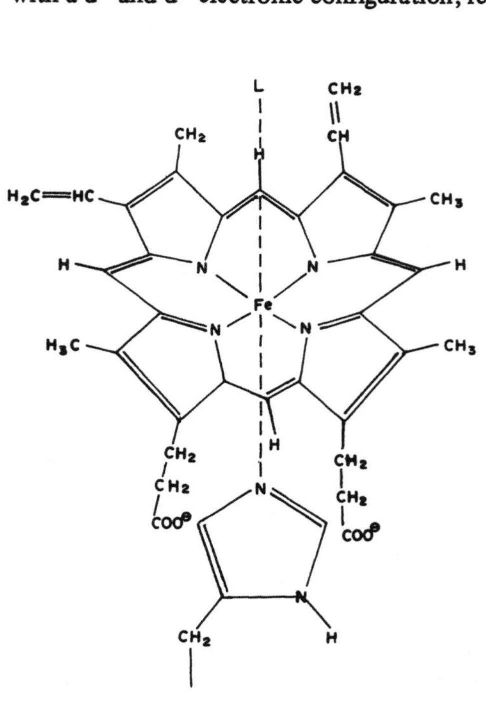

Fig. 4.8.4. Structure of the heme group indicating the positions of the polypeptide chain and the sixth ligand (L)

states can exist either in a high-spin [S = 2 for Fe(II) and 5/2 for Fe(III)] or in a low-spin state (S = 0 for Fe(II) and 1/2 for Fe(III)). Thus, there are four electronic states of iron. In hemeproteins, iron is coordinated to a His group of a protein chain, in addition to the four nitrogen atoms of the porphyrin ring. Hence, the minimum coordination number of iron is 5. This situation occurs in deoxy-Hb where Fe(II) is in a high-spin state. The sixth coordination number of iron can be saturated by oxygen or other ligands (such as H_2O, OH, CO, CN, etc.). The geometry of the heme group differs slightly according to the state of iron. In high-spin Fe(II) states, for example (as in deoxy-Hb), the central hole of porphyrin is unable to hold the metal ion and it is displaced by about 0.75 Å out of the heme plane. Thus, the sixth ligand, which approaches the metal ion from the other side, finds itself at a large distance from the ion. On the other hand, in low-spin states, the metal ion is in the plane of the porphyrin ring and such a situation is ideally suited to hexacoordination.

Extensive NMR investigations have been carried out on hemoproteins [37, 43, 48, 1017–1034]. These studies have attempted to compare the solution conformation of these proteins with the crystal structures, monitor changes in the molecular conformation and in the process of biological action, and to elucidate the electronic structure. The work is too extensive to be covered in this review, and only some of the recent work is discussed here. More detailed reviews on this subject have appeared earlier [555, 556, 1020]. The spectra of this class of proteins are unusual because of:

(a) The ring current effects of the porphyrin rings are 3 to 4 times larger than those produced by aromatic amino acid side chains [1064].

(b) In paramagnetic complexes, the nuclei of the polypeptide chains near the heme group experience conformation-dependent pseudocontact shifts and broadening of NMR lines. In addition, the NMR lines arising from the heme group may be shifted due to contact interactions.

As a result of the factors described above, heme proteins give rise to a number of well resolved resonances which can be used for unravelling the solution conformations of these systems and their electronic structures. Some typical applications are outlined below:

(a) *Myoglobin* is a single-chain protein with one heme group which reversibly binds to oxygen. The Fe(II) ion in the heme group alternates between high-spin (S = 2) and low-spin (S = 0) states both in the free and liganded protein. NMR studies on myoglobin have thrown light on the tertiary structure, ligand binding and electronic structure around the heme iron [555, 556].

Azide, imidizole and hydroxide complexes of Fe(III) Mb and Fe(III) Hb exhibit a temperature-dependent spin equilibrium between high (S = 5/2) and low (S = 1/2) states. The chemical shifts follow a non-curie law behavior in such cases [1019]. Evidence of restricted rotation of one methyl group in the heme side chain of ferric horse cyanomyoglobin has been obtained [1017]. The barrier to rotation is 14.8 kcal/mol and arises from short van der Waals contacts in the environment of the heme group.

(b) *Hemoglobin:* Extensive investigations have been carried out on deoxy and oxyhemoglobin from horse and human [555, 556]. The hemoglobin molecules consist of four polypeptide chains (α_1, α_2, β_1 and β_2), each containing one heme group. The four chains are held together by salt bridges and hydrophobic interactions. Upon oxygen binding, a displacement occurs in the relative positions of the four chains causing changes in the quaternary structure of the protein. This, in turn, increases the oxygen affinity of hemoglobin. Thus, the oxygen binding capability of hemoglobin exhibits cooperativity. The oxygen affinity is also controlled by the *p*H (Bohr's effect), concentration of inorganic phosphates and carbon dioxide.

NMR spectroscopy has helped in the detection of tertiary and quaternary changes in the structure of hemoglobin during oxygen binding [43, 48, 1022, 1023], in the understanding of the mechanism of the interactions between ligands and hemoproteins [1021, 1026, 1027], of the Bohr effect [1025], and in studies of structure-function relationships of mutant hemoglobins [1026]. For half-ligated hemoglobin, two quaternary forms of Hb have been observed and it has been concluded that the interconversion between the two forms is responsible for the cooperativity in Hb-ligand interactions [1022, 1023]. Further, at neutral *p*H and in the presence of 2,3-diphosphoglycerate, NMR shows that oxygen binds to the α-hemes in preference to β-hemes. Thus, oxygenation of hemoglobin can be described by a model that allows the α-hemes to bind first the ligand [1024].

(c) *Cytochrome-c* is an important electron carrier in biological oxidations. It contains a single polypeptide chain with 103 to 109 amino acid residues, and one heme group. The iron oxidation state alternates between +2 and +3. The ^{1}H-NMR spectra of reduced and oxidized cytochrome c exhibit a number of well resolved lines. These features have been used to study the electron exchange between the two states of cytochrome c, the ligand binding and the relation between electronic structure and biological activity [37, 1028, 1030–1033].

4.8.5.10 Collagen

Collagen is a structural protein containing the repeating unit (Gly-X-Y) and a large percentage of Pro and Hyp residues. It imparts tensile strength to connective tissues such as skin, bone and tendon. The mechanical strength and biological function of collagen arise from its rope-like structure. This structure consists of a left-handed triple helix stabilized by interchain hydrogen bonds. The Gly-CH_2 group resides in the core of the triple helix while the side chains of residues X and Y adopt positions on the surface of the supercoiled triple helix.

NMR studies on fibrous and denatured collagen have been carried out using ^{1}H- and ^{13}C-NMR [53, 1035]. Cross polarization studies on ^{13}C labelled collagen indicate two types of molecular motions in this molecule. A rod-like reorientational diffusion motion along the longitudinal axis of the molecule and a torsional reorientation. Thus, rapid reorientation of the helix backbone occurs in the fibrils. The formation of the fibrillar structure does not require a unique set of intermolecular interactions at the surfaces of collagen helices [53].

4.8.5.11 α-Chymotrypsin

This enzyme has a molecular weight of 24.800 and reveals a high specificity for cleavage of peptide bonds at the carbonyl group of Phe, Tyr and Trp. Aromatic amino acids and their derivatives are competitive inhibitors of the enzyme. A number of studies using ^{19}F labelled substrates have been reported. It is possible to prepare modified analogs of the enzyme where amino acid residues near the active site can be labelled. The ^{19}F chemical shifts have been utilized to obtain information on structure-function relationships of such systems [45, 1036–1038].

4.9 Protein-Nucleic Acid Interaction

4.9.1 Specificity

The binding of protein and nucleotide fragments has attracted the attention of NMR spectroscopists in recent years (see Table 4.9.1). For example, it has been shown in a study on the association of 5'-nucleoside monophosphates with tryptamine and 5'-methyltryptamine that the following order of association affinity holds: G > A > T > U > C [1074]. Non-specific interactions between proteins and nucleic acids can occur whenever electrostatic or other physicochemical forces can bind the two classes of molecules with more or less the same efficiency. The interaction may then be subjected only to the

Table 4.9.1. NMR Studies on protein-nucleic acid interactions and nucleoproteins

Nucleus	System and information	Ref.
^{1}H	Adenylic acid amides, D- and L-amino acid residues	[1065]
^{1}H	Aromatic amines–polyamides: conformational changes	[1066]
^{1}H	Sex hormones: nucleobase interactions	[1067]
^{13}C	Native histone complexes, thymus histone	[1068]
^{1}H	Nucleotide–amino acid interactions: conformation of complexes	[1069]
^{31}P	Histone-nucleotides: conformation	[1070]
^{1}H, ^{31}P	Dinucleoside triphosphate–DNA polymerase: T_1	[1071]
^{1}H	Dipeptide amides–DNA: intercalation bending helix	[1072]
^{1}H	Proline–DNA: minor groove interactions	[1073]
^{1}H	NAD-spin labelled alcohol dehydrogenase	[177]
^{1}H	Phosphoglycerate–kinase–ADP/ATP-lanthanides: active sites	[1043]
^{1}H	Nucleotides-aromatic amino acid derivatives	[1074]
^{1}H	Dipeptide Arg-Glu–nucleic acid bases	[1075]
^{1}H	Protein–gene 5 interactions	[1076]
^{1}H, ^{19}F, ^{31}P	Protein–nucleotide complexes: multinuclear NMR	[1077]
^{1}H	Protein–nucleic: recognition model	[1078, 1079]
^{1}H	D-, L-amino acid– nucleotide complexes: differences in structure	[1080]
^{1}H	Model studies of nucleic acids–protein interactions	[1081]
^{1}H	Role of tyrosine in specific recognition	[1082]
^{1}H	Guanine interaction with carboxylate ions	[1083]
^{1}H	Hydrogen bonding, amino acid side chains and nucleic acid bases	[1084]
^{1}H, ^{13}C	Interactions of uracil with peptide backbones	[1085]
^{1}H	Amino acids–nucleic acids: conformational changes	[1086]
^{19}F	5-Fluorouridylate-lactobacillus Casei thymidylate synthetase	[1095]

steric consideration as dictated by the stereochemical constraints of proteins and nucleic acids. Examples of non-specific binding are histones or polyamines with DNA.

A more important aspect from the biological viewpoint is the selective recognition by which sequences of bases in polynucleotides can be recognized by specific side chains of polypeptide chains. Here, the binding constant for a specific site differs by a factor of 10^4 to 10^6 from the rest of the molecule. A selective recognition of nucleic acids by proteins is important in biological processes for the entire mechanism of information transfer including replication, transcription, translation, repressor mechanisms and differentiation. Among the more important examples of specific base sequence binding are repressors to operators, RNA polymerase to promotors, ribosomal proteins with RNAs, and DNA to restriction endonucleases. The interaction of aminoacyl-tRNA-synthetase with t-RNA probably involves base-sequences which are located at different positions on t-RNA. Certain proteins have structural rather than base specificity. For example, proteins involved in replication, recombination and repair of DNA, bind strongly to single-stranded nucleic acids [1082, 1086]. Our knowledge about the molecular events that occur during replication and translation is rather sketchy. Little is known about the nature of forces which bring about specificity in protein-nucleic acid interactions and about the structure and dynamics of such complexes. A review on the work in this field till 1971 has been written by von Hippel and Mc Ghee [1087].

4.9.2 Nature of Intermolecular Forces

The protein-nucleic acid complexes may be stabilized by

(a) Electrostatic interactions between the charged functional groups in proteins (Lys, Arg, protanated His etc.) and nucleic acids (phosphate group). Although electrostatic interactions involving point charges are radial in nature the directional character to such interactions can be imparted by the backbone structure of polynucleotides which governs the position and orientation of charged phosphate groups.

(b) Stacking interactions involving nucleic acid bases and aromatic side chains (Trp, Phe, His, Tyr) in proteins. Such interactions can be easily monitored utilizing the upfield shifts of the proton resonances of aromatic amino acid residues. A structural specificity is present in such interactions since stacking is strongly favored in single strands of nucleic acids [1082, 1086, 1088]. Such a specificity for single-stranded structures has been observed for both Tyr- and Trp-containing peptides. The stacking interaction is independent of the sequence of peptides [1082]. The nature of interaction between aromatic peptides and nucleic acid depends on the configuration of the peptides [1072].

(c) Hydrogen bonding between the functional groups in peptides and nucleotide fragments [1075, 1083–1085, 1087]. Since the hydrogen bonds are directional such interactions are highly specific in nature, especially when a particular functional group is involved in two hydrogen bonds [1087, 1089]. The amino acid side chains which can participate in such interactions include the acidic groups of Glu and Asp, the amide groups of Gln and Asn, the imidazole ring of His, and the guanidinium group of Arg. Evidences of interactions of the peptide backbone with nucleic acid bases have been reported in recent years [1085, 1090].

(d) Hydrophobic interactions involving aliphatic amino acid side chains and nucleic acid bases. Again, such interactions are specific only to the extent of structural complementarity.

(e) Interactions mediated by a metal ion or solvent: Such a situation is prevalent in certain enzymes involved in biological oxidation-reduction reactions where nucleotides act as cofactors. In NMR experiments, the divalent metal ions are replaced by Mn(II) in order to study the geometry of bound substrates and cofactor. For example, the conformations of enzyme-bound coenzymes and substrates have been extensively studied utilizing the effect of Mn(II) or another paramagnetic reagent on the proton and ^{31}P relaxation rate [177, 1043, 1071]. This technique determines the distances between Mn(II) and the substrate and allows the exploration of active site regions of enzyme which cannot be studied directly by NMR. The conformation of deoxynucleoside triphosphate on DNA polymerase obtained by this method [1071] shows that the preorientation of such substrates can facilitate the incorporation of complementary nucleotides and promotes pyrophosphate departure.

(f) Formation of a covalent linkage between the peptide and nucleotide fragments leading to formation of nucleoproteins. During the translation processes, the peptide chain is synthesized at the 3′-end of t-RNA. Characteristics of the conformation of nucleotidyl-amino acid complexes have been studied [1065, 1069] by ^{1}H-NMR.

During the interaction of natural proteins with DNA and RNA more than one type of interactions listed above may be operative.

4.9.3 Applications

Interactions of histones with DNA are of interest from the view of integrity of chromatin. Current models of the chromatin structure are based on a periodic repeating unit in which a segment of DNA containing about 160 base pairs is folded along with eight histone molecules two each of F2A1, F3, F2A2 and F2B. Such monomers are linked by the continuous DNA molecule. Conformational studies have been carried out both on histones and their complexes with nucleotides [1068, 1070]. Histones in the group F2A1 and F3 contain randomly coiled *N*-terminal peptides which are ideally equipped to bind to DNA [1068] via electrostatic interactions involving the positively charged basic residues of the proteins and negatively charged backbone of the polynucleotide. The histones are probably held together through hydrophobic interactions [1068]. Conformational changes induced in poly(A) upon binding with aromatic amines have been studied [1066]. A review on NMR studies of histones is available [1091].

Gene 5 of the filamentous bacteriophages fd, fl or M13 codes for a small single-stranded DNA-binding protein [1076, 1077]. Studies have been carried out on the gene 5 protein single-stranded DNA complex using ^{1}H-, ^{19}F- and ^{31}P-NMR. It appears that the positively charged residues Lys and Arg interact with phosphate groups of the nucleotide backbone via formation of salt bridges. The phosphodiester conformation does not change significantly as a result of protein binding. The protein-Tyr residues bind through intercalation.

5 Polysaccharides

5.1 Structures of Polysaccharides and Carbohydrates

Polysaccharides and carbohydrates are polymers containing some simple sugars as repeating units. The most common sugars are D-glucose, D-mannose, D-fucose, D-galactose and a few substituted sugars such as D-*N*-acetylglucosamine, D-*O*-acetylsugar etc. Condensation of the sugars can occur in several different ways, depending on the hydroxy groups involved in glycosidic linkage. One of the hydroxy groups always participating in the condensation is that at C1. The linkages are termed as 1 → 2, 1 → 3, 1 → 4, and 1 → 6, depending on whether two sugar units are joined at 1 and 2, 1 and 3, 1 and 4 or 1 and 6 positions respectively (for numbering see Fig. 5.1.1). In some carbohydrates, phosphate groups are also involved in the linkages. Several books on primary structures of carbohydrates (see for example [1096]) are available.

The determination of the three-dimensional structure of a carbohydrate involves the following steps:

(a) Identification of the configuration of each residue (D or L, α or β),
(b) determination of conformation of each residue,
(c) identification of the type of linkage between neighboring residues,
(d) determination of the conformation at the various linkages.

5.2 Conformations of Monosaccharides

Recently, NMR and theoretical studies on the conformation of sugar rings in mono- and oligosaccharides have been reviewed [1097]. We will therefore restrict our discussion of monomer conformations to a minimum. The monomers exist as six-membered rings (commonly known as pyranose rings) which, in principle, can have six boat and two chair conformations [1098] (Fig. 5.2.1). Of these, the conformations generally found in crystal structures and in solutions are the two chair conformations designated as *C1* and *1C*.

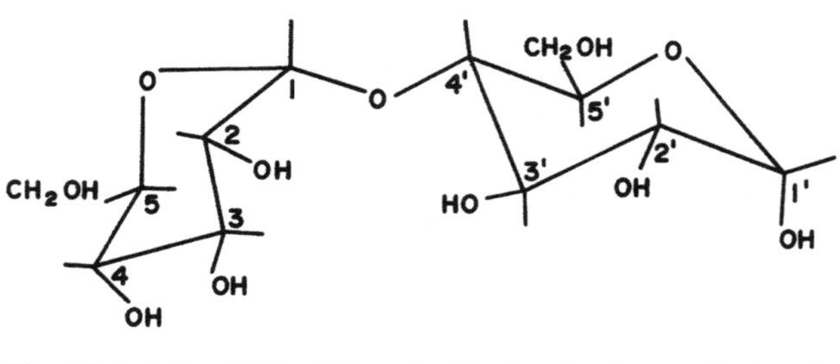

Fig. 5.1.1. A 1 → 4 linked disaccharide showing the numbering of atoms in the pyranose ring

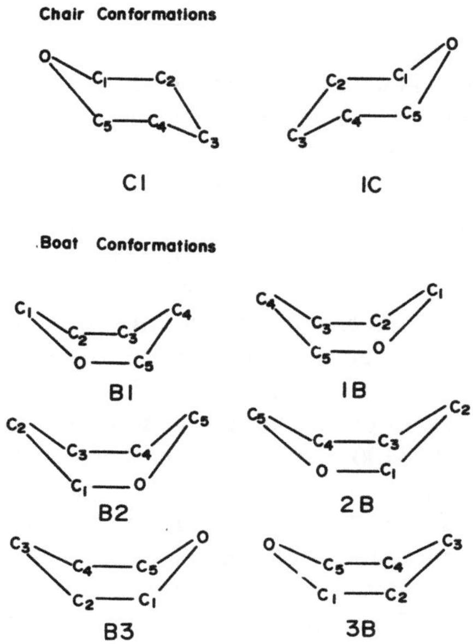

Fig. 5.2.1. Conformations of the pyranose ring

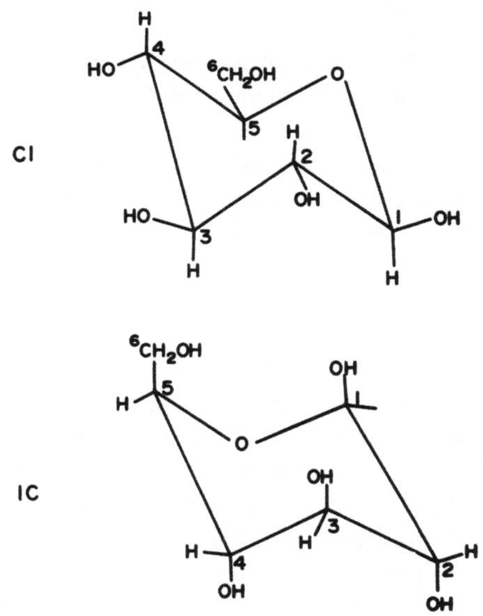

Fig. 5.2.2. *C1* and *1C* conformations of β-D-glucose showing the orientation (axial or equitorial) of different substituents

It may be seen from Fig. 5.2.2 that, in the chair conformations, alternate ring atoms C1, C3, C5 and the atoms C2, C4, and O lie in two parallel planes. When the plane containing the oxygen is above the other plane, the numbering of the atom is clockwise for *C1*(D) conformation and vice versa for *1C*(D) conformation [1097]. The bonds extending from a carbon atom are said to be equatorial when they are approximately in the plane containing the carbon atom, and are axial otherwise. The two chair forms differ not only in the spatial orientation of the ring atoms but also in the orientations of the side groups. The interconversion from one chair form to another changes the equatorial positions to axial positions and vice versa.

Since the early work of Lemieux et al. [63] high-resolution NMR has been extensively used in conjunction with theoretical studies on the determination of monosaccharide conformations [1099–1117]. It is generally observed that the chemical shifts and three-bond $^1H-^1H$ coupling constants in sugars depend on whether the protons are in axial or equatorial positions [63, 1099–1100]. Equatorially oriented protons appear at lower fields compared to the chemically similar but axially oriented protons. This difference in chemical shifts arises because of the long-range shielding effect associated with the diamagnetic anisotropy of C–C single bonds. Lemieux and Stevens [1099, 1100] have formulated empirical rules for the estimation of long-range shieldings that occur with changes in conformation and configuration of aldopyranoses and their derivatives. In the case of free sugars, these correlations have been established by using chemical shifts of β-D-xylose and β-D-glucose as reference points. The chemical shift of any proton can be estimated by the application of the following empirical rules:

(a) To obtain the chemical shift of an equatorially oriented proton add 0.6 ppm (δ scale) to the corresponding value of the reference compound (glucose or xylose).

(b) For the chemical shift of an axially oriented proton 1) add 0.3 ppm for each neighboring axial OH group and 2) add 0.35 ppm for each axial OH group whose direction is opposite to that of the proton in question.

In the case of substituted pyranose rings, the electronegativities of the substituents also influence the chemical shifts hence the above rules are not applicable. However, they are still useful for the assignment of the resonances in an NMR spectrum.

For conformational analysis, the more useful parameter is the three-bond coupling which is directly related to the dihedral angles involved. The most commonly used relation is given by [1101]

$$J_{H,H} = (7.8 - \cos\phi + 5.6 \cos 2\phi)(1 - M\,\Delta\chi) \tag{5.2.1}$$

where $\Delta\chi$ is the difference in the electronegativities of the substituent and hydrogen; M is usually taken as 0.1 but also depends on conformation (Chapter 2). It is clear that protons in axial-axial positions have a much larger coupling constants than the protons with axial-equatorial or equatorial-equatorial orientations. This effect of stereospecificity on M [117–119] should be taken into account in the calculation of component coupling constants for *C1* and *1C* conformations.

In pentopyranose derivatives two-bond $^1H-^1H$ coupling constants have also been found to be conformation-dependent [1102]. The H–C5–H coupling constant is numerically smaller for the *C1*(D) conformation than for the *1C*(D) conformation.

Using the above principles, solution conformations of several pyranoses have been obtained [1109–1110]; these are given in Table 5.2.1. It is seen that in most cases the *C1* conformation is the predominant one. In a few cases, *C1* and *1C* conformations coexist in dynamic equilibrium. The preference for the *C1*(D) conformation in aldohexopyranoses is probably due to the tendency of the bulky hydroxymethyl group and

Table 5.2.1. Favored conformations of D-aldopyranoses [1097] in aqueous (or DMSO) solutions

Aldopyranose	Conformation favored by NMR
α-D-allose	*C1*
β-D-allose	*C1*
α-D-altrose	*1C* ⇌ *C1*
β-D altrose	*C1*
α-D-galactose	*C1*
α-D-galactose*	*C1*
β-D-galactose	*C1*
α-D-glucose	*C1*
α-D-glucose*	*C1*
β-D-glucose	*C1*
β-D-glucose*	*C1*
α-D-gluose	*C1*
β D-gulose	*C1*
α-D-idose	*1C* ⇌ *C1*
β-D-idose	*C1*
α-D-mannose	*C1*
β-D-mannose	*C1*
α-D-talose	*C1*
β-D-talose	*C1*
α-D-arabinose	*1C*
β-D-arabinose	*1C* ⇌ *C1*
α-D-lyxose	*1C* ⇌ *C1*
β-D-lyxose	*C1*
α-D-ribose	*1C* ⇌ *C1*
β-D-ribose	*1C* ⇌ *C1*
α-D-xylose	*C1*
α-D-xylose*	*C1*
β-D-xylose	*C1*

* refers to NMR studies in DMSO solutions

the C5 atom to assume an equatorial position because otherwise, it would be in a sterically unfavorable arrangement. In L sugars, the 1C conformation is expected to be the most favored conformation.

In α- and β-D-mannose, β-talose and β-D-lyxose, the observed J_{12} coupling constants 1.0–1.7 Hz, 1.0 Hz, 1 Hz, and 1.5 Hz, respectively, are much smaller than that expected for the C1 conformation (3.5 Hz) on the basis of the Karplus-type relations. Similar discrepancies have been noted in the conformational studies of certain sugar derivatives [1101–1102, 1111]. The reasons for such a behavior are not yet clearly understood.

Another important aspect of the pyranose ring conformation is the orientation of the hydroxy groups or the substituents at the ring atoms. This information can be obtained from H—C—O—H coupling constants. Dimethyl sulfoxide (DMSO) is a suitable solvent for obtaining such coupling constants since in D_2O the hydroxy protons are exchanged by D_2O. The use of DMSO as a solvent may cast some doubt on their relevance to a biological system. However, it has been shown that α and β-D-glucoses, α-D-galactose and α-D-xylose exist in *C1*(D) conformations in both aqueous and DMSO

Table 5.2.2. Coupling constants (Hz) of hydroxy protons in some monosaccharides in DMSO solutions [1112–1113]

Monosaccharide	C1–H		O1–H doublets		O2–H O3–H doublets	O6–H triplet
	axial	equatorial	equatorial	axial		
α-D-glucose		3.0		4.5	5.0 4.5 6.5	5.0
β-D-glucose	6.5		6.5		3.5	5.5
α-D-galactose		3.0		4.5	5.0 7.0 4.0	5.0
α-D-xylose		3.5		4.5	4.0 3.5 6.5	
Methyl α-D-glucopyranoside					5.0 3.5 6.5	5.5

solutions [1117]. Thus, the results of studies performed in DMSO solutions may be supposed to apply to aqueous solutions as well. Based on such data, Jaffrey et al. [1118] suggested that the O–H bond at the anomeric carbon atom exists in both *anti* and *gauche* conformations of the β-anomer, whereas for the α anomer only the *gauche* conformation is allowed. Since $J_{anti} > J_{gauche}$, one expects larger values of coupling constants for the β anomer than for the α anomer and the observed results do show such a behavior (Table 5.2.2).

5.3 Conformations of Polysaccharides

Attempts to determine polysaccharide conformations by NMR spectrometry have been rather limited. Only recently have ^{1}H and ^{13}C NMR been applied to the conformational analysis of some polysaccharides. The general approach has been to examine the trends in chemical shifts and coupling constants [1119–1148].

Tables 5.3.1 and 5.3.2 compile the ^{1}H chemical shifts and coupling constants of the anomeric protons in different carbohydrates [1112–1114, 1123–1126]. For comparison, data on α- and β-D-glucose are also included. It is seen that in D_2O solutions, the anomeric proton signal appears at about (5.8 ± 0.3) ppm in all the α-linked compounds, which is close to the value observed for α-D-glucose in the *C1* conformation. The value of the $J_{H1,H2}$ coupling constant (3.0 ± 0.3 Hz) is also close to the value observed for α-D-glucose in the *C1* conformation. Similarly, the glycosidic anomeric proton signals in all the β-linked carbohydrates appear at 4.7 ± 0.1 ppm and $J_{H1,H2}$ = 7.0 ± 0.7 Hz.

Table 5.3.1. Chemical Shifts and Coupling Constants in Parantheses of Protons in Oligo and Polysaccharides [1097]

Name of compound	Anomeric proton at glycosidic linkage		Anomeric proton at reducing end	
	Equatorial	Axial	Equatorial	Axial
α, α′-Trehalose	5.91 (3.2)			
Turanose	5.3 (3.2)			
Methyl α-D-glucoside	4.75–4.8 (3.0 –3.3)			
D-maltose	5.4 (2.8–3.4)		5.2–5.25 (3.0–3.4)	4.3–4.7 (7.3–8.0)
Methyl β-maltoside	5.38 (3.2)			4.38 (7.6)
Sucrose	5.41–5.48 (3.0 –3.2)			
Maltrotriose	5.37 (3.5)		5.2 (3.7)	4.63 (7.3)
Raffinose	4.98 (2.7) 5.41 (2.9)			
Melibiose	4.96 (2.9)		5.21 (2.9)	4.96 (2.9)
α-D-glucose	5.27 (3.8)			
β-D-glucose		4.73 (8.1)		
Cyclohexa-amylose	5.05 (2.7)			
Cyclohexa-amylose	5.1–5.2 (2.7–3.5)			
Amylose	5.3			
Hydrolyzed starch	5.38 (3.0)			
Dextran B 512 E	4.91 (2.3)			
Dextran B 1355	4.93 (4.73)			
Methyl β-D-glucopyranoside	4.	4.35–4.46 (7.4–7.7)		
Cellobiose		4.5–4.624 (7.0–8.8)	4.22–5.3 (2.9 –3.3)	4.62–4.65 (7.4–8.0)
Lactose		4.42 (7.1)	5.22 (6.3)	4.64 (7.3)
Gentiobiose		4.5 (7.7)	5.22 (3.5)	4.66 (7.6)
Laminaran		4.75 (6.7)		
Crowngall-polysaccharide		4.88 (6.2)		
Digalactouronic acid	5.07			
Trigalacto-uronic acid	5.06			
Tetragalacto-uroni acid	5.06			

Table 5.3.2. Chemical Shifts and Coupling Constants (in Parenthesis) of C1–H Resonances of some Carbohydrates

	Equatorial	Axial
α, α′-Trehalose 1-O-(α-D-glucopyranosyl) α-D-Glucose	4.85 (3.0)	
β-Maltose-4-O-(α-D-glucopyranosyl)-β-D-glucose	4.99 (3.0)	4.3 (6.5, 6.5) pair of doublets
α-Melibiose-6-O-(α-D-galacotopyranosyl)-α-D-glucose	4.88–5.00	
β-Cellobiose-4-O-(β-D-galactopyranosyl)-β-D-glucose		4.4 (7.0, 7.0) pair of doublets
Amylose	5.1 (3.0)	
α-Cyclodextrin	4.81 (3.0)	
β-Cyclodextrin	5.83 (3.0)	
Linear dextran	4.6 (2.5)	
Maltotriose	(5.0)	
Isomaltotriose	4.9	
Maltotetraose	5.0	
Maltopentaose	5.0	
Maltohexaose	5.0	
Branced dextran	4.65–4.9	

These evidences indicate that the pyranose ring conformation in D-sugars is not significantly affected by polymerization. Also, methylation or acetylation of the hydroxy groups does not show marked effects. A comparison of the data in Tables 5.2.2 and 5.3.2 suggests that also in DMSO solutions the pyranose ring conformation is unaltered on polymerization. Similar results may be expected for L-sugars as well. In fact, in heparin containing alternating D and L repeating units, it is observed that the D residue has C1 conformation while the L residue exhibits *1C* conformation [1127–1128]. These results indicate that the pyronose ring conformation is least affected by changes of the surrounding environment. Thus, the magnitudes of $J_{H1,H2}$ can be directly used to identify the configuration and conformation of the pyranose ring [1131–1134].

Pasika and Cragg [1129–1130] have examined the ^{1}H NMR spectra of dextrans which are composed of glucose residues joined by α 1 → 6 glycosidic linkages. Depending on the source of dextran, the polymer may be branched to a greater or lesser extent. It has been observed that the ^{1}H spectrum of a branched dextran contains a peak which does not appear in the spectra of linear dextrans. This additional peak perhaps arises from non 1 → 6 linkages which form the branch points in branched dextrans [1129]. The intensity ratio of H1 resonances at 1 → 6 linkages to those at non 1 → 6 linkages has been used as a quantitative measure of the degree of branching of the polymer. Several similar studies on heparins and related compounds have been reported [1147–1148].

Since it has been established (Sect. 5.2) that the pyranose conformations in DMSO and D_2O are similar, ^{1}H NMR of polysaccharides in DMSO solutions is a potential tool for the understanding of the conformations around the glycosidic linkages. The glycosidic rotations determine the proximity and orientations of the hydroxy groups of neighboring pyranose rings and thus control the formation of intramolecular hydrogen bonds. Such bonds, if present, can be seen in the NMR spectra in DMSO solution. NMR studies [1112–1113] have revealed that some of the non-anomeric hydroxy protons appear on the low-field side (5.2–5.7 ppm) compared to their counterparts in monosaccharides. The low-field signals in $\alpha(1 \rightarrow 4)$ linked glucose derivatives have been assigned to the hydrogen-bonded hydroxy protons at C2 and C3. Cycloamyloses show stronger intramolecular hydrogen bonds than amylose and maltose indicating that the glycosidic conformation in the former is more rigid than in the latter. In the case of polysaccharides involving phosphodiester linkages, ^{31}P coupling with anomeric protons can be used to obtain information on the glycosidic dihedral angles.

The main difficulty arising in the application of ^1H NMR to carbohydrates is that the lines are broad and exhibit considerable overlapping. Consequently, much of the desired information on shifts and coupling constants lacks accuracy. In recent years, ^{13}C-NMR has become a very valuable technique in the elucidation of carbohydrate structures. Several conclusions have been drawn which are useful for the interpretation of data on complex carbohydrates [1149–1166]:

(a) The anomeric carbons resonate at the lowest field (90 to 98 ppm for pyranoses),

(b) the signals of the CH_2OH groups (C6) in hexapyranoses and of the ring CH_2 groups (C5) in pentopyranoses generally occur at the highest field positions (60 to 62 ppm and 61 to 68 ppm, respectively),

(c) the resonances of the ring carbons attached to hydroxy groups and those of C5 in hexapyranoses occur in the range 65 to 76 ppm,

(d) in 6-deoxysugars, the methyl groups resonate at the highest field positions (16 to 18 ppm); the resonances of ring deoxycarbons occur at lower fields (30 to 40 ppm),

(e) methylation of a sugar hydroxy group causes the resonances of the neighboring ring carbon atoms to shift 6–10 ppm downfield,

(f) carbon atoms bearing axial hydroxy groups generally resonate at higher field than the corresponding C-atoms with equatorial OH groups,

(g) the change of pyranose hydroxy group orientation from equatorial to axial, causes (due to 1,3-diaxial interaction) the resonances of γ carbon atoms to shift upfield,

(h) signal assignments are possible by spectral comparison of different analogs.

(i) proton-off resonance spectroscopy allows a distinction to be made between primary (quartet), secondary (triplet), tertiary (doublet), and quaternary (singlet) carbons,

(j) the signals of different anomers can be assigned by recording the ^{13}C-NMR spectrum both before and after attainment of mutarotational equilibrium. The more abundant isomer gives rise to stronger intensities in the spectrum,

(k) the carbon resonances of adjacent hydroxy groups in *cis* positions are identified by large downfield shifts in the presence of boric acid. Three-bond ^{13}C—^{31}P coupling constants in phosphocarbohydrates suggest that the phosphate groups preferentially adopt a *trans* position with respect to C2 of the ring implying that these carbohydrates are linearly extended chains,

(1) the chemical shifts are sensitive to the form of the sugar,

(m) the carbon atoms involved in linkages are shifted considerably downfield and the extent of shift is sensitive to overall conformation.

Following these guidelines serogroup A of meningococcal polysaccharide has been shown to be essentially a $1 \rightarrow 6$ linked homopolymer of partially *O*-acetylated 2-acetamido-2-deoxymannosyl phosphate and the serogroup X polysaccharide to be a 1 → 4 linked homopolymer of 2-deoxyglycosyl phosphate [1118, 1122]. The ^{13}C spectrum of serogroup X polysaccharide shows five couplings of ^{31}P with C1, C2, C3, C4, and C5. The chemical shift of C6 is similar to that of the monomer. These results indicate that the linkage is 1 → 4 via a phosphodiester group [1122].

In 1 → 3 β-D-glucans it is observed that the ^{13}C line widths are extremely large (~150 Hz) and that the glycosidic and C4 carbons are shifted considerably downfield as compared to those in the random-coil conformation. The downfield shifts are more than those in the case of (1 → 4) β-D-glucan which is a linear carbohydrate [1163–1165]. These results suggest that (1 → 3) β-D-glucan contains portions of ordered helices which exhibit a very rigid conformational structure. The downfield shifts of C4 signals have been ascribed to the presence of O4′–O5 intramolecular hydrogen bonds [1165].

Lentinan and schizophyllan are two (1 → 3) β-D-glucans which form polysaccharide gels. Lentinan has two branches for every 5-D-glucopyranosyl residues while schizophyllan has one branch for every 3 or 4 residues at the O6 position. In contrast to the linear (1 → 3) β-D-glucan (curlan), all of the ^{13}C signals due to the (1 → 3) linked β-D-glucosyl residues are completely suppressed. This complete loss of peak areas is explained in terms of a higher degree of cross-linking than that of linear glucans [1135]. The cross-links involve physical association of the helical segments such as the double or triple helices. These helical forms were converted at 0.2 M NaOH into randomly coiled forms, thereby restoring the peak areas.

^{13}C-NMR has also been used to study the composition of glucans synthesized by enzymes of the cariogenic organism *streptococcus mutans* [1136].

5.4 Glycoproteins and Peptide-Carbohydrate Interactions

Glycoproteins occur in a living cell as hormones, immunoglobulins, integral parts of mammalian membranes, etc. These play important roles in several functions of the cellular surface such as cell association, contact phenomena, growth control, recognition of foreign or tumour cells etc.

The general features of glycoprotein structures have been described in several reviews [1167–1168]. Glycoproteins consist of a carbohydrate portion and a protein portion joined by covalent linkages. The carbohydrate portion of glycoproteins is made up of sugars such as D-galactose, D-mannose, D-glucose (rarely), L-fucose, *N*-acetylglucosamine, *N*-acetylgalactosamine, and sialic acids. L-arabinose and D-xylose have also been found in some cases. The carbohydrate units encompass a wide range of polymerization ranging from a disaccharide to complex heterosaccharides consisting of 20–30 monosaccharide residues. These are linked through the C1 position of reducing terminal sugar residues to the side chain of an amino acid in the polypeptide portion of the glycoprotein. Three types of linkages have been found:

(a) Carbohydrates linked *N*-β-glycosidically to the amide group of the aspargine residues,

(b) carbohydrates linked *O*-β-glycosidically to the hydroxy groups of serine, threonine, hydroxylysine or hydroxyproline residues, and

(c) *S*-glycosidic linkages between cystein residues and carbohydrates.

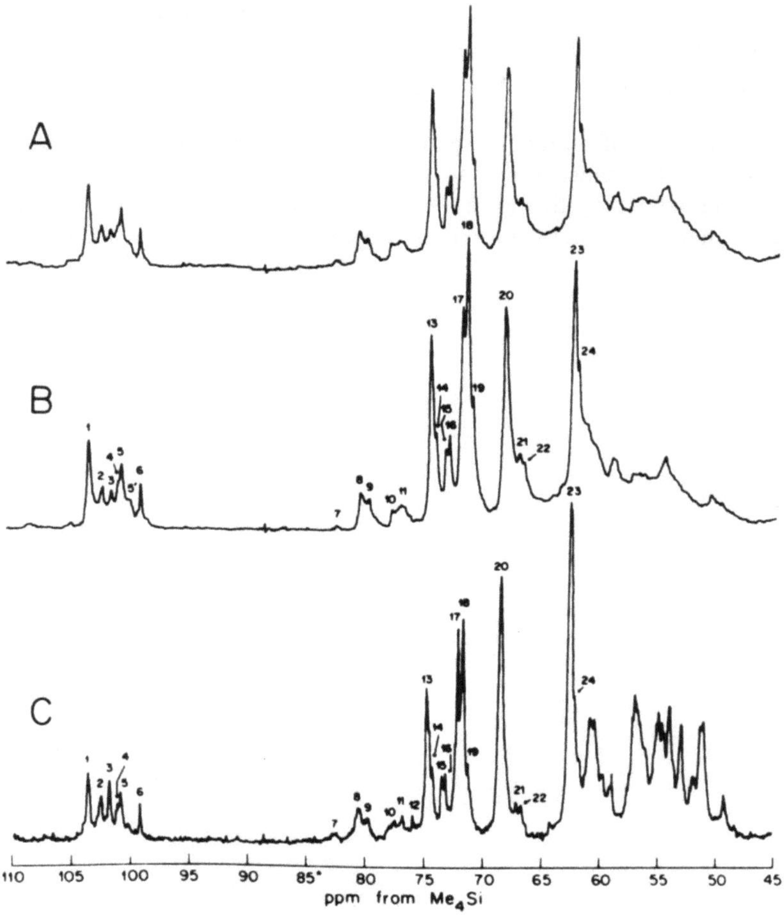

Fig. 5.4.1. Carbohydrate region of ^{13}C (67.9 MHz)-NMR spectrum of glycoamylase (reprinted with the permission from Ref. [1171], (A.) native protein from Boehringer, (B) native glycoprotein from SIGMA, (C) denatured glycoprotein from SIGMA, Peaks 7–24 are non-anomeric carbons

The NMR characteristics of glycoproteins have not yet been extensively studied. Recently, two PMR reports have appeared on the glycoprotein hormone porcine lutropin [1169–1170]. In these reports the interest is mainly centered on the identification of histidine C2–H resonances. Two structural studies have also been reported using ^{13}C-NMR spectroscopy [1171–1172]. It has been observed in the case of glycoprotein glucoamylase from Aspergillus niger [1171] that the ^{13}C resonances of all anomeric carbons and most non-anomeric carbons of carbohydrate residues do not overlap the resonances of amino acid residues. Figure 5.4.1 shows the carbohydrate region of the ^{13}C spectrum of a glycoprotein. These ^{13}C resonances are not significantly altered by denaturation or protein folding, suggesting that the carbohydrate groups are exposed to solvent. Integrated intensities and the chemical shifts indicate that the molecule contains 120 carbohydrate residues, out of which 80 are α-D-mannopyranoses and 10 are α-D-glucopyranoses. It has been suggested that all residues are involved in *O*-glycosidic linkages.

In the case of peptidophosphogalactomannan (pPGM) produced from pencillium charlesii, ^{13}C chemical shifts indicate that all the galactosyl residues are in the furanosyl form and linked together by β(1 → 5)-*O*-glycosidic linkages. ^{13}C-NMR data further reveal that the mannopyranosyl residues are involved in α(1 → 2)-*O*-glycosidic linkages [1172].

6 Lipids and Molecular Organization in Membranes

6.1 Biomembranes

A living cell is enclosed by a boundary which confines the interior of the cell and separates it from the external environment. This boundary is called membrane. Within the cell itself, there is an elaborate compartmentalization; nucleus, mitochondria, etc. Each of these compartments has a membrane surrounding it. The biological membranes possess many common structural features.

Biological membranes are essentially constituted of lipids, proteins and carbohydrates. The relative proportion of these components varies, depending on the function of the membrane. The lipids can be classified into several categories, i.e. phospholipids, triglycerides, cholesterol, glycolipids. Among these, phospholipids generally occur most abundantly in biomembranes. Molecular fromulae of some common lipids are shown in Fig. 6.1.1. The lipids are amphipathic since they have two long hydrocarbon chains and a short polar chain, attached to the glycerol moiety. In the biological membranes, the lipids form a bilayer matrix and the proteins are embedded in it as discrete particles. A schematic diagram of the structure of the membrane is shown in Fig. 6.1.2. (For details, the reader may consult several books and reviews [1173–1187].) Some of the proteins span the membrane while others do not. Depending on the ease with which proteins can be isolated from the membrane, they are classified as structural and peripheral proteins. Structural proteins need stronger treatments than peripheral proteins for their extraction.

The knowledge of the organization of these components in membranes is of paramount importance for the understanding of the membrane functions. It is believed that the proteins assume a conformation such that the hydrophobic portions lie buried in the lipid matrix while the exterior polar groups remain immersed in water.

NMR has been extensively used for the characterization of both the structure and organization of the lipids in biomembranes [1185, 1188–1190]. On the other hand, studies on membrane proteins are lacking due to the complex nature of these molecules. Some NMR investigations on myelin proteins discussed in Chapter 4 have been reported. It is worth recalling that these proteins show considerable segmental motions even when incorporated into liposomes (model systems for membrane studies). The conformations of these proteins are *p*H-dependent and tend to aggregate at neutral *p*H values. This implies that *p*H fluctuations in the solutions bathing the membrane can enhance the lateral diffusion of the proteins in the membrane.

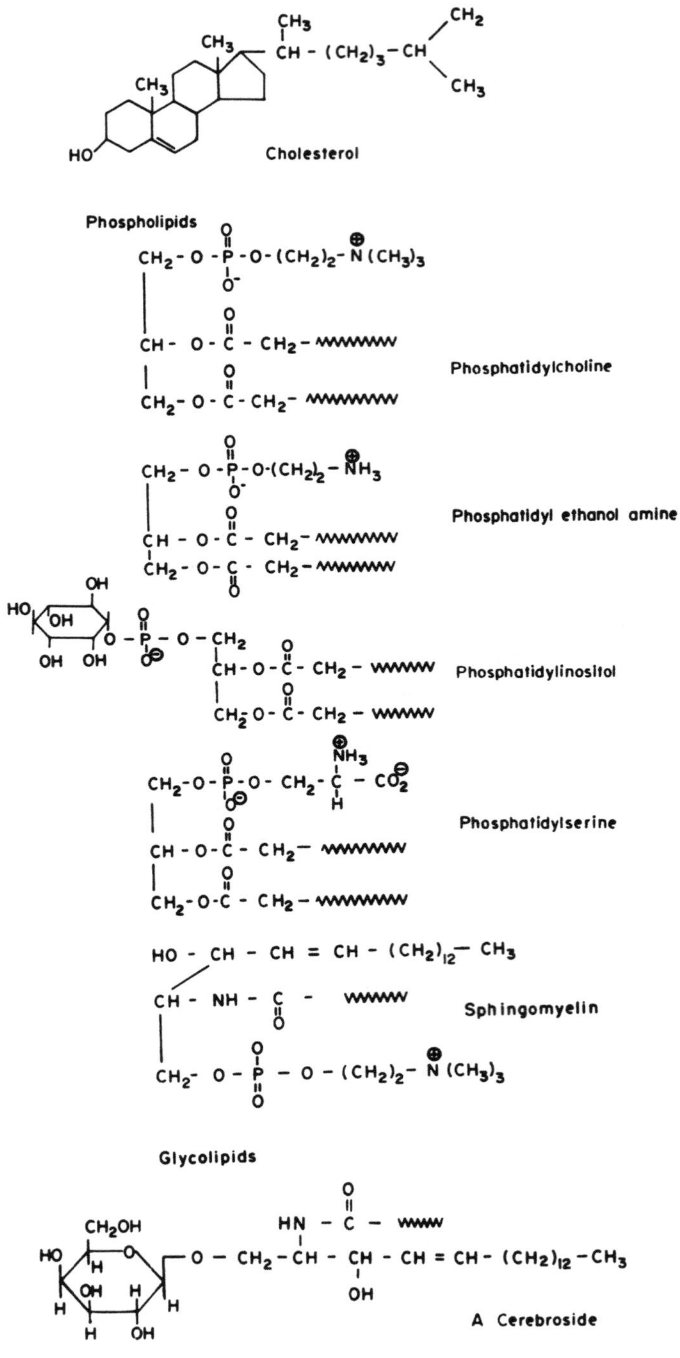

Fig. 6.1.1. Molecular formulae of common lipids found in biomembranes. The zig-zag lines indicate hydrocarbon chains containing 14–18 carbon atoms

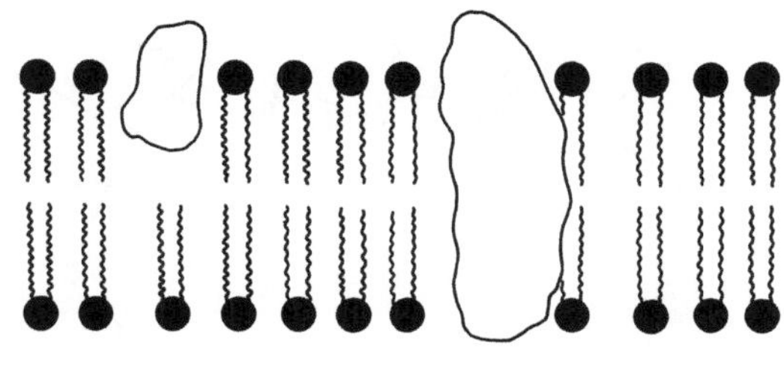

Fig. 6.1.2. Schematic diagram of a membrane structure. The proteins are embedded in the lipid bilayers and can be classified into integral and peripheral proteins (reprinted with the permission from Ref. [1186])

6.2 General Properties of Phospholipids

An important property of the phospholipid molecules is their amphipathic character. When mixed with water, the polar group tends to immerse in water while the non-polar hydrocarbon chains tend to keep away from it. Thus, depending on the properties of lipid and water, the lipids organize in different fashions, forming various structures. Secondly, the lipids display a large conformational flexibility owing to free rotations around the different single bonds in the molecule. The conformational flexibility is responsible for the dynamic nature of a biological membrane. It is extremely difficult to crystallize these molecules; thus, there have hitherto been reported only four crystal structure reports on lipids [1191–1194].

The rates of various rotational motions in a lipid molecule depend on temperature. It has been observed that lipids are capable of existing in two different states, namely gel and liquid crystalline states [1173]. In the gel state, the motions are more or less frozen while in the liquid-crystalline state, the mobility is relatively higher. The transition from the gel to the liquid crystalline state is smooth and is characterized by a temperature known as gel-to-liquid crystalline phase transition temperature (T_c). The temperature T_c depends on the charge on the polar group, length of the hydrocarbon chain, presence of unsaturation in the hydrocarbon chains etc. As a result, in a membrane where several types of lipids coexist, the physical states of the different lipids may be different and they may form islands of varying fluidities [1174, 1186].

6.3 Conformations of Phospholipids

Although phospholipids exhibit a very large conformational mobility, there are evidences to show that certain preferred conformations exist in the polar part of the molecule. For example, most of the crystal structure data on fragments of lipids show only a certain set of conformations [1195]. Intermolecular interactions in a membrane impose further restrictions on the allowed conformations of the polar head group and hydrocarbon chains.

Theoretical and NMR techniques have been the main contributors to the understanding of phospholipid conformation [1196–1215, 1220–1224]. Figure 6.3.1 shows the different dihedral angles which determine the conformation of a phospholipid molecule ([1195] see for notations). The dihedral angles, Θ_1, and Θ_3 determine the conformation in the glycerol moiety while the angles α_1–α_6, $\beta_1-\beta_n$ and $\gamma_1-\gamma_m$ dictate the relative orientations of the three chains and the structure in each chain.

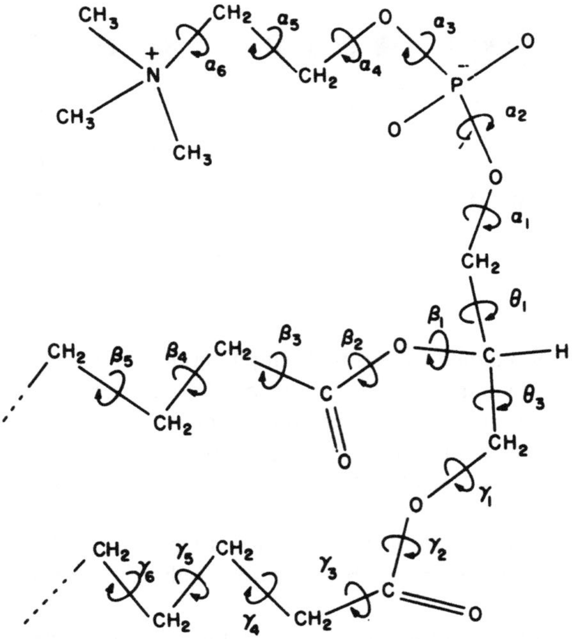

Fig. 6.3.1. Schematic formulae of lecithin showing different dihedral angles which fix the conformation of phospholipids (reprinted with the permission from Ref. [1186])

6.3.1 Conformations of the Glycerol Moiety

The conformations with respect to the angles Θ_1 and Θ_3 have been determined from the three-bond $^1H-^1H$ coupling constants. There are four such coupling constants, H2–H1, H2–H1′, H2–H3 and H2–H3′ which can be used to explore the preferred values of Θ_1 and Θ_3 (Fig. 6.3.2). The expected coupling constants for different staggered conformations calculated from the electronegativity rules of Abragam and Gatti [99] are also given in the figure. Comparing these with the observed time-averaged coupling constants, one finds that either (a) the conformation is different from the staggered conformations or (b) all the staggered conformations coexist in solution and the observed coupling constant is a statistical average over these conformations. The first alternative can be ruled out on the following grounds. Theoretical calculations show that the staggered conformations are the most stable ones. In none of the crystal structures [1191–1194] conformations other than staggered ones have been observed. Thus, considering the second alternative, the populations of the different staggered conformations can be obtained, using the component coupling constants and the following three relations,

$$J_{obs} = J_{g^-}P_{g^-} + J_tP_t + J_{g^+}P_{g^+} \tag{6.3.1}$$

$$J'_{obs} = J'_{g^-}P_{g^-} + J'_tP_t + J'_{g^+}P_{g^+} \tag{6.3.2}$$

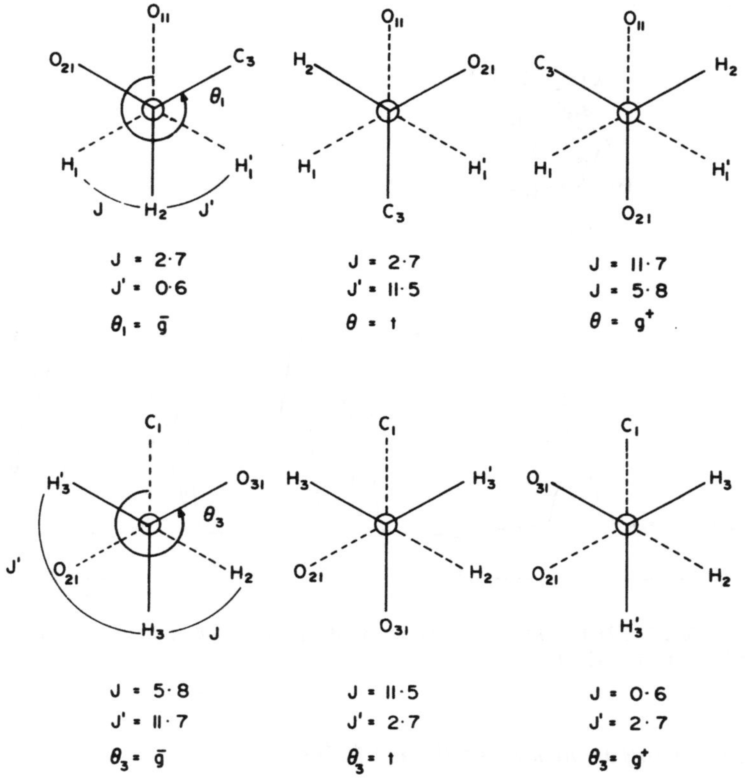

Fig. 6.3.2. Newman projection diagrams of the angles θ_1 and θ_3

$$1 = P_{g^-} + P_t + P_{g^+} \tag{6.3.3}$$

where P_{g^+}, P_{g^-} and P_t are the probabilities of g^+, g^- and t conformations (Fig. 6.3.2), respectively. J_{obs} and J'_{obs} are the observed three-bond coupling constants of H2 with H1 and H1′ (H3 and H3′), respectively. The assignment of the H1 and H'_1 protons or the H_3 and H'_3 protons in the NMR spectrum (Fig. 6.3.3) is not known unambiguously. Thus, there are two alternative assignments in each case (Θ_1 or Θ_3), which amounts to interchanging J_{obs} and J'_{obs} in Eqs. (6.3.1) and (6.3.2). Consequently, there are two solutions for population distributions with respect to both Θ_1 and Θ_3. Such an analysis in the case of dipalmitoyl lecithin [1203] (Fig. 6.3.3) shows that, with respect to Θ_1, $P_{g^-}:P_t:P_{g^+} = 31:41:28$, or 37:20:43, depending on how J and J′ are assigned. Similarly, with respect to Θ_3, $P_{g^-}:P_t:P_{g^+} = 0:60:40$ or 50:0:50. Thus, it is clear that, with respect Θ_1, all the three staggered conformations are substantially populated, while for Θ_3 only two of the three staggered conformations are populated in conformity with results of energy minimization techniques [1196–1221]. Similar conclusions have been reached in the case of triglycerides [1202].

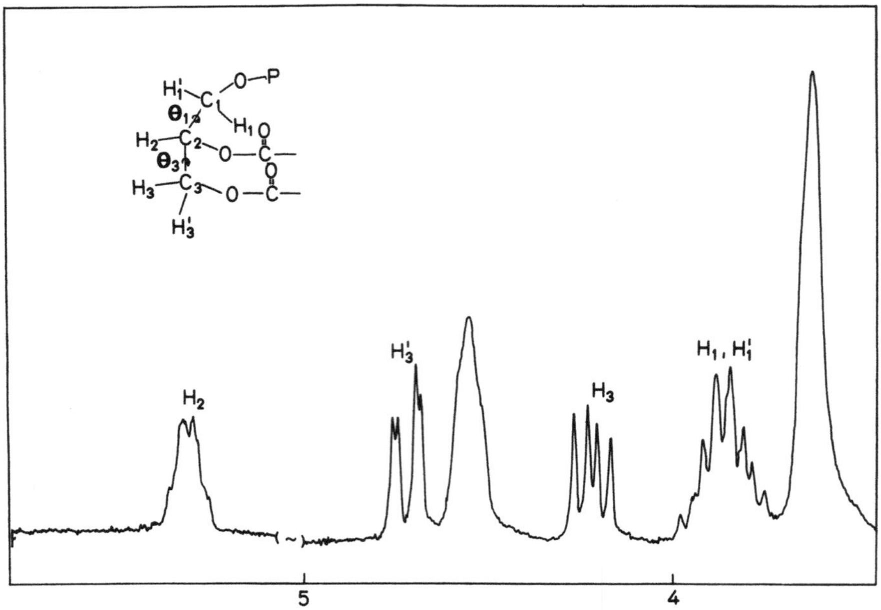

Fig. 6.3.3. 270 MHz ^{1}H-NMR (glycerol part of dipalmitoyl-lecithin showing chemical non-equivalence of H1, H1′ and H3, H3′

6.3.2 Conformations in the α-Chains

The conformation of the α-chain is determined by the torsion angles α_1 to α_6. Crystal structure reports on several fragments of phospholipids [1195] show that the phosphodiester group (O–P–O) prefers the *gauche-gauche* (either g^-, g^- or g^+, g^+) conformation (Chapter 3). It is possible that g^-, g^- and g^+, g^+ conformations coexist in solution in a dynamic equilibrium.

Information on the other dihedral angles α_1, α_4 and α_5 has been obtained from ^1H–^1H, ^1H–^{31}P and ^{13}C–^{31}P coupling constants. $^{31}P\{^1H\}$ NOE has also been used to determine the proximity of the $N(CH_3)$ group with respect to the phosphate group.

There are in all four ^1H–^{31}P coupling constants, two in the glycerol-CH_2OP group and the other two in the choline-CH_2OP group. The two protons in each group are non-equivalent. The values of all ^3J(H–P) coupling constants lie in the range 6–7 Hz in D_2O, chloroform and CD_3OD solutions [1205–1206]. Using these values, the population distributions among the three staggered conformations can be calculated. The component coupling constants necessary for this purpose are obtained from the usual relation (Eq. (3.4.1)) and these values for the staggered conformations are shown in Fig. 6.3.4. It has been found that the *trans* conformation contributes around 55–60% to the conformational equilibrium. Similar results have been obtained using ^{13}C–^{31}P coupling constants [1205]. The observed coupling constants for the glycerol C–C–O–P and choline C–C–O–P fragments are 7–8 Hz and 6 Hz, respectively. Using the data in Table 2.1.1, the component coupling constants for the three staggered conformations

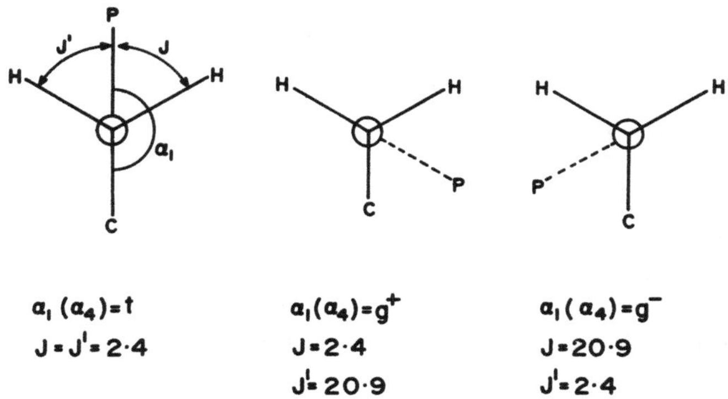

Fig. 6.3.4. Newman projection diagrams for α_1

are 10.1 and 2.07 Hz respectively for the *trans* and *gauche* conformations. It follows that an observed coupling constant of 6 Hz corresponds to ~50% of *trans* conformation.

Akutsu and Kyogoku [1212] used ^{1}H- and ^{31}P-NMR to investigate the conformation of *O*-phosphorylcholine (OPC), *O*-phosphorylethanolamine (OPE), L-α-glycerophosphoryl-phosphatidylcholine and phosphatidylethanolamine from E. coli in CH_3OH and $CHCl_3$ solutions. They observed that in every case the system prefers a *gauche* conformation; however, in the choline moiety, the dihedral angle around the C—C bond is distorted from the usual *gauche* angle of 60° to a larger value. The dihedral angle of OPC reveals greater flexibility compared to OPE. The ^{31}P coupling constants show that the *trans* conformation dominates in the P—O—C—C systems.

6.3.3 Conformations in the β and γ Chains

Conformations in the hydrocarbon chains are determined by the dihedral angles β_1—β_n for the β chain and γ_1—γ_m for the γ chain. Theoretical calculations [1196—1199] and carbon-13 T_1 measurements [1215] indicate a large flexibility in the hydrocarbon chains although the *trans* conformation is favored over *gauche* conformations. The conformational flexibility is somewhat restricted near the glycerol moiety. Thus, the relative orientation of the two chains is controlled mainly by the dihedral angles β_1, γ_1, Θ_3, β_2, γ_2 and, to some extent, by β_3 and γ_3.

Information on β_1 and γ_1 may be obtained using the three-bond coupling constants involving carboxy-^{13}C and ^{1}H: H3′ and H3 for γ_1 and H_2 for β_1 (Fig. 6.3.5). Figure 6.3.6 shows the carbonyl regions of the ^{13}C spectra of a triglyceride (glyceryl trivalerate) and a phospholipid (dipalmitoyl-lecithin) in $CDCl_3$ solutions. Analysis of these spectra yields $J_{C-H_2} \sim 3.5$ Hz, and $J_{C-H_3} = J_{C-H_3'} = 2.8$ Hz [1203]. Lemieux et al. [78] have established a calibration curve for ^3J(H—C—O—C) and obtained dihedral angles from which the latter can be estimated for a rigid conformation. Esperson and Martin [1216] have estimated $J_g = 1.3$ Hz and $J_t = 9.8$ Hz from amino acids. Qualitatively, it is clear that the observed ^{13}C—H3′ and ^{13}C—H3 coupling constants are consistent with a value

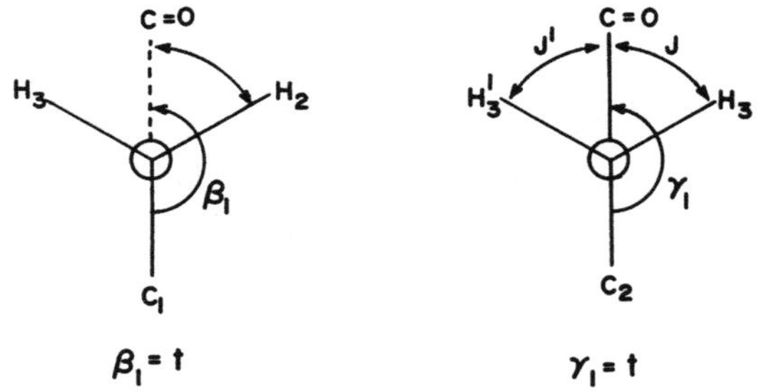

Fig. 6.3.5. Newman projection diagrams for β_1 and γ_1

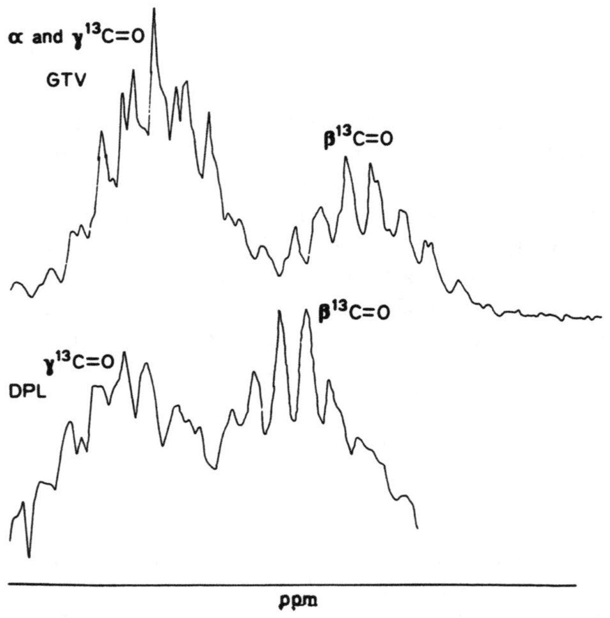

Fig. 6.3.6. The carbonyl region of the ^{13}C-NMR spectrum at 68 MHz of dipalmitoyllecithin (DPL) and glyceryl trivalerate (GTV). The fine structure is due to long-range ^1H—^{13}C coupling constants (reprinted with the permission from Ref. [1186])

of γ_1 (torsion angle C—C—O—C) close to 180°. Conclusions regarding β_1 from NMR data are less definite since there is only one three-bond coupling constant available. From the Lemieux calibration curve for ^{13}C—^1H coupling constants [78], a coupling constant of 3.5 Hz corresponds to dihedral angles of 90, 150, 260, and 340° for β_1. Conformations with β_1 values of 260 and 340° are sterically disallowed and hence can be ruled out. On the other hand, β_1 values of 90 and 150° correspond to the minimum energy conformations predicted by theory [1201–1202]. Thus, it is probable that these two conformations coexist in solutions in a dynamic equilibrium.

No information on β_2, γ_2 and β_3, γ_3 can be obtained from the NMR data, due to the lack of suitable coupling constants. However, theoretical [1197, 520] and X-ray crystallographic studies [1191–1195, 1219] on carboxylic ester groups indicate that β_2 and γ_2 predominantly assume *trans* conformation while β_3 and γ_3 reveal a marginal preference for the *trans* conformation.

One of the important questions with regard to phospholipid conformation is, whether the conformations observed in a free molecule or in an isotropic environment would be retained in a membrane. The determination of the conformations of the lipid molecules, when organized in a membrane, is difficult to perform for the following reasons. Firstly, the NMR spectrum of an organized system like a membrane is very broad and the coupling constant information is lost. Secondly, owing to the lack of proper understanding of "hydrophobic interactions", the various factors affecting the conformations cannot be properly discussed by theoretical methods. Fluidity of the membrane further complicates the matter. Nevertheless, in one case, convolution difference spectroscopy has been successfully applied to obtain the coupling constant information corresponding to a certain fragment of the lipid molecule [1217]. These coupling constants indicate limited conformational flexibility in the polar part of the molecule, as is the case in free lipid molecule. Further, the fact that phospholipid conformations in different solvents like D_2O, $CDCl_3$ and CD_3OD are nearly identical suggests that, also in the membrane the same conformations may be preferred, but with a different population distribution. Recent ^{1}H- and ^{13}C-NMR experiments support this view [1218]. It has been observed that the preferred conformation of the polar group of dihexanoyl-phosphatidylcholine is the same in both the monomeric and aggregated micellar form, suggesting that the conformation is primarily determined by the energetics of the compound.

6.4 Membrane Organization and Fluidity

6.4.1 Fluidity

The fluidity of the membrane arises due to three different types of motions of its components. These include (a) rotational motions around various single bonds (b), lateral diffusion on the surface of the membrane and (c) flip-flop motion from one surface of the membrane to the other [1182–1183]. The extent of these motions is of course different for the gel and liquid crystalline phases. In the gel phase, dipole-dipole interactions are not averaged out and hence the NMR lines are extremely broad. On the other hand, in the liquid crystalline phase, some of the lines (methyl and methylene groups) are fairly sharp and can be monitored for studying the different motions in the membrane. Thus, the resulting information often refers to the fluid portion of the membrane. In the following sections we will discuss NMR results on different motions.

6.4.1.1 Rotational Motions

Several NMR parameters have been used to study the rotational motions in lipid molecules.

(a) T_1: Quantitative information on the rates of rotational motions can be obtained from carbon-13 T_1 measurements on the lipid molecules. Levine et al. [1214–1215] measured T_1 values of ^{13}C nuclei at various positions along the chains within bilayers (Fig. 6.4.1). It has been found that T_1 values increase with increasing distance of the ^{13}C nuclei from the glycerol moiety. Several types of motions such as tumbling motions of the whole vesicles, rotation of alkyl chains as rigid units about their longitudinal axis and rotation about carbon-carbon single bonds contribute to T_1. It has been suggested that rotation about carbon-carbon single bonds makes a dominant contribution to the observed values [1214–1215]. A fit to the T_1 data could be obtained by allowing similar rates of motion for methylene groups close to the backbone and a sharp increase toward the terminal methyl group. Typical rates of motion are 10^7–10^8 s^{-1}. The mobility of the polar group has also been studied by ^{31}P relaxation time measurements on different lipid-water and lipid-water-cholesterol systems [1221]. The observed T_2 values are much larger than expected for rigid polar groups [1221] but not as large as one would expect for free motions. The results have been interpreted in terms of anisotropic motions in the head group.

(b) $^{31}P\{H\}$NOE: Yeagle et al. [1222] have observed that relaxation mechanisms of ^{31}P nuclei are predominantly dipolar in nature. From the observed NOE enhancements and from considerations of tumbling motions (correlation time τ_r) and rotational motions of the polar group (correlation time τ_c), τ_c has been estimated to be about 1.4×10^{-9} s.

(c) ^2H quadrupole splitting: ^2H quadrupole splitting is sensitive to the extent of randomness or degree of order in the motions around the ^2H nucleus [Eq. (2.10.1), p. 36], it increases with increasing degree of order. Seelig et al. [1210] have observed that the quadrupole splitting is larger than that expected for a free rotation of the polar group but much smaller than that expected for a rigid conformation.

(c) ^{31}P chemical shift anisotropy: Assuming free rotation around α_2, α_1 and Θ_1, chemical shift anisotropy has been calculated to be 51 ppm whereas the experimental

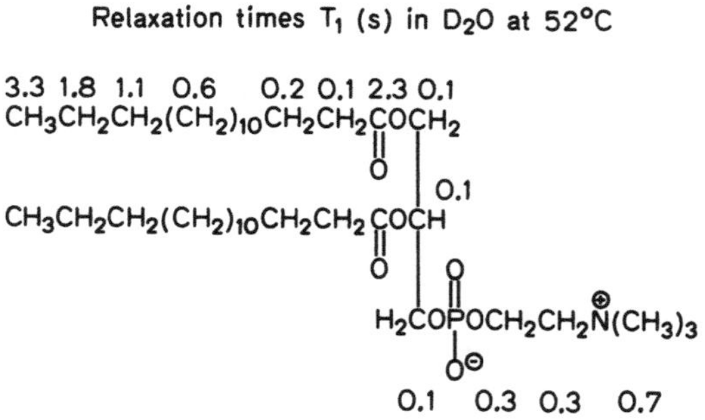

Fig. 6.4.1. ^{13}C spin-lattice relaxation time (T_1) of dipalmitoyl-lecithin in lipid bilayers (reprinted with the permission from Ref. [1214])

value is 47 ppm [1210, 1223] for phosphatidylcholine bilayers. These results suggested that the polar head group shows restricted mobility. A quantitative fit to the data is obtained by assuming a two-state model where the phosphorylcholine moiety jumps between two conformations [1210].

^{31}P spectra of dipalmitoyl-lecithin (unsonicated) below and above the gel-to-liquid crystalline phase transition temperature reveal small differences indicating that the polar group motion is not markedly affected by the phase transition and that hydrocarbon chain melting is the main cause of phase transition [1184].

6.4.1.2 Lateral Diffusion on the Surface of the Membrane

Lateral diffusion of the lipid molecules is reflected in the spin-spin interactions between the neighboring lipid molecules. Metcalfe and coworkers [1221–1225] have noted an increase in the proton relaxation time T_1 for dipalmitoyl-lecithin (DPL) samples upon the addition of increasing amounts of DPL deuterated in the alkyl chains. This implies that interchain relaxation plays an important role in fully protonated bilayers, i.e. dipolar interactions are not completely averaged out. From a consideration of the different types of motions contributing to relaxation mechanisms, it has been concluded that lateral diffusion of the lipid molecules, in the plane of the bilayer dominates the T_2 values of the protons. These values determined from line width measurements yielded a lower limit for the lateral diffusion constant. The diffusion constant in lecithin vesicles was estimated to be 0.9×10^{-8} cm^2 s^{-1}.

6.4.1.3 Flip-Flop Motions

The flip-flop motion is extremely slow on the NMR time-scale and hence this motion has not been studied by NMR.

6.4.2 Determination of Organization

In view of the fluidity discussed above, the term "organization in the membrane" refers to an average of the different possible organizations of the membrane components. With regard to organization, the following questions may be asked.

(a) What is the average orientation of the polar group at the interface?
(b) What are the different types of arrangements of the polar groups?
(c) Is there any order in the organization and if so what is the extent of it?
(d) Is there any relation between order and fluidity in the membrane?
(e) Is the organization identical on both sides of the membrane or is it different?

Some of these questions have been answered by ^{31}P, ^{1}H, and ^{2}H-NMR measurements [185–188, 1221, 1223, 1226–1230, 1238].

6.4.2.1 ^{31}P-NMR

Since there is only one ^{31}P nucleus in the entire phospholipid molecule, ^{31}P-NMR spectroscopy is naturally the method of choice for studying the conformation and orientations of the polar groups at the lipid-water interface.

The high-resolution NMR of bilayer membranes and vesicles is complicated by the fact that the environments of the spin systems are anisotropic, due to absence of random tumbling motions in these systems. As a consequence, the various dipolar interactions are not averaged to zero and the chemical shifts depend on the relative orientations of the magnetic field and the molecular axis. The angular dependence of the chemical shift can be expressed in terms of the angle between the magnetic field direction and the normal to the bilayer surface. In the last few years, techniques have been developed to obtain high-resolution NMR spectra of lipid bilayers. In these experiments [1223, 1229, 1230] the anisotropy of the ^{31}P chemical shift tensor has been measured. By combining this information with the knowledge of orientations of the tensor interactions with respect to molecular bonds, it has been possible to determine the molecular orientations in different samples. Figure 6.4.2 shows the ^{31}P spectra of the DPPC (dipalmitoyl phosphatidyl choline) bilayer for different orientations of the bilayer normal

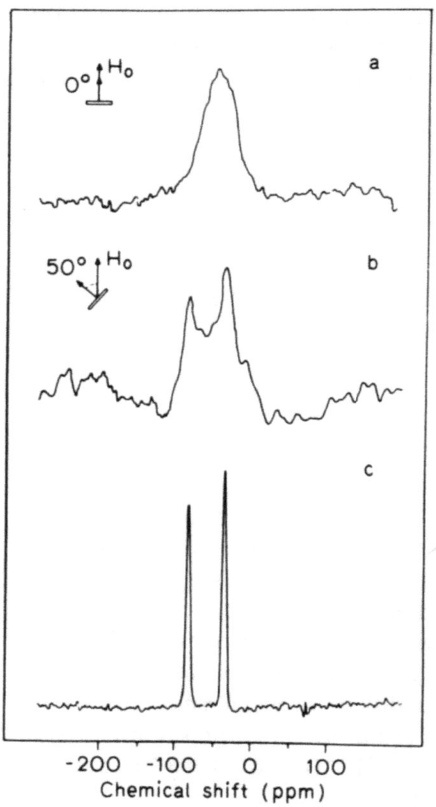

Fig. 6.4.2. High-field (6.8 T) ^{31}P-NMR spectrum illustrating the angular dependence of $^{31}PO_4$ chemical shifts.
(a) Bilayer normal parallel to H_0, $\sigma_{zz} = -46$ ppm, (b) bilayer normal at 50° with respect to H_0, $\sigma_{zz} = -80$ and -30 ppm, (c) Barium diethyl phosphate single-crystal spectrum (reprinted with the permission from Ref. [1228])

with respect to the magnetic field direction. Comparing these spectra with the ^{31}P spectrum of a single crystal of barium diethyl phosphate, one concludes that there is considerable disorder in the bilayer systems. From the study of the angular-dependent ^{31}P chemical shifts, it has been inferred that the O–P–O plane (where the oxygen atoms are the non-esterfied oxygen atoms of the phosphodiester) is tilted at an angle of $47 \pm 5°$ with respect to the bilayer normal. This results in an extension of the choline moiety parallel to the bilayer surface [1228].

An interesting result has been obtained by $^{31}P\{H\}$ NOE experiments on phosphatidylcholine vesicles [1222]. For $^{31}P\{H\}$, under extreme narrowing conditions, a maximum increase in intensity from 1.0 to 2.24, i.e. 124%, is expected. This is considered as a 100% nuclear Overhauser effect enhancement (NOEE). In the sonicated vesicles of phosphatidycholine, 40% NOEE was observed. Deuteration of the *N*-methyl protons does not change the $^{31}P\{^1H\}$ spectrum but reduces the NOEE from 40 to 10% suggesting that the *N*-methyl protons are the dominant contributors to ^{31}P-NOEE. This suggests that the positively charged $N(CH_3)_3$ group spends some time near the negatively charged phosphate groups, and the choline phosphate dipole is not statically oriented perpendicularly to the bilayer surface.

In the case of phosphorylcholine in D_2O solutions, the observed NOEE is 30% which is 1/4 of the maximum possible NOEE. In solutions where rapid isotropic tumbling motions occur, intermolecular dipolar interactions are averaged out and the observed NOEE arises from intramolecular dipolar relaxations. If one assumes that 40% NOEE is the maximum possible NOEE in phosphatidylcholine vesicles, then deuteration of the $\overset{\oplus}{N}(CH_3)_3$ group removes the contribution of methyl protons and the total NOEE is reduced to 1/4 of its original value indicating that 1/4 of NOEE is due to intramolecular interactions of methylene groups with phosphate residues while the rest arises from intramolecular interactions between $\overset{\oplus}{N}(CH_3)_3$ groups and the phosphate moiety. Thus, the surface of the bilayer consists of an interlocking set of intermolecular electrostatic associations of the positively charged $\overset{\oplus}{N}(CH_3)_3$ group with the negatively charged phosphate groups of the neighboring lipid molecules.

6.4.2.2 *²H-NMR*

Several workers have made use of ^2H-NMR (DMR) for studying the orientations of the polar and hydrocarbon chains and also to determine the extent of order in a membrane [185–187, 1227, 1243–1244]. Lipids are deuterated selectively at different sites in the hydrocarbon chains and also in the polar group. The ^2H quadrupole splittings ($\Delta\nu$) are then obtained from the DMR spectra. Figure 6.4.3 illustrates some typical DMR spectra. The separation of the two absorption peaks gives the residual quadrupole splitting $\Delta\nu$. The lines are quite broad suggesting that they are composed of two signals with slightly different quadrupole couplings. The doublets are indeed resolved for some of the deuterated DPL bilayers indicating that the two fatty acyl chains are not equivalent.

The deuterium quadrupole splittings have been used to determine the orientations and the extent of order in the hydrocarbon chains. The quadrupole splitting for a C–D vector depends on the orientation of the vector with respect to the direction of the normal to the bilayer and the anisotropy of the motions. For a fully ordered system, the

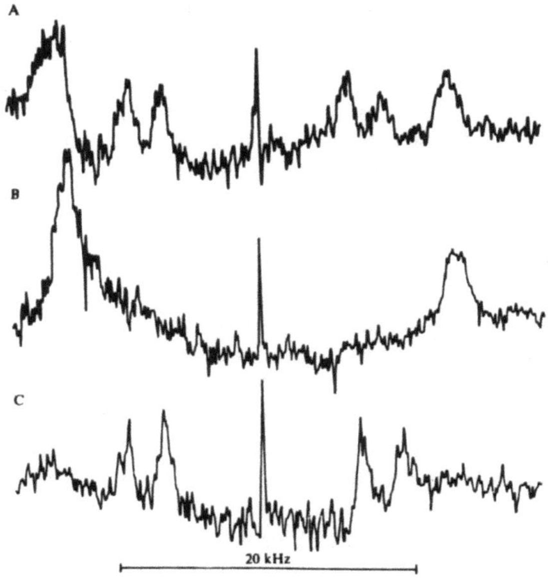

Fig. 6.4.3. 41.4 MHz ^{2}H-NMR spectrum of bilayers of dipalmitoyl-3-Sn-phosphatidylcholine at 50 °C.
(A) 1,2-[2'-d_2]-dipalmitoyl-3-Sn-phosphatidylcholine, (B) 1-[2'-d_2] palmitoyl-2-palmitoyl-3-Sn-phosphatidylcholine, (C) 1-palmitoyl-2-[2'-d_2]-palmitoyl-3-Sn-phosphatidylcholine (reprinted with the permission from Ref. [1188], copyright: Cambridge University Press)

quadrupole splitting attains a maximum and decreases as the randomness in the orientation of the vector increases. The extent of order represented by an order parameter (S_{CD}) is related to the quadrupole splitting by the relation

$$\Delta\nu = 3/4 \left(\frac{e^2qQ}{h}\right) S_{CD} \tag{6.4.1}$$

where $\left(\frac{e^2qQ}{h}\right)$ is the quadrupole splitting constant (ν_Q). For paraffin hydrocarbon chains this value is about 170 KHz. Often, another order parameter, namely the segmental order parameter, S_{mol}, is used. This parameter refers to the orientation of the segment (C–C bond) with respect to the bilayer normal. For an axially symmetric motion of the CD segment around the bilayer normal, the two order parameters are related as

$$S_{mol} = -2S_{CD} \tag{6.4.2}$$

Measurements of $\Delta\nu$ in different selectively deuterated lipid dispersion thus permit the estimation of the order at different points in the molecule. These values, in the case of L-α-dipalmitoyl-lecithin are shown in Fig. 6.4.4 [1188]. It is clear that as one goes toward the center of the bilayer, the order parameter decreases. Further, temperature variation studies on different phospholipid bilayers reveal that the order parameters in the two hydrocarbon chains are not identical [185]. One of the chains has a consistently lower order parameter as compared to the others (Fig. 6.4.5) indicating that two chains are inequivalent, in accordance with earlier observations.

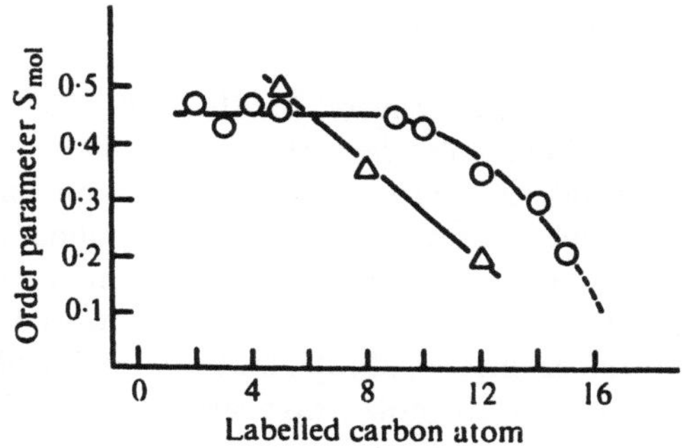

Fig. 6.4.4. Order parameters along the hydrocarbon chains in dipalmitoyl lecithin bilayers – o-o Deuterium data, – △-△ spin label data (reprinted with the permission from Ref. [1188], copyright: Cambridge University Press)

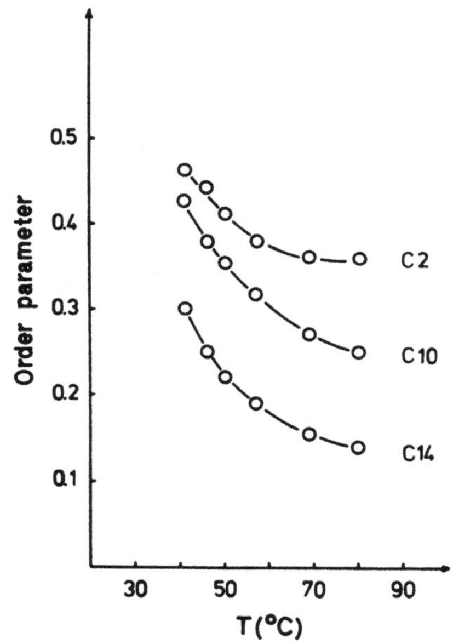

Fig. 6.4.5. Temperature variation of order parameter S_{mol} of DPL bilayers labelled at three different carbon atoms of the fatty acyl chains (reprinted with the permission from Ref. [187])

Information on the order in the membrane has also been obtained by spin-label ESR technique [1232]. Important differences between the results from ^{2}H-NMR and spin-label techniques have been observed [187]. This is attributed to the perturbation of the bilayer structure by the bulky spin labels.

It is worth taking a digression at this stage and clarify some of the misconceptions about the order and fluidity in the membrane. The two are often intuitively correlated by saying that a decrease in the order parameter implies an increase in the rate of rotational motions and vice versa. There are cases where fast rotational motions are associated with large-order parameters and conversely; in some cases, slow motions are associated with small-order parameters [1232–1233]. It has been suggested that order parameters can be better related with the coherence in rotational motions rather than with the rate of motions [1234–1235].

6.4.2.3 Other Studies

NMR has provided valuable results [1222–1236] on the question of whether the organization of lipid molecules is the same on the two sides of biomembranes. In the case of phosphatidylcholine bilayers, ^{31}P spectra obtained with proton-noise decoupling exhibit two phosphorus resonances separated by about 5 Hz. The addition of Eu(III) causes a shift of one of the resonances upfield by 1 ppm or more. This has been interpreted as an indication of different organizations on the two sides, the upfield resonance corresponding to the inner surface [1222]. Similarly, the ^{1}H-NMR spectrum of phosphatidylcholine in water at 220 MHz shows two peaks for the $\overset{\oplus}{N}(CH_3)_3$ group which have been assigned to groups internal and external to the vesicles. The addition of ions increases the separation between the two peaks [1236]. However, these interpretations are not unambiguous. It is possible that two different types of organizations are present on the same surface and the two observed resonances reflect these two organizations of the phosphatidylcholine molecules [1187, 1201, 1203].

The model that emerges from the above discussion on conformational and organizational features of the lipid bilayer is illustrated in Figs. 6.4.6 and 6.4.7. The phospholipid polar groups form an interlocked network which is stabilized by electrostatic and lipid-water hydrogen-bond interactions. The polar dipoles exhibit rotational motions about the bilayer normal and thus the organization is dynamic. The hydrocarbon chains display considerable mobility which increases toward the interior of the bilayer. The regu-

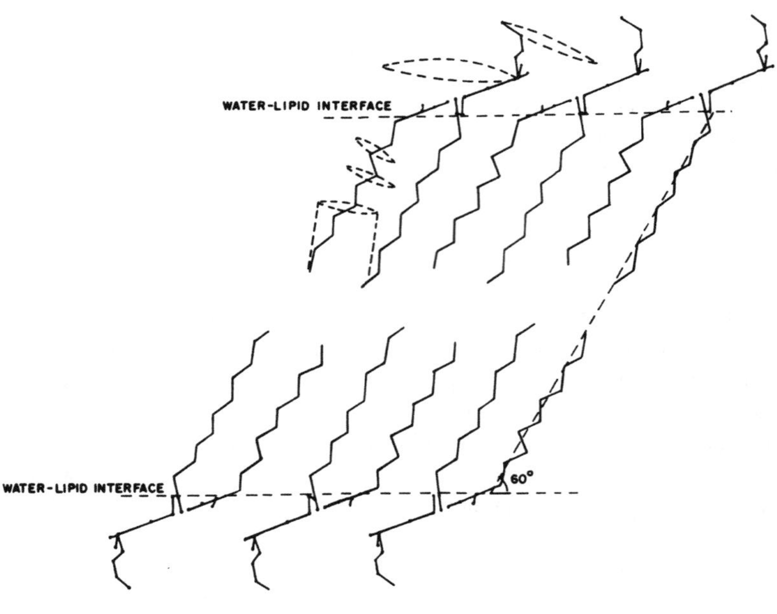

Fig. 6.4.6. Regular organization of phospholipid molecules in a bilayer. The different rotational motions are also indicated

larity of the organization or the order decreases as one moves toward the interior of the bilayer. There are more than one possible modes of organizations of lipid polar groups on the surface. The organizations may be different on the two sides of the bilayer.

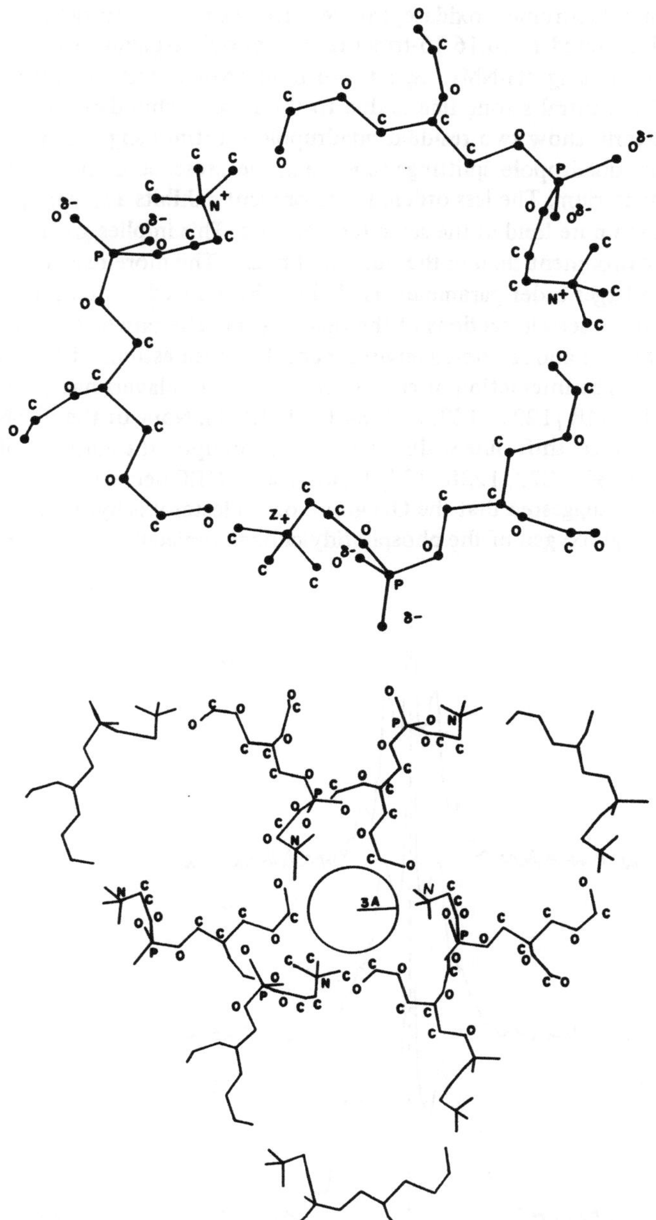

Fig. 6.4.7. Projection of the polar groups in an irregular organization onto the membrane surface. The interlocking of the polar groups is illustrated

6.5 Lipid-Protein and Lipid-Cholesterol Interactions

NMR studies on the interaction of lipids with membrane proteins and cholesterol have been reported [1222, 1237–1238]. Dahlquist et al. [1237] have studied the interaction of cytochrome c oxidase, isolated from beaf yeast mitochondria, with specifically deuterated 1-(16,16,16-trideuteropalmitoyl)-2-palmitoleoylphosphatidylcholine molecules using ^{2}H-NMR. Spectra of lipid protein particles are shown in Fig. 6.5.1 [1237]. The central strong line is due to the natural abundance of ^{2}H in water. The spectra clearly show two residual quadrupole splittings suggesting the presence of two residual quadrupole splittings suggesting the presence of two different environments for deuterium. The less ordered environment exhibits a splitting which is smaller than that for a pure lipid at the same temperature. This implies greater motional freedom in this environment than in the pure lipid phase. The more restricted environment corresponding to larger order parameters is likely to be induced by the protein, i.e. the protein immobilizes certain sections of the lipid matrix. The rate of exchange of the lipid between the free and restricted environments has been estimated to be less than 10^3 s^{-1}.

The interaction of cholesterol with lipid bilayers has been studied by ^{31}P- and ^{2}H-NMR [1221–1222, 1238–1241, 1244]. None of the ^{31}P-NMR parameters such as chemical shift, line width and T_1 changes upon the addition of cholesterol to the bilayer [77–80, 1222, 1238–1241]. However, NOEE decreases sharply from 40 to 10%. It has been suggested that the OH group of cholesterol is hydrogen bonded to the ester carbonyl oxygen of the phosphatidylcholine molecules. In an earlier study [1239] it has

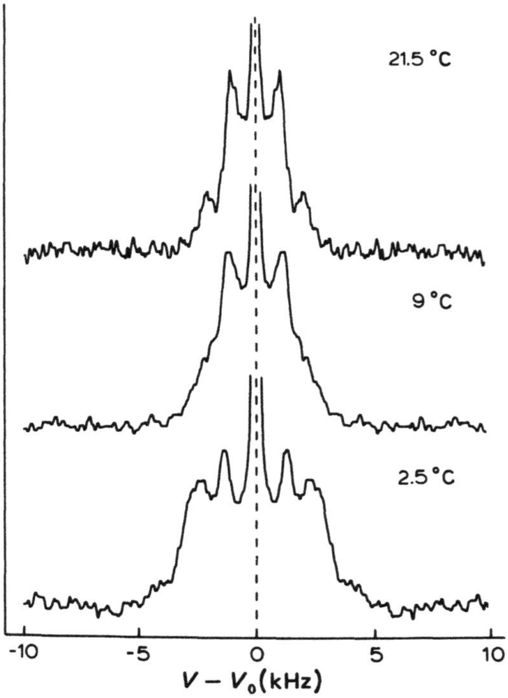

Fig. 6.5.1. ^{2}HMR (13.8 MHz) spectra at various temperatures of cyctochrome oxidase-lipid particles (reprinted with the permission from Ref. [1237])

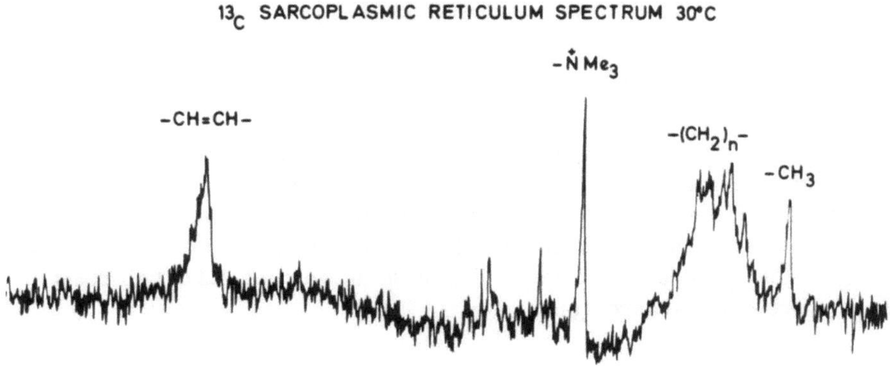

Fig. 6.5.2. ^{13}C-NMR (67.8 MHz) spectra of vesicles prepared from sarcoplasmic reticulum membranes (reprinted with the permission from Ref. [1242])

been reported that the C=O group experiences a substantial shift upon the addition of cholesterol to the bilayer. However, this carbonyl shift could not be reproduced [1184]. Phosphorus-31 T_2 measurements on lipid cholesterol systems indicate that cholesterol increases the fluidity of the head group [1221]. ^{2}H-NMR studies [1244] with specifically labelled lipids on lipid-cholesterol systems reveal a complex behavior. The addition of cholesterol causes an increase in the quadrupole splitting in the hydrocarbon chains but a decrease of this splitting in the polar chain. This suggests that cholesterol increases the order in the interior but reduces it at the surface of the membrane. No change in deuterium T_1 has been observed as a result of the addition of cholesterol [1244].

There have been very few NMR experiments on intact biological membranes [1242, 1245]. Lee et al. [1242] obtained ^{13}C spectra of vesicles prepared from sarcoplasmic reticulum (Fig. 6.5.2). A comparison of the relative intensities of the ^{13}C spectra of membranes and extracted lipids reveals that only 3/4 of the lipids contributes to the NMR spectrum of the membrane. This correlates with the proportion of the membrane which is fluid. Proton T_1 suggests that lateral diffusion in the membrane is similar to that in lipid bilayers. Very recently, ^{13}C-NMR spectra of some intact bacterial membranes have been obtained [1245]. The membranes were enriched with ^{13}C via biosynthetic pathways. ^{2}H- and ^{31}P-NMR have also been used to study lipid-protein interactions [1246–1247]. It has been observed that the binding of lipid with sarcoplasmic reticulum Ca(II), Mg(II)-ATPase results in a decrease in the T_1 and T_2 values of the ^{31}P nucleus suggesting involvement of polar head groups in the binding with the protein [1247].

7 Acknowledgements

We are grateful to the following colleagues for the critical reading and constructive suggestions on the manuscripts: Dr. E. D. Becker, Dr. H. T. Miles, Dr. K. B. Roy, Dr. C. Fisk, Dr. C. L. Khetrapal, Dr. M. M. Dhingra, Dr. A. Saran, and Mr. N. V. Kumar. We are also grateful to the National Institutes of Health, Bethesda, USA, for the various facilities which they provided in the preparation of this manuscript and to publishers for the permission to reproduce figures.

8 References

1. Blundell, T. L., Johnson, L. N.: Protein crystallography. New York: Academic Press 1976
2. Wüthrich, K.: NMR in biological research: peptides and proteins. Amsterdam: North Holland, Publ. 1976
3. Dwek, R. A.: Nuclear magnetic resonance in biochemistry: applications to enzyme systems. Oxford: Clarendon Press 1973
4. Knowles, P. F., Marsh, D., Rattle, H. W. E.: Magnetic resonance of biomolecules. London: John Wiley and Sons 1976
5. James, T. L.: Nuclear Magnetic Resonance in Biochemistry. New York: Academic Press 1975
6. Dwek, R. A., et al. (eds.): NMR in biology. London: Academic Press 1977
7. Sadler, P. J. In: Progress in medicinal chemistry. Ellis, G. P., West, G. B. (eds.). Amsterdam: North Holland, Publ. 1975
8. Rattle, H. W. E.: Prog. Biophys. Mol. Biol. *28*, 3 (1974)
9. Bovey, F. A.: High resolution NMR of macromolecules. New York: Academic Press 1972
10. Casy, A. F.: PMR spectroscopy in medicinal and biological chemistry. London: Academic Press 1971
11. Chemical Society (London). Specialist Periodical Report: Natural macromolecules
12. Hanack, M: Conformation theory. New York: Academic Press 1965
13. Winnewisser, G., Winnewisser, M., Gordy, W.: J. Chem. Phys. *49*, 3465 (1968)
14. Hunt, R. A., et al.: J. Chem. Phys. *42*, 1931 (1965)
15. Van Wart, H. E., Scheraga, H. A.: J. Phys. Chem. *80*, 1812 (1976)
16. Mizushima, S.: The structure of molecules and internal rotation. New York: Academic Press 1954
17. Binsch, G.: Top. Stereochem. *3*, 97 (1968)
18. Flory, P. J.: Statistical mechanics of chain molecules. New York: Interscience Publishers 1969
19. Bartell, L. S., Kohl, D. A.: J. Chem. Phys. *39*, 3097 (1963)
20. Pullman, B. In: Quantum mechanics of molecular conformations. Pullman, B. (ed.). London: John Wiley and Sons 1976
21. Hopfinger, A. J.: Conformational properties of macromolecules. New York: Academic Press 1973
22. Patel, D. J., Canuel, L. L., Bovey, F. A.: Biopolymers *14*, 987 (1975)
23. Jardetzky, O., Markley, J. L.: Magn. Reson. Biol. Res. Rep., 3rd Int. Conf. New York: Gordon and Breach 1971, p 35
24. Nagayama, K., et al.: Biochem. Biophys. Res. Comm. *86*, 218 (1979)
25. Mantsch, H. H., Saito, H., Smith, I. C. P.: Progr. NMR Spectrosc. *11*, 211 (1977)
26. Egan, W., Shindo, H., Cohen, J. S.: Ann. Rev. Biophys. Bioeng. *6*, 383 (1977)
27. Komorosky, R. A., Peat, I. R., Levy, G. C.: Top. Carbon-13 NMR Spectrosc. *2*, 179 (1976)
28. Oldfield, E., Norton, R. S., Allerhand, A.: J. Biol. Chem. *250*, 6381 (1975)
29. Oldfield, E., Norton, R. S., Allerhand, A.: J. Biol. Chem. *250*, 6368 (1975)
30. Oldfield, E., Allerhand, A.: J. Biol. Chem. *250*, 6403 (1975)
31. Dill, K., Allerhand, A.: J. Am. Chem. Soc. *99*, 4508 (1977)
32. Torchia, D. A., Van der Hart, D.: J. Mol. Biol. *104*, 315 (1976)
33. Randall, E. W., Gillies, D. G.: Progr. NMR Spectrosc. *5*, 136 (1971)
34. Levy, G. C., Lichter, R. L.: Nitrogen-15 nuclear magnetic resonance spectroscopy. New York: John Wiley and Sons 1977
35. Cary, L. W., Takita, T., Ohnishi, M.: FEBS Lett. *17*, 145 (1971)
36. Kricheldorf, H. R., Hull, W. E., Formack, V.: Biopolymers *16*, 1609 (1977)
37. Morishima, I., Inubushi, T.: EEBS Lett. *81*, 57 (1977)

38. Letcher, J. H., Van Wazer, J. R. In: Topics in phosphorus chemistry. Grayson, M., Griffiths, E. J. (eds.), vol. 5, p. 169. New York: Interscience Publishers
39. Hollis, D. P.: Bull. Magn. Reson. *1*, 27 (1979)
40. Hull, W. E., Sykes, B. D.: Biochemistry *15*, 1535 (1976)
41. Hull, W. E., Sykes, B. D.: J. Mol. Biol. *98*, 121 (1975)
42. Chaiken, I. M., et al.: J. Biol. Chem. *248*, 884 (1973)
43. Raftrey, M. A., Huestis, W. H., Millet, F.: Cold Spring Harbor Symp. Quant. Biol. *36*, 541 (1971)
44. Gally, H. U., Spencer, A. K., Armitage, I. M., Prestegard, J. H., Cronan, J. E. Jr.: Biochemistry *17*, 5377 (1978)
45. Lauterbur, P. C., Kaufman, B. V., Crawford, M. K.: Biomol. Struct. Funct. (Symp.) *1978*, 329
46. Lindman, B., Zeppezauer, M., Akeson, A.: Wenner-Gren Cent. Int. Symp. Ser. *18*, 603 (1972)
47. Carlstrom, D. E., Miller, W. G., Bryant, R. G.: J. Phys. Chem. *77*, 2759 (1973)
48. Collins, T. R., et al.: J. Am. Chem. Soc. *95*, 1649 (1973)
49. Crabbe, P.: ORD and CD in chemistry and biochemistry. New York: Academic Press 1972
50. Freeman, S. K.: Applications of Laser Raman Spectroscopy. New York: Wiley Interscil. Publ. 1974
51. Tobin, M. C.: Laser Raman Spectroscopy. New York: Wiley Intersci. Publ. 1971
52. Bailey, P. M.: Progr. Biophys. Molec. Biol. *27*, 1 (1973)
53. Jelinski, L. W., Torchio, D. A.: J. Mol. Biol. *133*, 45 (1979)
54. Allerhand, A.: Pure and Appl. Chem. *41*, 247 (1975)
55. Collins, D. M., et al.: Science *190*, 1047 (1975)
56. Blake, C. C. F.: Progr. Biophys. Mol. Biol. *25*, 83 (1971)
57. Ramachandran, G. N.: Conformation of biopolymers. New York: Academic Press 1967
58. Levinthal, C., et al.: Emerging concepts in interactive computer graphics. New York: Benjamin, W. A., Inc. 1968
59. Allerhand, A.: Methods Enzymol. *61*, 458 (1979)
60. Bystrov, V. F.: Progr. NMR Spectrosc. *10* (II), 41 (1976)
61. Lemieux, R., U.: Ann. N. Y. Acad. Sci. *222,* 915 (1973)
62. Feeney, J.: Proc. Roy. Soc. *345A*, 61, (1975)
63. Lemieux, R. U., et al.: J. Am. Chem. Soc. *80*, 6098 (1958)
64. Karplus, M.: J. Chem. Phys. *30*, 11 (1959)
65. Karplus, M.: J. Am. Chem. Soc. *85*, 2870 (1963)
66. Richarz, R., and Wüthrich, K.: Biopolymers, *17*, 2133 (1978)
67. Kopple, K. D., Wiley, G. R., Tauke, R.: Biopolymers *12*, 627 (1973)
68. Bothner-By, A. A.: Advan. Magn. Reson. *1*, 195 (1965)
69. Solkan, V. N., Bystrov, V. F.: Tetrahedron Lett. *1973*, 2261
70. Barfield, M., Gearhart, H. L.: J. Am. Chem. Soc. *95*, 641 (1973)
71. Govil, G.: Ind. J. Chem. *9*, 824 (1971)
72. Cung, M. H., Marraud, M., Neel, J.: Macromolecules *7*, 606 (1974)
73. Bystrov, V. F., et al.: Tetrahedron *29,* 673 (1973)
74. Ramachandran, G. N., Chandrasekaran, R., Kopple, K, D.: Biopolymers *10*, 2113 (1971)
75. Fraser, R. B., et al.: Can. J. Chem. *47*, 403 (1969)
76. Stolow, R. D., Gallo, A. A.: Tetrahectron Lett. *1968*, 3331
77. Solkan, V. N., Bystrov, V. F.: Izv. Akad. Nauk SSSR, Ser. Khim *6*, 1308 (1977)
78. Lemieux, R. U., Nagbhushan, T. L., Paul, B.: Can. J. Chem. *50*, 773 (1972)
79. Schwarez, J. A., Perlin, A. S.: Can. J. Chem. *50*, 3667 (1972)
80. Wasylishan, R., Schafer, T.: Can. J. Chem. *50*, 2710 (1972)
81. Karplus, S., Karplus, M.: Proc. Nat. Acad. Sci. USA *69*, 3204 (1972)
82. Kopple, K. D., Ahsan, A., Barfield, M.: Tetrahedron Lett. *1978*, 3519
83. Govil, G.: Mol. Phys. *21*, 953 (1971)
84. Gopinathan, M. S., Narasimhan, P. T.: Mol. Phys. *21*, 1141 (1971)
85. Ihring, A. M., Smith, S. L.: J. Am. Chem. Soc. *92*, 759 (1970)
86. Geissner-Prettre, C., Pullman, B.: J. Theor. Biol. *72*, 751 (1978)

87. Bottner-By, A. A., Cox, R. H.: J. Phys. Chem. *73*, 1830 (1969)
88. Mantsch, H. H., Smith, I. C. P.: Biochem. Biophys. Res. Commun. *46*, 808 (1972)
89. Kondo, N. S., Danyluk, S. S.: Biochemistry *15*, 756 (1976)
90. Harr, W., et al.: Proc. Natl. Acad. Sci. USA *72*, 4948 (1975)
91. Son Tran-Dinh, Fermandjian, S.: J. Phys. (Paris) Colloq. *8*, 45 (1973)
92. Marshall, J. L., Miller, D. E.: Proc. Natl. Acad. Sci. USA *72*, 4948 (1975)
93. Dodrell, D., et al.: J. Am. Chem. Soc. *96*, 1241 (1974)
94. Lapper, R. D., Mantsch, H. H., Smith, I. C. P.: J. Am. Chem. Soc. *94*, 6243 (1972)
95. Schleich, T., et al.: In: Structure and conformation of nucleic acids and protein nucleic acid interactions. Sundaralingam, M., Rao, S. T. (eds).: Baltimore: University Press 1975
96. Smith, I. C. P., et al.: In: Conformation of biological molecules and polymers. Bergman, E. D., Pulman, B. (eds.). New York: Academic Press 1973
97. Govil, G., Smith, I. C. P.: Biopolymers *12*, 2589 (1973)
98. Govil, G., Bernstein, H. J.: J. Chem. Phys. *48*, 285 (1968)
99. Abraham, R. J., Gatti, G.: J. Chem. Soc. B *1969*, 961
100. Abraham, R. J., Pachler, K. G. R.: Mol. Phys. *7*, 165 (1964)
101. Pachler, K. G. R.: J. Magn. Reson. *8*, 183 (1972)
102. Feeney, J., Hansen, P. E., Roberts, G. C. K.: J. Chem. Soc. Chem. Commun. *1974*, 465
103. Solkan, V. N., Bystrov, V. F.: Tetrahedron Lett *1973*, 2261
104. Ostlund, N. S., Prunski, M. J.: J. Magn. Reson. *15*, 549 (1974)
105. Pachler, K. G. R.: J. Chem. Soc. Perkin Trans. II *1972*, 1936
106. Hoffman, R. A., Kellerhals, H., Lustig, E.: NMR basic principles and progress *5*, 1 (1971)
107. Egli, H., Voegeli, U., von Philipsborn, W.: Jerusalem Symp. Quant. Chem. Biochem. *11*, 247 (1978)
108. La Mar, G. N., et al.: Biochem. Biophys. Res. Commun. *82*, 19 (1978)
109. Renugopalkrishnan, V., et al.: Int. J. Quant. Chem. Quant. Biol. Symp. *4*, 97 (1977)
110. Mohanakrishnan, P., Easwaran, K. R. K.: Org. Magn. Reson. *12*, 196 (1979)
111. Barfield, M. Hruby, V. J., Meraldi, J. P.: J. Am. Chem. Soc. *98*, 1308 (1976)
112. Sarma, R. H., et al.: J. Am. Chem. Soc. *95*, 6457 (1973)
113. Davies, D. B., Khaled, M. A.: Tetrahedron Lett. *1973*, 2829
114. Davies, D. B., Khaled, M. A., Urry, D. W.: J. Chem. Soc. Perkin Trans. II, *1977*, 1294
115. Sternhell, S.: Quart. Rev. *23*, 236 (1969)
116. Vicar, J., et al.: Biopolymers *18*, 1021 (1979)
117. Martin, R. B.: J. Phys. Chem. *83*, 2404 (1979)
118. Barfield, M., Chakrabarti, B.: Chem. Revs. *69*, 757 (1969)
119. Bovey, F. A.: Chem. Biol. Pept., Proc. 3rd Am. Pept. Symp. Meienhofer, J. (ed.), p. 3. New York: Ann Arbor, Ann Arbor Sci. Publ. 1972
120. Bovey, F. A. In: Chemistry and biology of peptides. Proc. 3rd Amer. Peptide Symp. Meienhofer, J. (ed.). New York: Ann Arbor, Ann Arbor Sci. Publ. 1972
121. Thiery, C., et al.: Biochim. Biophys. Acta. *494*, 293 (1977)
122. Higashijima, T., Tsutomu, T., Mitsuo, M. T.: FEBS Lett. *57*, 175 (1975)
123. Herwats, L., Laszlo, P., Wiertz, J.: Bull. Soc. Chim. Belg. *85*, 137 (1976)
124. Urry, D. W., Mitchell, L. W., Ohnishi, T.: Proc. Natl. Acad. Sci. USA. *71*, 3265 (1974)
125. Thomas, W. A.: Ann Rep. NMR Spectrosc. *6B*, 1 (1976)
126. Mc Connell, H. M.: J. Chem. Phys. *27*, 226 (1957)
127. Pople, J. A.: Proc. Roy. Soc. *A239*, 550 (1957)
128. Pople, J. A.: J. Chem. Phys. *24*, 1111 (1956)
129. Johnson, C. E., Bovey, F. A.: J. Chem. Phys. *29*, 1012 (1958)
130. Giessner-Prettre, C., Pullman, B.: J. Theor. Biol. *27*, 87 (1970)
131. Giessner-Prettre, C., Pullman, B.: Biochem. Biophys. Res. Commun. *70*, 578 (1976)
132. Giessner-Prettre, C., et al.: Biopolymers *15*, 2277 (1976)
133. Bundi, A., Wüthrich, K.: Biopolymers *18*, 299 (1979)
134. Anteunis, M. J. O.: Tetrahedron Lett. *1977*, 1535
135. Siemion, I. Z., Wieland, T., Pook, K. H.: Angew. Chem. *87*, 712 (1975)
136. Voelter, W., Oster, O., Breitmeier, E.: Z. Naturforsch. *28*, 370 (1973)

137. Thomas, W. A., Williams, M. K.: J. Chem. Soc. Chem. Commun. *1972*, 994
138. London, R. E., et al.: Biochemistry *17*, 2277 (1978)
139. Deber, C. M., Fossel, E. T., Blout, E. R.: J. Am. Chem. Soc. *96*, 4015 (1974)
140. Dorman, D. E., Torchia, D. A., Bovey, F. A.: Macromolecules *6*, 80 (1973)
141. Chao, Y. Y. H., Bersohn, R.: Biopolymers *17*, 2761 (1978)
142. Bundi, A., Wüthrich, K.: FEBS Lett. *77*, 11 (1977)
143. Uesugi, S., Tanaka, S., Ikehara, M.: Eur. J. Biochem. *90*, 205 (1978)
144. Madison, V., et al.: J. Am. Chem. Soc. *96*, 6725 (1974)
145. Gorenstein, D. G., et al.: Biochemistry *15*, 3796 (1976)
146. Labotka, R. J., Glonck, T., Myers, T. C.: J. Am. Chem. Soc. *98*, 3699 (1976)
147. Prado, F. R., et al.: J. Am. Chem. Soc. *101*, 1737 (1979)
148. Gorenstein, D. G.: Jerusalem Sympos. Quant. Chem. Biochem. *11*, 1 (1978)
149. Herzfeld, J., Griffin, R. G., Haberkorn, R. A.: Biochemistry *17*, 2711 (1978)
150. Wehrli, F. W., Wirthlin, T.: Interpretation of carbon-13 NMR spectra. London: Heyden and Son 1976
151. Breitmaier, E., Spohn, K., Berger, S.: Angew. Chem. Internat. Engl. Edit. *14*, 144 (1975)
152. Bloembergen, N., Purcell, E. M., Pound, R. V.: Phys. Rev. *73*, 679 (1948)
153. Abragam, A.: The principles of nuclear magnetism. London: Oxford University Press 1973
154. Imoto, T., Akasaka, K., Hatano, H.: Chem. Lett. *1974*, 73
155. Akasaka, K., et al.: J. Magn. Reson. *18*, 328 (1975)
156. Abasaka, K., Imoto, T., Hatano, H.: Chem. Phys. Lett. *21*, 398 (1973)
157. Birdsall, N. J. M., et al.: J. Chem. Soc. Chem. Commun. *1973*, 757
158. Smith, I. C. P., Deslauriers, R., Schaumburg, K. In: Peptides: chemistry, structure and biology. Proc. 4th Am. Pept. Symp. Walter, R., Meienhofer, J. (eds.), p. 97. Michigan: Ann Arbor Science 1975
159. Freeman, R., Hill, H. D. W., Kaptein, R.: J. Magn. Reson. *7*, 327 (1972)
160. Hart, P. A., Davis, J. P.: Jerusalem Symp. Quant. Chem. Biochem.: *5*, 297 (1973)
161. Son Tran-Dinh, Chachaty, C., Gueron, M.: J. Phys. (Paris), Colloq. *8*, 57 (1973)
162. Balaram, P., Bothner-By, A. A., Dadok, J.: J. Am. Chem. Soc. *94*, 4015 (1972)
163. Noggle, J. H., Schirmer, R. E.: The nuclear Overhauser effect. Chemical applications New York: Academic Press 1971
164. Bell, R. A., Saunders, J. K.: Top. Stereochem. *7*, 1 (1973)
165. Alger, J. R., Prestegard, J. H.: J. Magn. Reson. *27*, 137 (1977)
166. Morris, A. T., Dwek, R. A.: Quart. Rev. Biophys. *10*, 421 (1977)
167. Campbell, I. D., et al.: Proc. Rare Earth Res. Conf. 10th *2*, 791 (1973)
168. Levine, B. A., Williams, R. J. P.: Proc. Roy. Soc. London, Ser. A. *345*, 5 (1975)
169. Dobson, C. M., Williams, R. J. P., Xavier, A. V.: J. Chem. Soc., Dalton Trans. *1974*, 1762
170. Dobson, C. M., Levine, B. A.: New Tech. Biophys. Cell Biol. *3*, 19 (1976)
171. Sievers, R. A.: Nuclear magnetic resonance shift reagents. New York: Academic Press 1973
172. Solomon, I.: Phys. Rev. *99*, 559 (1955)
173. Inagaki, F., Tasumi, M., Miyazawa, T.: Biopolymers *17*, 267 (1978)
174. Kopple, K. D., Schamper, T. J.: J. Am. Chem. Soc. *94*, 3644 (1972)
175. Eisinger, J., Shulman, R. G., Blumberg, W. E.: Nature *192*, 963 (1961)
176. Burton, D. R., et al.: Prog. NMR Spectrosc. *13*, 1 (1979)
177. Sloan, D. L., Mildvan, A. S.: Biochemistry *13*, 1711 (1974)
178. Iizuka, T., Morishima, I.: Biochim. Biophys. Acta *400*, 143 (1975)
179. Sullivan, G. R.: J. Am. Chem. Soc. *98*, 7162 (1976)
180. Diehl, P., Khetrapal, C. L.: NMR, basic principles and progress *1*, 1 (1969)
181. Khetrapal, C. L., Ramprasad, S.: J. Ind. Inst. Sci. *59*, 53 (1977)
182. Khetrapal, C. L., et al.: NMR Basic Princip. Progr. *9*, 1 (1975)
183. Khetrapal, C. L., Kunwar, A. C., Easwaran, K. R. K.: Pramana Supp. *1*, 483 (1975)
184. Khetrapal, C. L., Kunwar, A. C., Ramaprasad, S.: Mol. Crys. Liq. Crys. Lett. *34*, 123 (1976)
185. Seelig, J., Browning, J. L.: FEBS Lett. *92*, 41 (1978)
186. Seelig, J., Neiderberger, W.: J. Am. Chem. Soc. *96*, 2069 (1974)

187. Seelig, A., Seelig, J.: Biochemistry *13*, 4839 (1974)
188. Seelig, A., Seelig, J.: Biochem. Biophys. Res. Commun. *57*, 406 (1974)
189. Pitner, T. P., et al.: Pept.: Chem., Struct. Biol., Proc. Am. Pept. Symp. 4th; p. 159, Ann Arbor Science 1975
190. Glickson, J. D., Dadok, J., Marshall, G. R.: Biochemistry *13*, 11 (1974)
191. Sugihara, T., Imanishi, Y., Higashimura, T.: Biopolymers *14*, 733 (1975)
192. Kaptein, R., Dijkstra, K., Nicolay, K.: Nature *274*, 293 (1978)
193. Krishna, N. R., et al.: Biophys. J. *26*, 345 (1979)
194. Moore, P. B., Laughrea, M.: Nucl. Acids Res. *6*, 2353 (1979)
195. Kopple, K. D., Schamper, T. J.: Chem. Biol. Peptides. Am. Pept. Symp. 3rd., Ann Arbor Science 1972, p. 75
196. Meienhofer, J.: Chem. Biol. Pept., 3rd Am. Pept. Symp., Ann. Arbor Science 1972, p. 81
197. Glickson, J. D., Lenkinsky, R. M., Agresti, D. G. In: Peptides, 5th Am. Pept. Symp. Goodman, M., Meienhofer, J. (eds.) New York: John Wiley and Sons 1977
198. Kessler, H., et al.: In: Peptides, 5th Am. Pept. Symp. Goodman, M., Meienhofer (eds.), p. 343. New York: John Wiley and Sons 1977
199. Pease L. G., Watson, C. In: Peptides, 5th Am. Pept. Symp. Goodman, M., Meienhofer, J. (eds.), p. 346. New York: John Wiley and Sons 1977
200. Tso, P. O. P. (ed.): Basic principles in nucleic acid chemistry, vols. I and II. New York, London: Academic Press 1975
201. Bloomfield, V. A., Crothers, D. M., Tinoco, I.: Physical chemistry of nucleic acids. New York: Harper and Row 1974
202. Duchesne, J. (ed.): Physicochemical properties of nucleic acids. London: Academic Press 1973
203. Watson, J. D. and Crick, F. H. C.: Nature *171*, 737 (1953)
204. Rodley, G. A., et al.: Proc. Nat. Acad. Sci. USA. *73*, 2959 (1976)
205. Sasisekharan, V., Pattabhiraman, N.: Curr. Sci. *45*, 779 (1976)
206. Crick, F. H. C., Wang, J. C., Bauer, W. Q.: J. Mol. Biol. *129*, 449 (1979)
207. Quigley, G. J., et al.: Nucl. Acids Res. *2*, 2329 (1975)
208. Sussman, J. L., Kim, S. H.: Biochem. Biophys. Res. Commun. *68*, 89 (1976)
209. Stout, C. D., et al.: Nucl. Acids Res. *3*, 111 (1976)
210. Ladner, J. E., et al.: Nucl. Acids Res. *2*, 1629 (1975)
211. Pullman, B., Saran, A.: Progr. Nucl. Acid. Res. Mol. Biol. *18*, 216 (1976)
212. In: Conformation of Biological Molecules and Polymers 5th Jerusalem Symp. Quant. Chem. Biochem. Bergmann, E. D., Pullman, B. (eds.), p. 815. New York: Academic Press 1973
213. Sundaralingam, M.: Biopolymers *7*, 821 (1969)
214. Arnott, S.: Progr. Biophys. Mol. Biol. *21*, 284 (1970)
215. Sasisekharan, V. In: 5th Jerusalem Symp. Quant. Chem. Biochem. Bergmann, E. D., Pullman, B. (eds.), p. 247. New York: Academic Press 1973
216. Olson, W. K., Flory, P. J.: Biopolymers *11*, 1 (1972)
217. Guschlbauer, W.: Nucleic acid structure. Berlin, Heidelberg, New York: Springer Verlag 1976
218. Davies, D. B.: Jerusalem Symp. Quant. Chem. Biochem. *11*, 509 (1978)
219. Tso, P. O. P., et al.: Ann. N. Y. Acad. Sci. *222*, 290 (1973)
220. Davies, D. B.: Progr. Nucl. Magn. Reson. Spectrosc. *12*, 135 (1978)
221. Son, T.-D.: Analusis *4*, 280 (1976)
222. Tso, P. O. P. In: Fine structure of proteins and nucleic acids. Fasman, G. D., Timasheff (eds.), p. 49. New York: Marcel Dekker 1970
223. Imoto, T., Akasaka, K., Hatano, H.: Chem. Lett. *1974*, 73
224. Akasaka, K., et al.: J. Mag. Reson. *17*, 413 (1975)
225. Donohue, J., Trooblood, K. N.: J. Mol. Biol. *2*, 363 (1960)
226. Davies, D. B.: Stud. Biophys. *55*, 29 (1976)
227. Dea, P., et al.: In Proc. 1st Int. Conf. Stable Isotop. Chem. Biol. Med., p 84 Springfield Va 1973
228. Uesugi, S., Ikehara, M.: J. Am. Chem. Soc. *99*, 3250 (1977)
229. Govil, G., et al.: Biopolymers *20*, 573 (1981)
230. Suck, D., Sanger, W.: J. Am. Chem. Soc. *94*, 6520 (1972)
231. Schweizer, M. P., Kreishman, G. P.: Biopolymers *10*, 159 (1973)

232. Schweizer, M. P., Witkowsky, J. T., Robins, R. K.: J. Am. Chem. Soc. *93*, 277 (1971)
233. Hart, P. A., Davis, J. P.: J. Am. Chem. Soc. *93*, 753 (1971)
234. Nanda, R. K., et al.: Can. J. Chem. *52*, 371 (1974)
235. Egan, W., Forsen, S., Jacobus, J.: Biochemistry *14*, 735 (1975)
236. Son Tran-Dinh, Chachaty, C.: Biochim. Biophys. Acta. *500*, 405 (1977)
237. Neuman, J. M., et al.: Biochim. Biophys. Acta. *479*, 427 (1977)
238. Bock, K. H., Laurance, D.: J. Carbohydr. Nucleos. Nucleot. *4*, 83 (1977)
239. Son, T.-D., et al.: J. Am. Chem. Soc. *99*, 3267 (1977)
240. Bobruskin, I. D., et al.: Mol. Biol. (Moscow) *13*, 118 (1979)
241. Chachaty, C., Morange, M.: Eur. J. Biochem. *62*, 45 (1976)
242. Giessner-Prettre, C., Pullman, B.: J. Theor. Biol. *65*, 189 (1977)
243. Giessner-Prettre, C., Pullman, B.: J. Theor. Biol. *65*, 171 (1977)
244. Giessner-Prettre, C., Pullman, B.: 11th Jerusalem Symp. Quant. Chem. Biochem. Bergmann, B. D., Pullman, B (eds.). 161 (1978)
245. Prado, F. R., Giessner-Prettre, C., Pullman, B.: J. Theor. Biol. *74*, 259 (1978)
246. Ikehara, M., Tada, H.: 11th Jerusalem Symp. Quant. Chem. Biochem. Bergmann, B. D., Pullman, B. (eds.). 455 (1972).
247. Jordan, F., Niv, H.: Biochim. Biophys. Acta. *476*, 265 (1977)
248. Cadet, J., Ducolomb, R., Taieb, C.: Tetrahedron Lett. *1975*, 3455
249. George, A. L., et al.: Can. J. Chem. *56*, 1170 (1978)
250. Thewalt, U., Bugg, C. E., Marsh, R. E.: Acta Cryst. *B26*, 1089 (1970)
251. Nagashima, N., Iitaka, Y.: Acta. Cryst. *B24*, 1936 (1968)
252. Ikehara, M., Muraoka, M.: J. Carbohydr. Nucleos., Nucleot. *4*, 1 (1977)
253. Akhrem, A. A., Mikhailopulo, I. A., Abramov, A. I.: Org. Mag. Reson. *12*, 247 (1979)
254. Sundaralingam, M., Jensen, L. H.: J. Mol. Biol. *13*, 930 (1965)
255. Govil, G., Saran, A.: J. Theor. Biol. *33*, 399 (1971)
256. Altona, C., Sundaralingam, M.: J. Am. Chem. Soc. *94*, 8205 (1972)
257. Petzold, D. R., Schwarz, D.: Stud. Biophys. *55*, 93 (1976)
258. Davies, D. B., Danyluk, S. S.: Biochemistry *14*, 543 (1975)
259. Altona, C., Sundaralingam, M.: J. Am. Chem. Soc. *95*, 2333 (1973)
260. Fang, K. N., et al.: J. Am. Chem. Soc. *93*, 6647 (1971)
261. Jaworski, A., Ekiel, I., Shugar, D.: J. Am. Chem. Soc. *100*, 4357 (1978)
262. Rabezenko, A., Jankowski, K., Devies, D. B.: Stud. Biophys. *67*, 33 (1978)
263. Guschlbauer, W., Son, T.-D.: Nucl. Acids Res. Symp. Chem. Nucleic Acid Comp. *1*, 85 (1975)
264. Davies, D. B., Danyluk, S. S.: Biochemistry *13*, 4417 (1974)
265. Leudemann, H. D., et al.: Biophys. struct. Mech. *1*, 121 (1975)
266. Lee, C. H., Evans, F. E., Sarma, R. H.: J. Biol. Chem. *250*, 1290 (1975)
267. Sarma, R. H., et al.: J. Am. Chem. Soc. *96*, 7337 (1974)
268. Sundaralingam, M. In: Structure and comformation of nucleic acids and protein nucleic acid interactions. Sundaralingam, M., Rao, S. T. (eds.) Baltimore: University Press 1974
269. Govil, G., et al.: Nucleic Acids Res. *4*, 2573 (1977)
270. Silverton, J., Miles, H. T.: Private communication
271. Robins, M. J., Mac Coss, M., Wilson, J. S.: J. Am. Chem. Soc. *99*, 4660 (1977)
272. Hruska, F. E. In: Conformation of biological molecules and polymers. Bergmann, E. D., Pullman, B. (eds.). New York: Academic Press 1973
273. Hruska, F. E.: Can. J. Chem. *49*, 2111 (1971)
274. Hruska, F. E.: J. Am. Chem. Soc. *93*, 1795 (1970)
275. Blackburn, B. J., et al.: J. Am. Chem. Soc. *95*, 2873 (1973)
276. Lapper, R. D., Mantsch, H. H., Smith, I. C. P.: J. Am. Chem. Soc. *95*, 2878 (1973)
277. Lapper, R. D., Smith, I. C. P.: J. Am. Chem. Soc. *95*, 2880 (1973)
278. Tewari, R.: Ph.D. Thesis, University of Bombay 1975
279. Tewari, R., Nanda, R. K., Govil, G.: Biopolymers *13*, 2015 (1974)
280. Cozzone, P. J., Jardetzky, O.: Biochemistry *15*, 4860 (1976)
281. Smith, I. C. P., et al.: Jerusalem Symp. Quant. Chem. Biochem. *5*, 381 (1973)
282. Nanda, R. K.: Ph.D. Thesis, University of Bombay 1975

283. Bolton, P. H., Kearns, D. R.: J. Am. Chem. Soc. *101*, 479 (1979)
284. Lee, C. H., Sarma, R. H.: J. Am. Chem. Soc. *97*, 1225 (1975)
285. Cheng, D. M., Sarma, R. H.: J. Am. Chem. Soc. *99*, 7333 (1977)
286. Son, T.-D., Guschlbauer, W.: Nucl. Acids Res. *2*, 873 (1975)
287. Remin, M., Shugar, D.: Biochem. Biophys. Res. Commun. *48*, 636 (1972)
288. Davies, D. B., Rabezenko, A.: J. Chem. Soc. Perkin Trans. II *1975*, 1703
289. Hruska, F. E., et al.: Can. J. Chem. *52*, 497 (1974)
290. Govil, G. In: Conformation of biological molecules and polymers. Bergman, E. D., Pullman, B. (eds.). New York: Academic Press 1973
291. Arter, D. B., Schmidt, P.G.: Nucl. Acids Res. *3*, 1437 (1976).
292. Gorenstein, D. G., et al.: Biochemistry *15*, 3796 (1976)
293. Gueron, M., Shulman, R. G.: Proc. Nat. Acad. Sci. USA *72*, 3482 (1975)
294. Lee, C. H., Evans, F. E., Sarma, R. H.: FEBS Lett. *51*, 73 (1975)
295. Haasnoot, C. A. G., Altona, C.: Nucl. Acids Res. *6*, 1133 (1977)
296. Deslauriers, R., Smith, I. C. P.: Can. J. Chem. *51*, 833 (1973)
297. Davies, D. B., Danyluk, S. S.: Can. J. Chem. *48*, 3112 (1970)
298. Kitano, S., et al.: Biochem. Biophys. Res. Commun. *64*, 996 (1975)
299. Hruska, F. E., et al.: Can. J. Chem. *49*, 2449 (1971)
300. Freyne, E. J., et al.: J. Carbohydr. Nucleos., Nucleot. *3*, 113 (1976)
301. Esmans, E. L., et al.: J. Carbohydr. Nucleos., Nucleot. *3*, 93 (1976)
302. Lee, C. H., Sarma, R. H.: Biochemistry *15*, 697 (1976)
303. Paszye, S., Bujarski, J.: Bull. Acad. Pol. Sci. Ser. Sci. Chim. *23*, 477 (1975)
304. Lo, A., Shefter, E., Cochran, T. G.: J. Pharm. Sci. *64*, 1707 (1975)
305. Szarek, W. A., Vyas, D. M., Achmatowicz, B.: Tetrahedron Lett. *1975*, 1553
306. Cadet, J., Teoule, R.: J. Chem. Res. (S) *1977*, 296
307. Ekiel, I., et al.: Biochim. Biophys. Acta. *562*, 177 (1979)
308. Westhof, E., et al.: Z. Naturforsch. *30C*, 131 (1975)
309. Follmann, H., Gremels, G.: Eur. J. Biochem. *47*, 187 (1974)
310. Blandin, M., et al.: Biochim. Biophys. Acta. *361*, 249 (1974)
311. Dea, P., Schweizer, M. P., Freishman, G. P., Biochemistry *13*, 1862 (1974)
312. Matsuda, A., Nomoto, Y., Ueda, T., Chem. Pharm. Bull. *27*, 183 (1979)
313. Manor, P. C., et al.: Biochim. Biophys. Acta *340*, 472 (1974)
314. Remin, M., Shugar, D.: J. Am. Chem. Soc. *95*, 8146 (1973)
315. Wood, D. J., et al.: Can. J. Chem. *51*, 2571 (1973)
316. Slessor, K. N., Tracey, A. S.: Carbohydr. Res. *27*, 407 (1973)
317. Grey, A. A., Smith, I. C. P., Hrusha, F. E.: J. Am. Chem. Soc. *93*, 1765 (1971)
318. Belikova, A. M., Grineva, N. I., Kabashiva, G. N.: Izv. Sib. Otd. Akad. Nauk SSR, Ser. Khim. Nauk. *1972*, 94
319. Deslauriers, R., Lapper, R. D., Smith, I. C.P.: Can. J. Biochem. *49*, 1279 (1971)
320. Pless, R., Dudycz, L., Stolarski, R., Shugar, D.: Z. Naturforsch. *33C*, 902 (1978)
321. Luedemann, H. D., Westhof, E.: Jerusalem Symp. Qant. Chem. Biochem. *11*, 41 (1978)
322. Huynh, D. T., et al.: Nouv. J. Chim. *2*, 357 (1978)
323. Gaudemer, A., et al.: Org. Magn. Reson. *10*, 135 (1977)
324. Leclereq, F., Jumelet-Bach, J., Antonakis, K.: Carbohydr. Res. *62*, 73 (1978)
325. Ikehara, M., Limn, W., Fukui, T.: Chem. Pharm. Bull. *25*, 2702 (1977)
326. Plach, H., Westhof, E., Luedemann, H. D., Mengel, R.: Eur. J. Biochem. *80*, 295 (1977)
327. Luedemann, H. D., Westhof, E.: Z. Naturforsch. *32C*, 528 (1977)
328. Plochoka, D., Rabezenko, A., Davies, D. B.: Biochim. Biophys. Acta. *476*, 1 (1977)
329. Nottoli, E. M., Lambert, J. B., Letsinger, R. L.: J. Am. Chem. Soc. *99*, 3486 (1977)
330. De Kok, A. J., et al.: J. Chem. Soc. Perkin Trans. II *1977*, 487
331. Hruska, F. E., Wood, D. J., Singh, H.: Biochim. Biophys. Acta. *474*, 129 (1977)
332. Oppenheimer, N. J., Kaplan, N. O.: Biochemistry *15*, 3981 (1976)
333. Rabideau, P. W., Parikh, D. K., Watson, R. R.: J. Carbohydr. Nucleos. Nucleot. *5*, 535 (1979)
334. Hayashi, F., Akasaka, K., Hatano, H.: Nucl. Acids Res. Special Publ. No. *2*, 53 (1977)
335. Roussel, C., Chanon, M., Metzger, J.: FEBS Lett. *29*, 253 (1973)

336. Lee, C. G., Chang, C. J.: J. Carbohydr. Nucleos. Nucleot. *5*, 343 (1978)
337. Nottoli, E. M., Lambert, J. B., Letsinger, R. L.: Org. Magn. Reson. *9*, 499 (1977)
338. Dale, B. J., Jones, D. M.: Spectrochim Acta. *33A*, 887 (1977)
339. Evans, F. E., Sarma, R. H.: J. Am. Chem. Soc. *97*, 3215 (1975)
340. Hruska, F. E., et al.: FEBS Lett. *31*, 153 (1973)
341. Son, T.-D., et al.: Biochim. Biophys. Acta *281*, 289 (1972)
342. Davies, D. B.: Jerusalem Symp. Quant. Chem. Biochem. *11*, 71 (1978)
343. Cerletti, N., et al.: Helv. Chim. Acta. *61*, 2589 (1978)
344. Uesugi, S., Ikehara, M.: Chem. Pharm. Bull. *26*, 3040 (1978)
345. Dobson, C. M., et al.: Eur. J. Biochem. *88*, 259 (1978)
346. Barrio, J. R., et al.: J. Am. Chem. Soc. *101*, 1564 (1979)
347. Geraldes, C. F. G. C., Williams, R. J. P.: Eur. J. Biochem. *85*, 463 (1978)
348. Chung, H. L., Horwitz, J. F.: J. Carbohydr. Nucleos., Nucleot. *5*, 469 (1979)
349. Evans, F. E., et al.: Bioorg. Chem. *7*, 57 (1978)
350. Cozzone, P. J., Jardetzky, O.: Biochemistry *15*, 4853 (1976)
351. Reitz, G., Pfleiderer, W.: Chem. Ber. *108*, 2895 (1975)
352. Mac Coss, M., et al.: Carbohydr. Res. *62*, 203 (1978)
353. Geraldes, C. F. G. C., Williams, R. J. P.: Eur. J. Biochem. *85*, 471 (1978)
354. Govil, G., Smith, I. C. P. (unpublished)
355. Schweizer, M. P., Robins, R. K.: Jerusalem Symp. Quant. Chem. Biochem. *5*, 329 (1973)
356. Lavallee, D. K., Zettmann, A. H.: Proc. 11th Rare Earth Res. Conf. *1*, 218 (1974)
357. Lavellee, D. K., Coulter, C. L.: J. Am. Chem. Soc. *95*, 576 (1973)
358. Kainosho, M., Ajisaka, K.: J. Am. Chem. Soc. *97*, 6839 (1975)
359. Lee, C. H., Sarma, R. H.: J. Am. Chem. Soc. *98*, 3541 (1976)
360. Oppenheimer, N. J.: J. Carbohydr. Nucleos., Nucleot. *5*, 251 (1978)
361. Oppenheimer, N. J., Arnold, L. J. Jr., Kaplan, N. D.: Biochemistry *17*, 2613 (1978)
362. Ellis, P. D., et al.: J. Biol. Chem. *248*, 7677 (1973)
363. Blumenstein, M., Raftery, M. A.: Biochemistry *12*, 3585 (1973)
364. Williams, T. J., et al.: Arch. Biochem. Biophys. *172*, 490 (1976)
365. Czeisler, J. L., Hollis, D. P.: Biochemistry *14*, 2781 (1975)
366. Zens, A. P., et al.: J. Am. Chem. Soc. *97*, 2850 (1975)
367. Birdsall, B., et al.: J. Am. Chem. Soc. *97*, 2845 (1975)
368. Torreilles, J., Crastes de Paulet, A.: Biochim. Biophys. Acta. *362*, 375 (1974)
369. Torreilles, J., Crastes de Paulet, A.: Biochim. *55*, 1077 (1973)
370. Sarma, R. H., Mynott, R. J.: J. Am. Chem. Soc. *95*, 7470 (1973)
371. Sarma, R. H., et al.: Can. J. Chem. *51*, 1843 (1973)
372. Sarma, R. H.: J. Chem. Soc. Chem. Commun. *1973*, 140
373. Sarma, R. H., Mynott, R. J.: J. Am. Chem. Soc. *95*, 1641 (1973)
374. Sarma, R. H., Mynott, R. J.: J. Chem. Soc. Chem. Commun. *1972*, 975
375. Sarma, R. H., Mynott, R. J.: J. Chem. Soc. Chem. Commun. *1972*, 977
376. Oppenheimer, N. J., Arnold, L. J., Kaplan, N. O.: Proc. Nat. Acad. Sci. USA *68*, 3200 (1971)
377. Olda, T., et al.: Nucl. Acids Res., 4th Symp. Nucl. Acid Chem. *2*, 47 (1976)
378. Bose, K. S., Sarma, R. H.: Biochem. Biophys. Res. Commun. *66*, 1173 (1975)
379. Sarma, R. H., Mynott, R. J.: Jerusalem Symp. Quant. Chem. Biochem. *5*, 591 (1973)
380. Van, R., et al.: T. J. Bioorg. Chem. *6*, 403 (1977)
381. Zens, A. P., et al.: J. Am. Chem. Soc. *98*, 7559 (1976)
382. Arnold, L. J. Jr., et al.: Biochemistry *18*, 2787 (1979)
383. Hirano, S., Fukuda, T., Kondo, S.: Agr. Biol. Chem. *38*, 1515 (1974)
384. Kozlowski, H., Matczak-Jon, E.: Inorg. Chim. Acta. *32*, 143 (1979)
385. Fazakerley, G. V., Reid, D. G.: Eur. J. Biochem. *93*, 535 (1979)
386. Chaudhuri, P., Sigel, H.: J. Am. Chem. Soc. *99*, 3142 (1977)
387. Granot, J.: J. Am. Chem. Soc. *100*, 1539 (1978)
388. Mikhailov, S. N., et al.: Tetrahedron *32*, 2409 (1976)
389. Evans, F. E., Sarma, R. H.: Nature *263*, 567 (1976)
390. Dhingra, M. M., Sarma, R. H.: Biochemistry (in press).

391. Sarma, R. H., Dhingra, M. M.: Nature *278*, 582 (1979)
392. Dhingra, M. M., Sarma, R. H.: Proc. Int. Symp. Biomol. Struct., Conformation, Function and Evolution. Srinivasan, R. (ed.). Oxford: Pergamon Press 1979
393. Dhingra, M. M., Sarma, R. H.: Int. J. Quan. Chem. QBS*6*, 131 (1979)
394. Govil, G.: Biopolymers *15*, 2303 (1976)
395. Lee, C. H., et al.: Biochemistry *15*, 3627 (1976)
396. Altona, C. In: Structure and conformation of nucleic acids and protein – nucleic acid interaction. Proc. 4th Annu. Harry Steenbock Symp. Sundaralingam, M., Rao, S. T. (eds.), p. 613, Baltimore: University Park Press 1975
397. Ezra, F. S., et al.: Biochemistry *16*, 1977 (1977)
398. Kan, L. S., et al.: Biopolymers *12*, 2225 (1973)
399. Altona, C., et al.: Jerusalem Symp. Quant. Chem. Biochem. *11*, 87 (1978)
400. Cheng, D. M., Dhingra, M. M., Sarma, R. H.: Nucl. Acids Res. *5*, 4399 (1978)
401. Dhingra, M. M., et al.: J. Am. Chem. Soc. *100*, 4669 (1978)
402. Uesugi, S., et al.: J. Am. Chem. Soc. *94*, 5480 (1972)
403. Barry, C. D., et al.: Biochim. Biophys. Acta. *262*, 101 (1972)
404. Danyluk, S. S., Ainsworth, C. F., Maccoss, M.: Jerusalem Symp. Quant. Chem. Biochem. *11*, 111 (1978)
405. Peeling, J., Hruska, F. E., Loewen, P. C.: Can. J. Chem. *56*, 522 (1978)
406. Dhingra, M. M., Sarma, R. H.: Nature *272*, 798 (1978)
407. Hillen, W., Gassen, H. G.: Biochim. Biophys. Acta. *518*, 7 (1978)
408. Lee, C. H., Tinoco, I. Jr.: Biochemistry *16*, 5403 (1977)
409. Cheng, D. M., Sarma, R. H.: Biopolymers *16*, 1687 (1977)
410. Chachaty, C., et al.: Biophys. Chem. *6*, 151 (1977)
411. Topal, M. D., Warshaw, M. M.: Biopolymers *15*, 1755 (1976)
412. Uesugi, S., Limn, W., Ikehara, M.: Chem. Pharm. Bull. *26*, 3732 (1978)
413. Kondo, N. S., et al.: Biochemistry *11*, 1991 (1972)
414. Kondo, N. S., Ezra, F., Danyluk, S. S.: FEBS Lett. *53* 213 (1975)
415. Wood, D. J., Ogilvie, K. K., Hruska, F. E.: Can. J. Chem. *53*, 2781 (1975)
416. Altona, C., Van Boom, J. H., Haasnoot, C. A. G.: Eur. J. Biochem. *71*, 557 (1976)
417. Tazawa, S., et al.: Biochemistry *11*, 3544 (1972)
418. Evans, F. E., Lee, C. H., Sarma, R. H.: Biochem. Biophys. Res. Commun. *63*, 106 (1975)
419. Dhingra, M. M., et al.: Biochemistry *17*, 5815 (1978)
420. Suck, D., et al.: Nature, New Biol. *246*, 161 (1973)
421. Ezra, F. S., et al.: Nucl. Acids Res. *3*, 2549 (1976)
422. Dhingra, M. M., Sarma, R. H. In: Topics in nucleic acid structures. Neidle, S. (ed.). London: Pergamon Press 1979
423. Sarma, R. H., Dhingra, M. M., Feldman, R. J. In: Stereodynamics of molecular systems. Sarma, R. H. (ed.). Oxford: Pergamon Press 1979
424. Cheng, D. M., et al.: Biochemistry *19*, 2491 (1980)
425. Olson, W. K., Manning, S.: Biopolymers *15*, 2391 (1976)
426. Olson, W. K.: Biopolymers *14*, 1797 (1975)
427. Schleich, T., et al.: In: Structure and conformation of nucleic acids and protein-nucleic acid interactions. Proc. 4th Annu. Harry Steenbock Symp. Sundaralingam, M., Rao, S. T. (eds.), p. 223. Baltimore: University Park Press 1975
428. Nelson, J. H., et al.: J. Mol. Biol. *62*, 331 (1971)
429. Kreishman, G. P., Chan, S. I.: Biopolymers *10*, 159 (1971)
430. Komoroski, R. A.: Ph.D. Thesis, Indiana University, Bloomington 1973 (referred to in [27])
431. Ts'o, P. O. P., et al.: In: Structure and conformation of nucleic acids and protein-nucleic acid interaction. Proc. 4th Annu. Harry Steenbock Symp. Sundaralingam, M., Rao, S. T. (eds.), p. 193. Baltimore: University Park Press 1975
432. Yamada, A., Akasaka, K., Hatano, H.: Biopolymers *15*, 1315 (1976)
433. Miwa, M., et al.: Nucl. Acids Res. *4*, 3997 (1977)
434. Patel, D. J.: Biopolymers *18*, 553 (1979)
435. Davanloo, P., Armitage, I. M., Crothers, D. M.: Biopolymers *18*, 663 (1979)

436. Cross, A. D., Crothers, D. M.: Biochemistry *10*, 4015 (1971)
437. Patel, D. J., Biochemistry *14*, 3984 (1975)
438. Patel, D. J., Tonelli, A. E.: Biochemistry *14*, 3990 (1975)
439. Patel, D. J., Hilbers, C. W.: Biochemistry *14*, 2651 (1975)
440. Patel, D. J., Tonelli, A. E.: Biopolymers *13*, 1943 (1974)
441. Patel, D. J., Canuel, L. L.: Eur. J. Biochem. *96*, 267 (1979)
442. Arnott, S., Chandrasekaran, R., Selsing, E. In: Structure and conformation of nucleic acids and protein-nucleic acid interaction. Proc. 4th Annu. Harry Steenbock Symp. Sundaralingam, M., Rao, S. T. (eds.). Baltimore: University Park Press 1975
443. Hibers, C. W., Patel, D. J.: Biochemistry *14*, 2656 (1975)
444. Kan, L. S., et al.: Biochemistry *14*, 3278 (1975)
445. Gromova, E. S., et al.: Biorg. Khim. *1*, 1716 (1975)
446. Borer, P. N., Kan, L. S., Ts'o, P. O. P.: Biochemistry *14*, 4847 (1975)
447. Kallenbach, N. R., Daniel, W. E., Jr., Kaminker, M. A.: Biochemistry *15*, 1218 (1976)
448. Borer, P. M., Kan, L. S., Ts'o, P. O. P.: Biochemistry *14*, 4864 (1975)
449. Fritzsche, H.: Z. Chem. *12*, 1 (1972)
450. Patel, D. J.: Biochemistry *13*, 2396 (1974)
451. Early, T. A., et al.: Biochemistry *16*, 541 (1976)
452. Kearns, D. R.: Annu. Rev. Biophys. Bioeng. *6*, 477 (1977)
453. Kearns, D. R.: Progr. Nucl. Acid. Res. Mol. Biol. *18*, 91 (1976)
454. Patel, D. J.: Accounts. Chem. Res. *12*, 118 (1979)
455. Kallenbach, N. R., Berman, H. M.: Quart. Rev. Biophys. *10*, 138 (1977)
456. Crothers, D. M., Hilbers, C. W., Shulman, R. G.: Proc. Nat. Acad. Sci. USA *70*, 2899 (1973)
457. Patel, D. J., Canuel, L. L.: Proc. Nat. Acad. Sci. USA *73*, 674 (1976)
458. Katz, L., Penman, S.: J. Mol. Biol. *15*, 220 (1966)
459. Patel, D. J.: Biopolymers *15*, 533 (1976)
460. Viswamitra, M. A., et al.: Nature *273*, 687 (1978)
461. Klug, A., et al.: J. Mol. Biol. *131*, 669 (1979)
462. Shindo, H., Simpson, R. T., Cohen, J. S.: J. Biol. Chem. *254*, 8125 (1979)
463. Hosur, R. V., et al.: J. Mol. Struct. *72*, 261 (1981)
464. Patel, D. J., Cannel, L. L., Pohl, F. M.: Proc. Nat. Acad. Sci. USA *76*, 2508 (1979)
465. Kearns, D. R., Wong, Y. P.: J. Mol. Biol. *87*, 755 (1974)
466. Baan, R. A., et al.: Proc. Nat. Acad. Sci. USA *74*, 1028 (1977)
467. Zachav, H. G.: Angew. Chem. Int. Engl. Ed. *8*, 71 (1969)
468. Sueoka, N., Kano-Sueoka, T.: Prog. Nucl. Acids Res., Mol. Biol. *10*, 23 (1971)
469. Lewis, J. A., Ames, B. N.: J. Mol. Biol. *66*, 131 (1972)
470. Holley, R. W., et al.: Science *147*, 1462 (1966)
471. Levitt, M.: Nature *224*, 759 (1969)
472. Daniel, W. E., Jr., Cohn, M.: Biochemistry *15*, 3917 (1976)
473. Shulman, R. G., et al.: In: Structure and conformation of nucleic acids and protein-nucleic acid Interaction. Proc. 4th Annu. Harry Steenbock Symp. Sundaralingam, M., Rao, S. T. (eds.), p. 149. Baltimore: University Park Press 1975
474. Wong, Y. P., et al.: J. Mol. Biol. *74*, 403 (1973)
475. Kastrup, R. V., Schmidt, P. G.: Nucl. Acids Res. *5*, 257 (1978)
476. Robillard, G. T. In: NMR in biology. Dwek, R. A., Campbell, I. O., Richards, R. E., Williams, R. J. P. (eds.), p. 201, London: Academic Press 1977
477. Kyogoku, Y., et al.: Nucl. Acids Res. *4*, 585 (1977)
478. Tompson, J. G., et al.: Biochemistry *18*, 2079 (1979)
479. Kearns, D. R., et al.: Nature *247*, 541 (1974)
480. Kearns, D. R., Bolton, P. H. In: Biomolecular structure and function. Agris, P. E. (ed.). New York: Academic Press 1978
481. Reid, B. R., Hurd, R. E.: Accounts Chem. Res. *10*, 396 (1977)
482. Patel, D. J.: Annu. Rev. Phys. Chem. *29*, 337 (1978)
483. Kearns, D. R.: NATO Adv. Study Inst. Ser. *B22*, 375 (1977)
484. Kearns, D. R., et al.: Biochemistry *13*, 4736 (1974)

485. Crothers, D. M., et al.: J. Mol. Biol. *87*, 63 *(1974)*
486. Cole, P. E., Crothers, D. M.: Biochemistry *11*, 4368 (1972)
487. Hilbers, C. W., Shulman, R. G., Kim, S. H.: Biochem. Biophys. Res. Comm. *55*, 593 (1973)
488. Kearns, D. R., Wong, K. L., Wong, Y. P.: Proc. Nat. Acad. Sci. USA *70*, 3843 (1973)
489. Daniel, W. E., Cohn, M.: Proc. Nat. Acad. Sci. USA *72*, 2582 (1975)
490. Patel, D. J., Canuel, L. L.: Proc. Nat. Acad. Sci. USA *74*, 2624 (1977)
491. Patel, D. J., Canuel, L. L.: Proc. Nat. Acad. Sci. USA *73*, 3343 (1976)
492. Patel, D. J. In: Peptides, polypeptides and proteins. Blout, E. R., Bovey, F. A., Goodman, M. (eds.), p. 459. New York: Wiley Intersci. Publ. 1974
493. Krugh, T. R., Neely, J. W.: Biochemistry *12*, 4418 (1973)
494. Patel, D. J.: Biochemistry *13*, 2388 (1974)
495. Patel, D. J.: Biopolymers *15*, 533 (1976)
496. Reuben, J., Baker, B. M., Kallenbach, N. R.: Biochemistry *17*, 2915 (1978)
497. Barbet, J., et al.: Biochemistry *15*, 2642 (1976)
498. Heller, M. J., Tu, A. T., Maciel, G. E.: Biochemistry *13*, 1623 (1974)
499. Patel, D. J., Shen, C.: Proc. Nat. Acad. Sci. USA *75*, 2553 (1978)
500. Krugh, T. R., Wittlin, F. N., Cramer, S. P.: Biopolymers *14*, 197 (1975)
501. Kreishman, G. P., Chan, S. I., Bauer, W.: J. Mol. Biol. *61*, 45 (1971)
502. Krugh, T. R., Reinhardt, C. G.: J. Mol. Biol. *97*, 133 (1975)
503. Lee, C., Tinoco, I.: Nature *274*, 609 (1978)
504. Sobell, H. M., et al.: J. Mol. Biol. *114*, 333 (1977)
505. Berman, H. M., Neidle, S., Stodala, R. K.: Proc. Nat. Acad. Sci. USA *75*, 828 (1978)
506. Patel, D. J., Canuel, L. L.: Proc. Nat. Acad. Sci. USA *76*, 24 (1979)
507. Sobell, H. M., et al.: Cold Spring Harbor Symp. Quart. Biol. *42*, 87 (1978)
508. Patel, D. J., Canuel, L. L.: Proc. Nat. Acad. Sci. USA *74*, 5207 (1977)
509. Alden, C. J., Arnott, S.: Nucl. Acids Res. *2*, 1701 (1975)
510. Bolte, J., et al.: Biochemistry *18*, 4928 (1979)
511. Heller, M. J., Tu, A. T., Maciel, G. E.: Biochemistry *13*, 1623 (1974)
512. Patel, D. J.: Biochim. Biophys. Acta *442*, 98 (1976)
513. Urbanke, C., Romer, R., Maass, G.: Eur. J. Biochem. *33*, 511 (1973)
514. Jones, C. R., Kearns, D. R.: Biochemistry 14, 2660 (1975)
515. Boyer, P. D. (ed.): The Enzymes, vol. 1 onwards. London: Academic Press 1970 onwards
516. Schroeder, W. A.: The primary structure of proteins. New York: Harper and Row 1968
517. Leach, S. J. (ed.): Physical principles and techniques of protein chemistry. New York: Academic Press 1969
518. Fasman, G. D. (ed.): Poly(α-amino acids). New York: Marcel Dekker 1967
519. Neurath, H. (ed.): The proteins. Vol. 1. New York: Academic Press 1963
520. Ramachandran, G. N., Sasisekharan, V.: Adv. Protein Chem. *23*, 283 (1968)
521. Pullman, B., Pullman, A.: Adv. Protein. Chem. *28*, 347 (1974)
522. Edsall, J. T., et al.: Biopolymers *4*, 121 (1966)
523. Mizushima, S., Shimanouchi, T.: Adv. Enzymol. *23*, 1 (1961)
524. La Pranche, L. A., Roger, M. T.: J. Am. Chem. Soc. *86*, 357 (1964)
525. Govil, G.: J. Chem. Soc.(A) *1971*, 386
526. Urry, D. W., et al.: Pept. Chem., Struct. Biol., 4th Proc. Am. Pept. Symp. *1975*, 113
527. Young, P. E., et al.: Pept. Chem., Struct. Biol., 4th Proc. Am. Pept. Symp. *1975*, 187
528. Luisi, P. L., et al.: Pept. Chem., Struct. Biol., 4th Proc. Am. Pept. Symp. *1975*, 255
529. Levine, B. A., Williams, R. J.P.: Jerusalem, Symp. Quant. Chem. Biochem. *8*, 95 (1976)
530. Deslauriers, R., Smith, I. C. P.: Top. Carbon-13 Spectrosc. *2*, 1 (1976)
531. Deber, C. M., Madison, V., Blout, E. R.: Accounts Chem. Res. *9*, 106 (1976)
532. Bovey, F. A.: Pept. Polypept. Proteins. Proc. Rehovot. Symp., p. 248, 1974
533. Rattle, H. W.: Amino Acids, Pept., Prot. *6*, 223 (1975)
534. Feeney, J., Roberts, G. C. K., Burgen, A. S. V.: Jerusalem Symp. Quant. Chem. Biochem. *7*, 301 (1974)
535. Bovey, F. A.: J. Polym. Sci., Macromol. Rev. *9*, 1 (1974)
536. Urry, D. W.: Res. Develop. *24*, 20 (1973)

537. Bovey, F. A.: Accounts Chem. Res. *5*, 193 (1972).
538. Hruby, V. J.: Chem. Biochem. Amino Acids, Pept., Proteins *3*, 1 (1974)
539. Imanishi, Y., et al.: Kyoto Daigaku Nippon Kagakuseni Kenkyusho Koenshu *33*, 57 (1976)
540. Urry, D. W.: Enzym. Biol. Membr. *1*, 31 (1976)
541. Rattle, H. W. E.: Amino Acids, Pept., Proteins *8*, 199 (1976)
542. Rattle, H. W. E.: Amino Acids, Pept., Proteins *4*, 246 (1972)
543. Rattle, H. W. E.: Amino Acids, Pept. Proteins *7*, 190 (1976)
544. Wüthrich, K.: Pure Appl. Chem. *37*, 235 (1974)
545. Wüthrich, K.: Experientia *30*, 577 (1974)
546. Bystrov, V. F., Arseniev, A. S., Garilov, Yu. D.: J. Magn. Reson. *30*, 151 (1978)
547. Wüthrich, K., Wagner, G.: Trends Biochem. Sci. *3*, 227 (1978)
548. Nemethy, G.: Biochimie *57*, 471 (1975)
549. Roberts, G. C. K., Feeney, J.: Biochem. J. *125*, 88 (1971)
550. Bystrov, V. F.: 4th Int. Biophys. Congr. *2*, 746 (1973)
551. Wüthrich, K.: Naturwissenschaften *60*, 221 (1973)
552. Craig, L. C., Chen. H. C., Gibbons, W. A.: Polym. Prepr. Amer. Chem. Soc., Div. Polym. Chem. *12*, 901 (1971)
553. Wüthrich, K.: NATO Adv. Study, Inst. Ser., *B22*, 347 (1977)
554. Smith, I. C. P., Deslauriers, R.: Recent Progr. Horm. Res. *33*, 309 (1977)
555. Wüthrich, K.: Structure and Bonding *8*, 53 (1970)
556. Wüthrich, K. In: Probes of structure and function of macromolecules and membranes. Chance, B., Yonetani, T., Mildvan, A. S. (eds.), vol. 2. New York: Academic Press 1971
557. Fraser, R. D. B., et al.: Biopolymers *5*, 251 (1967)
558. Carrer, J. P., Blout, E. R.: Treatise on collagen, Vol. 1. New York: Academic Press 1967
559. Crawford, J. L., Lipscomb, W. N., Shellman, C. G.: Proc. Nat. Acad. Sci. USA *70*, 538 (1973)
560. Chou, P. Y., Fasman, G. D.: Biochemistry *13*, 222 (1974)
561. Pullman, B., Maigret, B.: Jerusalem. Symp. Quant. Chem. Biochem. *5*, 13 (1973)
562. Saran, A., Govil, G.: J. Chem. Soc.(A) *1971*, 3624
563. Venkatachalam, C. M.: Biopolymers *6*, 1425 (1968)
564. Bovey, F. A.: Polym. Prepr. Am. Chem. Soc., Div. Polym. Chem. *14*, 163 (1973)
565. Evans, C. A., Rabenstein, D. L.: J. Am. Chem. Soc. *96*, 7312 (1974)
566. Wüthrich, K., Grathwohl, C.: FEBS Lett. *43*, 337 (1974)
567. Davies, D. B., Khaled, M. A.: J. Chem. Soc. Perkin Trans. II *1973*, 1651
568. Dorman, D. E., Bovey, F. A.: J. Org. Chem. *38*, 2379 (1973)
569. Stimson, E. R., Zimmerman, S. S., Scheraga, H. A.: Macromolecules *10*, 1049 (1977)
570. Kricheldorf, H. R., Schilling, G.: Makromol. Chem. *178*, 3115 (1977)
571. Dechter, J. J., Clark, D. S., Mandelkern, L.: Macromolecules, *11*, 274 (1978)
572. Savrda, J.: Pept., Proc. 14th Eur. Pept. Symp. *1976*, 653
573. Rothe, M., Rott, H., Mazanek, J.: Pept., Proc. 14th Eur. Pept. Symp. *1976*, 309
574. Roques, B. P., et al.: Biopolymers *16*, 937 (1977)
575. Grathwohl, C., Wüthrich, K.: Biopolymers *15*, 2043 (1976)
576. Rothe, M., Rott, H.: Angew. Chem. *88*, 576 (1976)
577. Di Blesi, R., Verdini, A. S.: Biopolymers *18*, 735 (1979)
578. Voelter, V. W., Oster, O.: Chem.-Ztg. *96*, 547 (1973)
579. Voelter, V. W., Oster, O.: Org. Magn. Reson. *5*, 547 (1973)
580. Deslauriers, R., et al.: FEBS Lett. *31*, 59 (1973)
581. Dorman, D. G., Bovey, F. A.: J. Org. Chem. *38*, 2379 (1973)
582. Deslauriers, R., Walter, R., Smith, I. C. P.: Biochem. Biophys. Res. Commun. *53*, 244 (1973)
583. Grathwohl, C., et al.: FEBS Lett. *29*, 271 (1973)
584. Grathwohl, C., Wüthrich, K.: J. Mag. Reson. *13*, 217 (1974)
585. Brewster, A. I. R., et al.: Biochemistry *12*, 1643 (1973)
586. Toma, F., et al.: Biochim. Biophys. Acta. *534*, 112 (1978)
587. Gerig, J. T.: Biopolymers, *10*, 2435 (1971)
588. Ramachandran, G. N.: Jerusalem Symp. Quant. Chem. Biochem. *5*, 1 (1973)
589. Bystrov, V. F., et al.: Pure Appl. Chem. *36*, 19 (1973)

590. Gibbons, W. A., et al.: Pept. Chem., Struct. Biol., Proc. 4th Am. Pept. Symp., p. 127, 1975
591. Govil, G.: J. Ind. Chem. Soc. *48*, 731 (1971)
592. Pachler, K. G. R.: Spectrochim. Acta. *20*, 581 (1964)
593. Kainosho, M., Ajisaka, K.: J. Am. Chem. Soc. *97*, 5630 (1975)
594. Kopple, K. D., Go, A.: Pept. Chem., Struct. Biol., Proc. 4th Am. Pept. Symp., p. 139, 1975
595. Deslauriers, R., Smith, I. C. P., Rothe, M.: Proc. 4th Am. Peptide Symp., 1975
596. Niu, C. H., Pease, L. G., Blout, E. R.: Biopolymers *17*, 115 (1978)
597. Howarth, O. W.: J. Chem. Soc. Faraday Trans. II. *75*, 863 (1979)
598. Deslauriers, R., Smith, I. C. P., Walter, R.: J. Biol. Chem. *249*, 7006 (1974)
599. Deslauriers, R., et al.: Int. J. Pept. Protein Res. *13*, 473 (1979)
600. Takeda, M., Jardetzky, O.: J. Chem. Phys. *26*, 1346 (1957)
601. Dale, B. J., Jones, D. W.: Spectrochim. Acta. *31A*, 83 (1975)
602. Dale, B. J., Jones, D. W.: J. Chem. Soc. Perkin Trans. II. *1976*, 1190
603. Bartle, K. D., Jones, D. W., L'Amie, R.: J. Chem. Soc. Perkin Trans. II *1972*, 650
604. Bartle, K. D., Jones, D. W., L'Amie, R.: J. Chem. Soc. Perkin Trans. II *1972*, 646
605. Dale, B. J., Jones, D. W.: Proc. 5th Int. Wolltextil – Forschungskonf. *2*, 287 (1976)
606. Fermandjian, S., et al.: Biochim. Biophys. Acta. *399*, 313 (1975)
607. Tran, T., et al.: Biochim. Biophys. Acta. *492*, 245 (1977)
608. Horsley, W., Sternlicht, H., Cohen, J. S.: J. Am. Chem. Soc. *92*, 680 (1970)
609. Voelter, W., et al.: Z. Naturforsch *26*, 213 (1971)
610. Sternlicht, H., et al.: J. Am. Chem. Soc. *93*, 199 (1971)
611. London, R. E., et al.: J. Am. Chem. Soc. *100*, 3723 (1978)
612. Hansen, P. E., Batchelor, J., Feeney, J.: J. Chem. Soc. Perkin Trans. II *1977*, 50
613. Sugiyama, H., Shiraishi, H., Seto, S.: Chem. Lett. *1973*, 621
614. London, R. E.: J. Am. Chem. Soc. *100*, 2678 (1978)
615. Abraham, R. J., Thomas, W. A.: J. Chem. Soc. *1964*, 3739
616. Torchia, D. A.: Macromolecules *4*, 440 (1971)
617. Ellenberger, N., et al.: Chem. Phys. Lett. *27*, 419 (1974)
618. Pogliani, L., Ellenberger, M., Valat, J.: Org. Magn. Reson. *7*, 61 (1975)
619. Young, P. E., Madison, V., Blout, E. R.: J. Am. Chem. Soc. *95*, 6142 (1971)
620. Anteunis, M. J. O., et al.: Bull. Soc. Chim. Belg. *87*, 41 (1978)
621. Venkatachalam, C. M., Price, B. J., Krimm, S.: Biopolymers *14*, 1121 (1975)
622. De Tar, D. F., Luthra, N. P.: J. Am. Chem. Soc. *99*, 1232 (1977)
623. Deslauriers, R., Walter, R., Smith I. C. P.: J. Biol. Chem. *249*, 7006 (1974)
624. Blomberg, F., et al.: Org. Magn. Reson. *11*, 598 (1978)
625. Joshua, H., Deber, C. M.: Proc. 3rd Am. Pept. Symp., p. 67, 1972
626. Lung, M. T., Marraud, M., Neel, J.: Jerusalem Symp. Quant. Chem. Biochem. *5*, 69 (1973)
627. Sergeev, G. B., et al.: H. A. 1st Mater. uses Konf. Din. Stereokhim Konformatsionnomu Anal., p. 31, 1971
628. Kessler, H., Molter, M.: Angew. Chem. *86*, 553 (1974)
629. Nance, L. E., Schreiner, A. F., Frye, H. G.: Bioinorg. Chem. *3*, 135 (1974)
630. Martinelth, L. C., Honigberg, I. L., Sternson, L. A. Tetrahedron *29*, 1671 (1973)
631. Cung, M. T., Marraud, M., Neel, J.: Ann. Chim. *7*, 183 (1972)
632. Kozlowski, H., et al.: Org. Magn. Reson. *10*, 146 (1977)
633. Ptak, M., Heitz, A., Dreux, M.: Biopolymers *17*, 1129 (1978)
634. Forchioni, A., et al.: Biopolymers *16*, 2755 (1977)
635. Johnson, P. G., McCormik, D. B.: Biochemistry *12*, 3359 (1973)
636. Kwan, S. J., Rhee, C.: Taehan Hwahak Hoechi *21*, 171 (1977)
637. Kazlowski, H., Jezowska, M.: Chem. Phys. Lett. *47*, 452 (1977)
638. Goodman, M., Chen, Fu, Lee, Chi-yu: J. Am. Chem. Soc. *96*, 1479 (1974)
639. Neel, J.: Pure Appl. Chem. *31*, 201 (1972)
640. Idoux, J. P., Scandrett, J. M., Sikorski, J. A.: J. Am. Chem. Soc. *99*, 4577 (1977)
641. Casey, J. M., Martin, R. B.: J. Am. Chem. Soc. *94*, 6141 (1972)
642. Boussard, G., et al.: Biopolymers *16*, 1033 (1977)

643. Kainosho, M., et al.: Biochem. Biophys. Res. Commun. *64*, 425 (1975)
644. Gerig, J. T., McLeod, R. S.: J. Am. Chem. Soc. *98*, 3970 (1976)
645. Dale, B. J., Jones, D. W.: J. Chem. Soc. Perkin Trans. II *1976*, 91
646. Anteunis, M. J., et al.: Bull. Soc. Chim. Belg. *86*, 161 (1977)
647. Lala, A. K., Anteunis, M. J. O., Lala, K.: Biochim. Biophys. Acta. *453*, 133 (1976)
648. Bellocq, A. M., Dubien, M., Dupart, E.: Biochem. Biophys. Res. Commun. *65*, 1393 (1975)
649. Donzel, B., Rivier, J., Goodman, M.: Biopolymers *13*, 2631 (1974)
650. Deslauriers, R., Walter, R., Smith, I. C. P.: Biochem. Biophys. Res. Commun. *53*, 244 (1973)
651. Blagdon, D. E., Rivier, J., Goodman, M.: Proc. Nat. Acad. Sci. USA *70*, 1166 (1973)
652. Deslauriers, R., et al.: Biochemistry *13*, 3443 (1974)
653. Fermandjian, S., et al.: FEBS Lett. *28*, 156 (1972)
654. Walter, R., Bernal, I., Johnson, L. F.: Proc. 3rd Am. Pept. Symp., p. 131, 1972
655. Higashijima, T., et al.: Eur. J. Biochem. *89*, 543 (1978)
656. Deslauriers, R., Walter, R., Smith, I. C. P.: FEBS Lett. *37*, 27 (1973)
657. Howard, J. C., et al.: Macromolecules. *8*, 607 (1975)
658. Kopple, K. D., Go, A., Pilipauskar, D. R.: J. Am. Chem. Soc. *97*, 6830 (1975)
659. Takai, M., et al.: Chem. Pharm. Bull. *23*, 2556 (1975)
660. Rich, D. H., Mathiaparanam, P.: Tetrahedron, Lett. 4037 (1974)
661. Bundi, A., Wüthrich, K.: Biopolymers *18*, 285 (1979)
662. Fournie Zaluski, M. Cl.: Biochem. Biophys. Res. Commun. *79*, 1199 (1977)
663. Lam Thanh Hung, et al.: Pept., Proc. 14th Eur. Pept. Symp. *1976*, 609
664. Khaled, M. A., Renugopalkrishnan, V., Urry, D. W.: J. Am. Chem. Soc. *98*, 7547 (1976)
665. Bleich, H. E., Cutnell, J. D., Glasel, J. A.: Biochemistry *15*, 2455 (1976)
666. Kopple, K. D., Go, A.: Biopolymers *15*, 1701 (1976)
667. Hallenga, K., et al.: Biochim. Biophys. Acta *577*, 82 (1979)
668. Urry, D. W., Ohnishi, T.: Biopolymers. *13*, 1223 (1974)
669. Urry, D. W., Mitchell, L. W., Ohnishi, T.: Biochemistry *13*, 4083 (1974)
670. Urry, D. W., Long, M. M.: CRC Crit. Rev. Biochem. *4*, 1 (1976)
671. Feeney, J., et al.: J. Chem. Soc. Perkin Trans. II *1972*, 601
672. Urry, D. W., et al.: J. Mol. Biol. *96*, 101 (1975)
673. Urry, D. W., Cunningham, W. D., Ohnishi, T.: Biochemistry *13*, 609 (1974)
674. Keim, P., et al.: J. Biol. Chem. *248*, 7811 (1973)
675. Loucheux-Lefebvre, M. H., Helbecque, N.: Proc. 5th Am. Pept. Symp., p. 415, 1977
676. Garbay-Jaureguiberry, C., et al.: Biochem. Biophys. Res. Commun. *71*, 558 (1976)
677. Jones, C. R., et al.: Proc. 5th Am. Pept. Symp., p. 329, 1977
678. Combrisson, S., Roques, B. P., Oberlin, R.: Tetrahedron Lett. *1976*, 3455
679. Stimson, E. R., Meinwald, Y. C., Scheraga, H. A.: Biochemistry *18*, 1661 (1979)
680. Fischman, A. J., Riemen, M. W., Cowburn, D.: FEBS Lett. *94*, 236 (1978)
681. Jones, C. R., Garsky, V., Gibbons, W. A.: Biochem. Biophys. Res. Commun. *76*, 619 (1977)
682. Khaled, M. A., et al.: Biochem. Biophys. Res. Commun. *76*, 224 (1977)
683. Anteunis, M., et al.: Biochemistry *16*, 1462 (1977)
684. Renugopalkrishnan, V., Khaled, M. A., Urry, D. W.: J. Chem. Soc. Perkin Trans. II, 111 (1978)
685. Skrabal, P., et al.: Biopolymers *18*, 995 (1979)
686. Siemion, I. Z., Klis, W. A.: Rocz. Chem. *48*, 1941 (1974)
687. Urry, D. W., et al.: J. Am. Chem. Soc. *100*, 696 (1978)
688. Urry, D. W., Mitchell, L. W., Ohnishi, T.: Biochim. Biophys. Acta. *393*, 296 (1975)
689. Urry, D. W., Ohnishi, T.: Bioinorganic Chem. *3*, 305 (1974)
690. Goodman, M., Ribeiro, A. A., Naider, F.: Proc. Nat. Acad. Sci. USA *75*, 4647 (1978)
691. Wüthrich, K., DeMarco, A.: Helv. Chim. Acta. *59*, 2228 (1976)
692. London, R. E., et al.: Biochemistry *17*, 2270 (1978)
693. Deber, C. M., Adawadkar, P. D., Tom-Kun, J.: Proc. 5th Am. Pept. Symp., p. 357, 1977
694. Mehlis, B., et al.: Proc. 12th Eur. Pept. Symp., p. 328, 1973
695. Deslauriers, R., et al.: FEBS Lett. *62*, 50 (1976)
696. Deslauriers, R., et al.: Biochemistry *14*, 878 (1975)

697. Bleich, H. E., et al.: Biochemistry *12*, 4950 (1973)
698. Zimmer, S., et al.: Eur. J. Biochem. *29*, 80 (1972)
699. Fermandjian, S., Greff, D., Fromageot, P.: Proc. 3rd Pept. Symp., p. 545, 1972
700. Glickson, J. D., Cunningham, W. D., Marshall, G. R.: Biochemistry *12*, 3684 (1973)
701. Donzel, B., et al.: Proc. 4th Am. Pept. Symp., p. 165, 1975
702. Wessels, P. L., et al.: J. Chem. Soc. Perkin Trans. II *1973*, 1691
703. Donzel, B., Rivier, J., Goodman, M.: Biochemistry *16*, 2611 (1977)
704. Deslauriers, R., et al.: Biochemistry *14*, 4335 (1975)
705. Academia Sinica, Institute of Biophysics Peop. R. China Sci. Sin. *19*, 497 (1976)
706. Academia Simica, Institute of Biophysics. Peop. R. China Sheng Wu Hua Hsueh Yu Sheng Wu Wu Li Hsueh Pao *8*, 71 (1976)
707. Bundi, A., et al.: FEBS Lett. *64*, 126 (1976)
708. Jung, G., et al.: Biochim. Biophys. Acta *433*, 164 (1976)
709. Leibfritz, D., et al.: Tetrahedron *34*, 2045 (1978)
710. Nagraj, R., Shamala, N., Balram, P.: J. Am. Chem. Soc. *101*, 16 (1979)
711. Bystrov, V. F., et al.: Tetrahedron *25*, 493 (1969)
712. Tonelli, A. E., Bovey, F. A.: Macromolecules *3*, 410 (1970)
713. Tonelli, A. E., Brewster, A. I., Bovey, F. A.: Macromolecules *3*, 412 (1970)
714. Burgess, A. W., Momany, F. A., Scheraga, H. A.: Proc. Nat. Acad. Sci. USA *70*, 1456 (1973)
715. Reed, L. L., Johnson, P. L.: J. Am. Chem. Soc. *95*, 7523 (1973)
716. Ralston, E., De Coen, J. L., Walter, R.: Proc. Nat. Acad. Sci. USA *71*, 1142 (1974)
717. Smith, G. D., Griffin, J. F.: Science *199*, 1214 (1978)
718. Loew, G. H., Burt, S. K.: Proc. Nat. Acad. Sci. USA *75*, 7 (1978)
719. Saunders, D. J., Offord, R. E.: FEBS Lett. *26*, 286 (1972)
720. Kopple, K. D., Marr, D. H.: J. Am. Chem. Soc. *91*, 962 (1969)
721. Ziauddin, Kopple, K. D.: J. Org. Chem. *35*, 253 (1970)
722. Ziauddin, Kopple, K. D., Bush, C. A.: Tetrahedron Lett. *1972*, 483
723. Deslauriers, R., Grizonka, Z., Walter, R.: Biopolymers *15*, 1677 (1976)
724. Davies, D. B., Khaled, M. A.: J. Chem. Soc. Perkin Trans. II *1976*, 1238
725. Young, P.. E., Madison, V., Blout, E. R.: J. Am. Chem. Soc. *98*, 5365 (1976)
726. Liberek, B., Kitowska, A., Janiak, T.: Rocz. Chem. *49*, 2123 (1975)
727. Ciarkowski, J., et al.: Rocz. Chem. *49*, 2119 (1975)
728. Ptak, P.: Inf. Chim. *147*, 237 (1975)
729. Siemion, I. Z., Konopinska, D., Klis, W. A.: Rocz. Chem. *49*, 781 (1975)
730. Deslauriers, R., et al.: J. Am. Chem. Soc. *97*, 5093 (1975)
731. Andom, M., Sheinblatt, M.: Pept. Polypeptides Proteins. Proc. 2nd Rehovot. Symp., p. 293, 1974
732. Ptak, M., Heitz, A.: Org. Magn. Reson. *6*, 358 (1974)
733. Konopinska, D., Siemion, I. Z., Zabza, A.: Bull. Acad. Pol. Sci., Ser. Sci. Chim. *22*, 253 (1974)
734. Siemion, I. Z., Sucharda-Sobczyk, A.: Rocz. Chem. *46*, 1257 (1972)
735. Young, P. E., Madison, V., Blout, E. R.: J. Am. Chem. Soc. *95*, 6142 (1973)
736. Donzel, B., et al.: Helv. Chim. Acta. *55*, 947 (1972)
737. Ziauddin, Kopple, K. D., Bush, C. A.: Tetrahedron Lett. *1972*, 483
738. Vicar, J., Budesinsky, M., Blaha, K.: Collect. Czech. Chem. Commun. *38*, 1940 (1973)
739. Blaha, K., et al.: Tetrahedron Lett. *1972*, 1437
740. Dale, J., Titlestad, K.: Acta. Chem. Scand. *B*29, 353 (1975)
741. Titlestad, K.: Acta Chem. Scand. *B29*, 153 (1975)
742. Siemion, I. Z.: Org. Magn. Reson. *8*, 432 (1976)
743. Liberek, B., Bednarek, M.: Pol. J. Chem. *52*, 1099 (1978)
744. Haslinger, E.: Tetrahedron *34*, 685 (1978)
745. Liberek, B., et al.: Pept., Proc. 14th Eur. Pept. Symp. *1976*, 641
746. Exner, M. M., Kostelnik, R. J.: Biopolymers *16*, 1387 (1977)
747. Liberek, B., et al.: Rocz. Chem. *51*, 189 (1977)
748. Meraldi, J. P., et al.: Biopolymers *17*, 2401 (1978)
749. Deber, C. M., Torchia, D. A., Blout, E. R.: J. Am. Chem. Soc. *93*, 4893 (1971)

750. Dale, J., Titlestad, K.: J. Chem. Soc. Chem. Commun. *1972*, 255
751. Rich, D. H., Bhatnagar, P. K.: J. Am. Chem. Soc. *100*, 2212 (1978)
752. Rich, D. H., Bhatnagar, P. K.: Pept., Proc. 5th Am. Pept. Symp., p. 340, 1977
753. Grathwohl, C., et al.: Helv. Chim. Acta. *58*, 415 (1975)
754. Dale, J., Titleslad, K.: J. Chem. Soc. Chem. Commun. *1970*, 1403
755. Hassal, C. H., Moschidis, M. C., Thomas, W. A.: J. Chem. Soc. *B 1971*, 1757
756. Hassal, C. H.: Chem. Biol. Pept., Proc. 3rd Am. Pept. Symp., p. 153, 1972
757. Shimohigashi, Y., et al.: Bull. Chem. Soc. Japan. *51*, 584 (1978)
758. Wüthrich, K., Tun-Kyi, A., Schwyzer, R.: FEBS Lett. *25*, 104 (1972)
759. Williamson, K. L., Pease, L. G., Roberts, J. D.: J. Am. Chem. Soc. *101*, 714 (1979)
760. Bara, Y. A., et al.: Chem. Ber. *111*, 1045 (1978)
761. Pease, L. G., Watson, C.: J. Am. Chem. Soc. *100*, 1279 (1978)
762. Pease, L. G., Watson, C.: Pept., Proc. 5th Am. Pept. Symp., p. 346, 1977
763. Kessler, H., et al.: Pept., 5th Proc. Am. Pept. Symp., p. 343, 1977
764. Titlestad, K.: Acta Chem. Scand. *B30*, 753 (1976)
765. Demel, D., Kessler, H.: Tetrahedron Lett. *1976*, 2801
766. Meraldi, J. P., et al.: J. Phys (Paris), Collocq. 41, 1973
767. Titlestad, K., Groth, P., Dale, J.: J. Chem. Soc. Chem. Commun. *1973*, 646
768. Meraldi, J. P., et al.: Helv. Chim. Acta. *55*, 1962 (1972)
769. Cavalieri, P., et al.: Gazz. Chim. Ital. *108*, 509 (1978)
770. Mauger, A. B., Rzeszotarski, W. J., Ford, R. A.: Org. Magn. Reson. *5*, 231 (1973)
771. De Santis, P., Rizzo, R., Ughetto, G.: Tetrahedron Lett. *1971*, 4309
772. Lackner, H.: Tetrahedron Lett. *1975*, 1921
773. Torchia, D. A., et al.: J. Am. Chem. Soc. *94*, 616 (1972)
774. Sugihara, T., Imanishi, Y., Higashimura, T.: Biopolymers *15*, 1529 (1976)
775. Torchia, D. A., et al.: J. Am. Chem. Soc. *94*, 609 (1972)
776. Steinfeld, A., Lepore, U., Goodman, M.: Proc. 3rd Am. Pept. Symp., p. 159, 1972
777. Deber, C. M., et al.: Proc. 3rd Am. Pept. Symp., p. 39, 1972
778. Ivanov, V. T., et al.: Bioorg. Khim. *1*, 25 (1975)
779. Ivanov, V. T., et al.: Khim. Prir. Soed. *11*, 202 (1975)
780. Kopple, K. D.: Proc. 2nd Am. Pept. Symp., p. 241, 1972
781. Ivanov, V. T., et al.: Izv. Akad. Nauk. SSSR. Ser. Khim. *1974*, 2320
782. Blaha, K., Budesinsky, M.: Int. J. Pept. Protein Res. *5*, 399 (1973)
783. Kopple, K. D., et al.: J. Am. Chem. Soc. *95*, 6090 (1973)
784. Tonelli, A. E., Brewster, A. I. R.: Biopolymers *12*, 193 (1973)
785. Grathwohl, C., et al.: FEBS Lett. *29*, 271 (1973)
786. Pease, L. G., Deber, C. M., Blout, E. R.: J. Am. Chem. Soc. *95*, 258 (1973)
787. Schwyzer, R., et al.: Helv. Chim. Acta *55*, 2545 (1972)
788. Kopple, K. D., Ohmishi, M., Go, A.: Biochemistry *8*, 4087 (1969)
789. Portnova, S. L., et al.: Tetrahedron Lett. *1971*, 3085
790. Deber, C. M., et al.: Proc. Nat. Acad. Sci. USA. *69*, 1825 (1972)
791. Tonelli, A. E., Brewster, A. I.: J. Am. Chem. Soc. *94*, 2851 (1972)
792. Kopple, K. D., et al.: J. Am. Chem. Soc. *94*, 973 (1972)
793. Al-Hassan, J., Davies, J. S.: Tetrahedron Lett. *1978*, 3843
794. Imanishi, Y., Kimura, S.: Jerusalem Symp. Quant. Chem. Biochem. *11*, 273 (1978)
795. Khaled, M. A., et al.: J. Phys. Chem. *82*, 2743 (1978)
796. Bartman, B., Deber, C. M., Blout, E. R.: J. Am. Chem. Soc. *99*, 1028 (1977)
797. Hassall, C. H., Thomas, W. A., Moschidis, M. C.: J. Chem. Soc. Perkin Trans. I. *1977*, 2369
798. Llinas, M., et al.: J. Mol. Biol. *104*, 853 (1976)
799. Llinas, M., Wilson, D. M., Klein, M. P.: J. Am. Chem. Soc. *99*, 6846 (1977)
800. Llinas, M., Wilson, D. M., Neilands, J. B.: J. Am. Chem. Soc. *99*, 3631 (1977)
801. De Marco, A., Llinas, M.: Biochemistry *18*, 3846 (1979)
802. Llinas, M., Klein, M. P.: J. Am. Chem. Soc. *97*, 4731 (1975)
803. Llinas, M., Klein, M. P., Neilands, J. B.: Int. J. Pept. Protein Res. *4*, 157 (1972)

804. Khaled, M. A., Sugano, H., Urry, D. W.: Biochim. Biophys. Acta *577*, 273 (1979)
805. Kopple, K. D., Savrda, J.: Peptd., Proc. 11th Eur. Pept. Symp. *1973*, 400
806. Tsetlin, V. I., et al.: Pept., Proc. 12th Eur. Pept. Symp., p. 384, 1973
807. Tonelli, A. E.: Biochemistry *12*, 689 (1973)
808. Abdullaev, N. D., et al.: Zh. Org. Khim. *8*, 1853 (1972)
809. Hollosi, M., Radies, K., Wieland, T.: Int. J. Pept. Protein Res. *10*, 286 (1977)
810. Klis, W. A., et al.: Rocz. Chem. *51*, 1499 (1977)
811. Siemion, I. Z., et al.: Rocz. Chem. *51*, 1489 (1977)
812. Madison, V., Deber, C. M., Blout, E. R.: J. Am. Chem. Soc. *99*, 4788 (1977)
813. Naganawa, H., et al.: Agr. Biol. Chem. *40*, 2223 (1976)
814. Wüthrich, K., et al.: Proc. 1st Eur. Biophys. Congr. *1*, 93 (1971)
815. Meraldi, J. P., et al.: Proc. 4th Am. Pept. Symp., p. 803, 1975
816. Fric, I., et al.: Collect. Czech. Chem. Commun. *41*, 1704 (1976)
817. Glickson, J. D.: Proc. Am. Pept. Symp., p. 787, 1975
818. Hruby, V. J.: Struct. – Act. Relat. Protein polypept. Horm., Proc. 2nd Int. Symp., p. 458, 1972
819. Deslauriers, R., Walter, R., Smith, I. C. P.: Proc. 1st Int. Conf. Stable Isotop. Chem., Biol., Med., p. 64, 1973
820. Honig, B., et al.: J. Mol. Biol. *80*, 277 (1973)
821. Walter, R., Glickson, J. D.: Proc. Nat. Acad. Sci. USA *70*, 1199 (1973)
822. Brewster, A. I. R., et al.: Biochemistry *12*, 1643 (1973)
823. Lyerla, J. R., Jr., Freedman, M. H.: J. Biol. Chem. *247*, 8183 (1972)
824. Glickson, J. D., et al.: Proc. Nat. Acad. Sci. USA *69*, 2136 (1972)
825. Feeney, J., Roberts, G. C. K., Burgen, A. S. V.: Biochem. J. *125*, 62 (1971)
826. Hruby, V. J., et al.: J. Med. Chem. *22*, 7 (1979)
827. Live, D. H., et al.: J. Am. Chem. Soc. *101*, 474 (1979)
828. Hruby, V. J., et al.: J. Am. Chem. Soc. *101*, 202 (1979)
829. Blumenstein, M., Hruby, V. J.: Biochemistry *16*, 5169 (1977)
830. Convert, O., et al.: Biochemistry *16*, 5061 (1977)
831. Blumenstein, M., Hruby, V. J., Vishwanath, V.: Biochemistry *18*, 3552 (1979)
832. Boicelli, C. A., Bradbury, A. F., Feeney, J.: J. Chem. Soc. Perkin Trans. II, *1977*, 477
833. Meraldi, J. P., Hruby, V. J., Brewster, A. I. R.: Proc. Nat. Acad. Sci. USA *74*, 1373 (1977)
834. Kopple, K. D., et al.: Biochemistry *15*, 2945 (1976)
835. Brewster, A. I., Glasel, J. A., Hruby, V. J.: Proc. Nat. Acad. Sci. USA *69*, 1470 (1972)
836. Pitner, T. P., Urry, D. W.: J. Am. Chem. Soc. *94*, 1399 (1972)
837. Patel, D. J., Tonelli, A. E.: Biopolymers *15*, 1623 (1976)
838. Schwyzer, R., Ludescher, U.: Biochemistry *7*, 2519 (1968)
839. Kumar, N. G., et al.: J. Am. Chem. Soc. *97*, 4105 (1975)
840. Jones, C. R., Kuo, M., Gibbons, W. A.: J. Biol. Chem. *254*, 10307 (1979)
841. Kumar, N. G., et al.: Biochemistry *14*, 2197 (1975)
842. Kuo, M., et al.: J. Biol. Chem. *254*, 10301 (1979)
843. Allerhand, A., Komoroski, R. A.: J. Am. Chem. Soc. *95*, 8228 (1973)
844. Siemion, I. Z., Soboczyk, K., Picur, B.: Pol. J. Chem. *52*, 667 (1978)
845. Sogan, J., Craig, L. C., Gibbons, W. A.: J. Am. Chem. Soc. *96*, 3306 (1974)
846. Jones, C. R., Kuo, M., Gibbons, W. A. In: Proc. Symp. on Biomolecular Structure and Function. Agris, P. F. (ed.). p. 401. New York: Academic Press 1978
847. Zenin, S. V., et al.: Vestsi, Akad. Navuk, SSR, Ser. Khim. *1977*, 87
848. Miroshnikov, A. I., et al.: Bioorg. Khim. *3*, 180 (1977)
849. Weinstein, S., et al.: Proc. Nat. Acad. Sci. USA *76*, 4230 (1979)
850. Hawkes, G. E., Randall, E. W., Bradley, C. H.: Nature *257*, 767 (1975)
851. Khaled, M. A., et al.: Biochemistry *17*, 2490 (1978)
852. Davis, D. G., Gisin, B. F., Tosteson, D. C.: Biochemistry *15*, 768 (1976)
853. Servis, K. L., Patel, D. J.: Tetrahedron *31*, 1359 (1975)
854. Ivanov, V. T., et al.: Bioorg. Khim. *1*, 214 (1975)
855. Ivanov, V. T., et al.: Izv. Akad. Nauk SSSR. Ser. Khim *1974*, 2310

856. Patel, D. J., Tonelli, A. E.: Biochemistry *12*, 486 (1973)
857. Davies, D. B., Khaled, M. A.: J. Chem. Soc. Perkin Trans. II *1976*, 1327
858. Ivanov, V. T.: Ann. N. Y. Acad. Sci. *264*, 221 (1975)
859. Kumar, N. G., Urry, D. W.: Biochemistry *12*, 4392 (1973)
860. Pitner, T. P., Urry, D. W.: Biochemistry *11*, 4132 (1972)
861. Viglino, P., et al.: Org. Magn. Reson. *4*, 237 (1972)
862. Chapman, T. M., Golden, M. R.: Biochem. Biophys. Res. Commun. *46*, 2040 (1972)
863. Webb, L. E., Lin. C.: J. Am. Chem. Soc. *93*, 3818 (1971)
864. Roillet, J., Pullman, B., Marigut, B.: Biopolymers *10*, 221 (1971)
865. Venkatachalam, C. M.: Biochim. Biophys. Acta *168*, 397 (1968)
866. Takeuschi, T., Prockop, D. J.: Biochim. Biophys. Acta. *175*, 142 (1969)
867. Sobell, H. M., Jain, S. C., Sakore, T. D.: Nature, New Biology, *231*, 200 (1971)
868. Schwyzer, R., Ludescher, U.: Helv. Chim. Acta. *52*, 2033 (1969)
869. Karle, I. L., Gibson, J. W., Karle, J.: J. Am. Chem. Soc. *92*, 3755 (1970)
870. Karle, I. L., Karle, J.: Acta. Cryst. *16*, 969 (1963)
871. Chandrasekharan, R., et al.: Biochim. Biophys. Acta. *303*, 14 (1973)
872. Smith, I. C. P., Deslauriers, R., Walter, R.: Proc. 3rd Am. Pept. Symp., p. 29, 1972
873. Kotelchuck, D., Scheraga, H. A., Walter, R.: Proc. 3rd Am. Pept. Symp., p. 35, 1972
874. Rudko, A. D., Lovell, F. M., Low, B.: Nature New Biol. *232*, 18 (1971)
875. Urry, D. W., Walter, R.: Proc. Nat. Acad. Sci. USA *68*, 656 (1971)
876. Deslauriers, R., Smith, I. C. P., Walters, R.: J. Am. Chem. Soc. *96*, 2289 (1974)
877. Izumiya, N., et al.: Synthetic aspects of biologically active cyclic peptides – gramicidins and tyrocidines. New York: Halsted Press, John Wiley 1979
878. Ovchinnikov, Yu. A., Ivanov, V. T.: Tetrahedron *31*, 2177 (1975)
879. Ivanov, V. T.: Peptides.: Proc. 5th Am. Pept. Symp., p. 307, 1977
880. Ivanov, V. T., et al.: Biochem. Biophys. Res. Commun. *34*, 803 (1969)
881. Neupert, L. K., Dopler, M.: Helv. Chim. Acta. *58*, 432 (1975)
882. Sisido, M., Imanishi, Y., Higashimura, T.: Macromolecules *9*, 389 (1976)
883. Heitz, F., Cary, P. D., Crane-Robinson, C.: Macromolecules *8*, 745 (1975)
884. Goodman, M., et al.: Pept., Polypept. Proteins, Proc. Rehovot 2nd Symp., p. 126, 1974
885. Bradbury, E. M., et al.: Pept., Polypept., Proteins, Proc. Rehovot 2nd Symp., p. 190, 1974
886. Paolillo, L., et al.: Pept., Polypept., Proteins. Proc. Rehovot 2nd Symp., p. 177, 1974
887. Goodman, M., Toniolo, C., Naider, F., Pept. Polypept., Proteins. Proc. Rehovot 2nd Symp., p. 308, 1974
888. Goodman, M., Ueyama, N., Naider, F.: Biopolymers *14*, 901 (1975)
889. Toniolo, C.: Dynamic Aspects Conform. Changes Biol. Macromol. Proc. 23rd Annu. Meet. Soc. Chim. Phys., p. 87, 1973
890. Torchia, D. A., Lyerla, J. R., Jr.: Biopolymers *13*, 97 (1974)
891. Ferretli, J. A., Jernigan, R. L., Weiss, G. H.: J. Polym. Sci., Polym. Symp. *42*, Part 2, 1051 (1973)
892. Bradbury, E. M., et al.: Pure Appl. Chem. *36*, 53 (1973)
893. Tadokoro, S., Fujiwara, S., Ichihara, Y.: Chem. Lett. *1973*, 849
894. Paolillo, L., et al.: Macromolecules *6*, 831 (1973)
895. Bradbury, E. M., et al.: Polymer *14*, 303 (1973)
896. Allerhand, A., Oldfield, E.: Biochemistry *12*, 3428 (1973)
897. Lyerla, J. R., Barber, B. H., Freedman, M. H.: Can. J. Biochemistry *51*, 460 (1973)
898. Paolillo, L., et al.: J. Chem. Soc. Chem. Commun. *1972*, 335
899. Di Blassi, R., Verdini, A. S.: Biopolymers *13*, 2209 (1974)
900. Todokoro, S., Fujiwara, S., Ichihara, Y.: Chem. Lett (Jap). *1973*, 849
901. Di Blassi, R., Verdini, A. S.: Biopolymers *13*, 765 (1974)
902. Saito, H., Smith, I. C. P.: Arch. Biochem. Biophys. *158*, 154 (1973)
903. Suzuki, Y.: Biopolymers *14*, 1223 (1975)
904. Bradbury, E. M., et al.: J. Am. Chem. Soc. *95*, 1683 (1973)
905. Goodman, M., Toda, F., Ueyama, N.: Proc. Nat. Acad. Sci. USA *70*, 331 (1973)

906. Miller, W. G.: Macromolecules *6*, 100 (1973)
907. Miller, W. G.: Polym. Prepr., Am. Chem. Soc., Div. Polym. Chem. *12*, 558 (1971)
908. Torchia, D. A.: Macromolecules *5*, 566 (1972)
909. Paolillo, L., et al.: Biopolymers *11*, 2043 (1972)
910. Scheraga, H. A., Silverman, D. N.: Biochemistry *10*, 1340 (1971)
911. Gratzer, W. B., Beaven, G. H.: Biopolymers *11*, 689 (1972)
912. Bradbury, E. M., et al.: Polymer. *13*, 33 (1972)
913. Shoji, A., Kawai, T.: Kobunshi Kagaku *28*, 810 (1971)
914. Paolillo, L., et al.: Biopolymers *10*, 2555 (1971)
915. Tanaka, T., Wada, A., Suzuki, M.: J. Chem. Phys. *59*, 3799 (1973)
916. Nagayama, K., Wada, A.: Biopolymers *14*, 2489 (1975)
917. Ullman, R.: Biopolymers *9*, 471 (1970)
918. Bradbury, E. M., Crane-Robinson, C., Hartman, P. G.: Polymer *14*, 543 (1973)
919. Tigelaar, H. L., Flygare, W. H.: J. Am. Chem. Soc. *94*, 343 (1972)
920. Anand, N., et al.: Macromolecules *4*, 564 (1971)
921. Williams, R. K., Fyfe, C. A., Bruck, D.: High-Pressure Sci. Technol., 6th AIRAPT Conf. *1*, 695 (1979)
922. Saito, H., et al.: Biopolymers. *17*, 2587 (1978)
923. Andrew, E. R., et al.: Biopolymers *17*, 1913 (1978)
924. Mutter, M., Mutter, H., Bayer, E.: Pept., Proc. 5th Am. Pept. Symp., p. 403, 1977
925. Shoji, A., Kawai, T., Nishioka, A.: Makromol Chem. *179*, 611 (1978)
926. Gearhart, H. L., Sawyer, W. P., II Anal. Lett. *10*, 887 (1977)
927. Boni, R., Di Blassi, R., Verdini, A. S.: Macromolecules *8*, 140 (1975)
928. Heitz, F.: Macromolecules *10*, 1289 (1977)
929. Oshima, R., Kumanotani, Ju.: J. Polym. Sci. *17*, 379 (1979)
930. Shoji, A., Kawai, T., Nishioka, A.: Macromolecules *10*, 1292 (1977)
931. Pysh, E. S., Toniolo, C.: J. Am. Chem. Soc. *99*, 6211 (1977)
932. Price, C., et al.: Chem. Phys. Lett. *49*, 504 (1977)
933. Lader, H. J., Komoroski, R. A., Mandelkern, L.: Biopolymers *16*, 895 (1977)
934. Heitz, F., Cary, P. D., Crane-Robinson, C.: Macromolecules *10*, 526 (1977)
935. Sase, S., et al.: Biopolymers *16*, 95 (1977)
936. Aksenov, S. I., et al.: Stud. Biophys. *62*, 127 (1977)
937. Grathwohl, C., Wüthrich, K.: Biopolymers *15*, 2025 (1976)
938. Hiraoki, T., Hikichi, K.: Polym. J. *11*, 299 (1979)
939. Peganetli, T., et al.: Pept., Proc. 5th Am. Pept. Symp., p. 395, 1977
940. Ferretti, J. A., Ninham, B. W., Parsegian, V. A.: Macromolecules *3*, 34 (1970)
941. Ferretti, J. A., Jernigan, R. L.: Macromolecules *6*, 687 (1973)
942. Ferretti, J. A., Jermigan, R. L., Weiss, G. H.: J. Polym. Sci. Symp. *42*, 1051 (1973)
943. Kobayashi, Y., Kyogoku, Y.: J. Mol. Biol. *81*, 337 (1973)
944. Bovey, F. A.: Polymer conformation and configuration, New York. Polytechnic Press. 1969
945. Bovey, F. A.: High resolution NMR of macromolecules. New York: Academic Press 1972
946. Sheard, B., Bradbury, E. M.: Progr. Biophys. Mol. Biol. *20*, 187 (1970)
947. Bovey, F. A.: J. Poly. Sci. Macromol. Rev. *9*, 1 (1974)
948. Bradbury, E. M., et al.: Nature *217*, 812 (1968)
949. Goodman, M., Ueyama, N., Naider, F.: Biopolymers *14*, 901 (1975)
950. Milstein, J. B., Ferretti, J. A.: Biopolymers *12*, 2335 (1973)
951. Joubert, F. J., Lotan, N., Scheraga, H. A.: Biochemistry *9*, 2197 (1970)
952. Bradbury, E. M., Fenn, M. D.: Aust. J. Chem. *22*, 357 (1969)
953. Lumry, R., Legare, R., Miller, W. G.: Biopolymers *2*, 489 (1964)
954. Cummings, A. L., Eyring, E. M.: Biopolymers *14*, 2107 (1975)
955. Zona, R.: J. Am. Chem. Soc. *94*, 3646 (1972)
956. Wada, A., Tanaka, T., Kihara, H.: Biopolymers *11*, 587 (1972)
957. Hammes, G. G., Roberts, P. B.: J. Am. Chem. Soc. *91*, 1812 (1969)
958. Barksdale, A. D., Stuehr, J. E.: J. Am. Chem. Soc. *94*, 3334 (1972)

959. Tsuji, V., et al.: J. Am. Chem. Soc. *98*, 813 (1976)
960. Go, N.: J. Phys. Soc. Jap. *22*, 416 (1967)
961. Bradbury, E. M., et al.: J. Am. Chem. Soc. *93*, 5916 (1971)
962. Gelin, B. R., Karplus, M.: Proc. Nat. Acad. Sci. USA *72*, 2002 (1975)
963. Suezaki, Y., Go, N.: Int. J. Pept. Protein Res. *7*, 333 (1975)
964. Wüthrich, K., Wagner, G., Bundi, A.: Jerusalem Symp. Quant. Chem. Biochem. *11*, 201 (1978)
965. Dobson, C. M., Williams, R. J. P.: FEBS Lett. *56*, 362 (1975)
966. Wagner, G., Wüthrich, K.: Nature *275*, 247 (1978)
967. Aksenov, S. I.: Ravnovesn. Dynam. Nalivn. Strukt. Belka *1977*, 42
968. Aksenov, S. I., Filatov, A. V.: Mol. Biol. *12*, 522 (1978)
969. Aksenov, S. I., et al.: Biofizika *23*, 224 (1978)
970. Wüthrich, K., Wagner, G., Richarz, R.: Proc. FEBS Meet. *52*, 143 (1979)
971. Dobson, C. M. In: NMR in biology. Dwek, R. A., Campbell, I. D., Richards, R. E., Williams, R. J. P. (eds.)., p. 63. London: Academic Press 1977
972. Bradbury, J. H., Norton, R. S.: Int. J. Pept. Protein Res. *6*, 295 (1974)
973. Patt, S. L., Dolphin, D. S., Brian, D. In: Lysozyme. Proc. Lysozyme Conf. Osserman, E. F. (edit.)., p. 229, New York: Academic Press 1974
974. Bradbury, J. H., Brown, L. R.: Eur. J. Biochem. *40*, 565 (1973)
975. Bradbury, J. H., King, N. L. R., O'Shea, J. M.: Int. J. Pept. Protein Res. *4*, 257 (1972)
976. Millet, F., Raftery, M. A.: Biochem. Biophys. Res. Commun. *47*, 625 (1972)
977. Gagelli, E., Tiezzi, E., Velensin, G.: Chem. Phys. Lett. *1979*, 155
978. Campbell, I. D., Dobson, C. M., Williams, R. J. P.: Proc. Roy. Soc. *A345*, 41 (1975)
979. Glickson, J. D., Phillips, W. D., Rupley, J. A.: J. Am. Chem. Soc. *93*, 4031 (1971)
980. Campbell, I. D., Dobson, C. M., Williams, R. J. P.: Proc. Roy. Soc. *B189*, 485 (1975)
981. Wagner, G., Wüthrich, K., Tschesche, H.: Eur. J. Biochem. *86*, 67 (1978)
982. Brown, L. R., et al.: Eur. J. Biochem. *62*, 103 (1976)
983. Perkins, S. J., Wüthrich, K.: Biochem. Biophys. Acta *536*, 406 (1978)
984. Harina, B. M., et al.: J. Am. Chem. Soc. *100*, 4897 (1978)
985. Wagner, G., Tschesche, H., Wüthrich, K.: Eur. J. Biochem. *93*, 239 (1979)
986. Wagner, G., Kalb, A. J., Wüthrich, K.: Eur. J. Biochem. *93*, 249 (1979)
987. Wüthrich, K., et al.: In: NMR in biology. Dwek, R. A., Campbell, I. D., Richards, R. E., Williams, R. J. P. (eds.), p. 51. London: Academic Press 1977
988. Roberts, G. C. K., et al.: NMR in biology. Dwek, R. A., Campbell, I. D., Richards, R. E., Williams, R. J. P. (eds.), p. 95. London: Academic Press 1977
989. Feeney, J., et al.: In: NMR in biology. Dwek, R. A., Campbell, I. D., Richards, R. E., Williams, R. J. P. (eds.), p. 113. London: Academic Press 1977
990. Birdsall, B., et al.: Proc. Roy. Soc. *B196*, 251 (1977)
991. Feeney, J., et al.: Proc. Roy. Soc. *B196*, 267 (1977)
992. Feeney, J.: Jerusalem Symp. Quant. Chem. Biochem. *11*, 297 (1978)
993. Dower, S. K., Dwek, R. A.: Biomolecular structure and function. Agris, P. A. (ed.), p. 295. New York: Academic Press 1978
994. Gettino, P., Dwek, R. A.: NMR in biology. Dwek, R. A., Campbell, I. D., Richards, R. E., Williams, R. J. P. (eds.), p. 125. London: Academic Press 1977
995. Strausbach, P. H., et al.: Biochemistry *10*, 4342 (1971)
996. Klostgaard, J., et al.: Immunochemistry *14*, 107 (1977)
997. Liebes, L. F., Zand, R., Phillips, W. D.: Biochim. Biophys. Acta. *405*, 27 (1975)
998. Liebes, L. F., Zand, R., Phillips, W. D.: Biochim. Biophys. Acta. *427*, 392 (1976)
999. Littlemore, L. A. T.: Aust. J. Chem. *31*, 2387 (1978)
1000. Chapman, B. E., Moore, W. J.: Aust. J. Chem. *31*, 2367 (1978)
1001. Campagnani, A. T., Whitehead, D. L., Rowan, R.: Biomolecular structure and function. Agris, P. F. (ed.), p. 413. New York: Academic Press 1978
1002. Deber, C. M., Mascarello, M. A., Wood, D. D.: Biochemistry *17*, 898 (1978)
1003. Chapman, B. E., Moore, M. J.: Biochem. Biophys. Res. Commun. *73*, 758 (1976)
1004. Block, R. E., Brady, A. H., Joeffe, S.: Biochem. Biophys. Res. Commun. *54*, 1595 (1973)

1005. Levine, B. A., et al.: Calcium binding proteins calcium function. Proc. 2nd Int. Symp. Wasserman, R. H. (ed.), p. 9. New York: Elsevier North Holland 1977
1006. Nelson, D. J., et al.: Biomolecular structure and function. Agris, P. F. (ed.), p. 393. New York: Academic Press 1978
1007. Edward, B. F. P., Lee, L., Sykes, B. D.: Biomolecular structure and function. Agris, P. F. (ed.), p. 275. New York: Academic Press 1978
1008. Cave, A., et al.: FEBS Lett. *65*, 190 (1976)
1009. Nelson, D. J., Opella, S. J., Jardetzky, O.: Biochemistry *15*, 5552 (1976)
1010. Birdsall, W. J., et al.: Biochem. Soc. Trans. *7*, 702 (1979)
1011. Urry, D. W., Mitchell, L. W.: Biochem. Biophys. Res. Commun. *68*, 1153 (1975)
1012. Urry, D. W., et al.: Int. J. Pept. Protein Res. *7*, 367 (1975)
1013. Jardetzky, O.: Proc. Int. Biophys. Cong. 4th series, *4*, Part 2, 734 (1973)
1014. Shindo, H., Cohen J. S.: J. Biol. Chem. *251*, 2648 (1976)
1015. Rueterjans, H.: DECHEMA *71*, 219 (1973)
1016. Niu, C. H., et al.: J. Biol. Chem. *254*, 3788 (1979)
1017. Morishima, I., Iizuka, T.: Biochim. Biophys. Acta. *386*, 542 (1975)
1018. Benko, B., et al.: J. Colloid. Interface Sci. *52*, 444 (1975)
1019. Morishima, I., Iizuka, T.: J. Am. Chem. Soc. *96*, 5279 (1974)
1020. Wüthrich, K., Keller, R. M., Baumann, R.: Dyn. Aspects. Conform. Changes Biol. Macromol., Proc., 23rd Annu. Meet. Soc. Chim. Phys., p. 151, 1973
1021. Scheler, W., et al.: Mol. Biol. *7*, 670 (1973)
1022. Ogawa, S., Shulman, R. G.: J. Mol. Biol. *70*, 315 (1972)
1023. Mayer, A., et al.: J. Mol. Biol. *81*, 187 (1973)
1024. Lindstrom, R. R., Ho, C.: Proc. Nat. Acad. Sci. USA *69*, 1707 (1972)
1025. Gupta, R. K., Mildvan, A. S.: J. Biol. Chem. *250*, 246 (1975)
1026. Mansuy, D., et al.: Colloq. Inst. Nat. Sante, Rech. Med. *70*, 23 (1977)
1027. Moon, R. B., Richards, J. H.; J. Am. Chem. Soc. *94*, 5093 (1972)
1028. Morishima, I., et al.: Biochim. Biophys. Acta. *495*, 287 (1977)
1029. Morishima, I., Inubhushi, T.: J. Am. Chem. Soc. *100*, 3568 (1978)
1030. Wüthrich, K., Aviram, I., Schejter, A.: Biochem. Biophys. Res. Commun. *253*, 98 (1971)
1031. Gupta, R. K., Koenig, S. H., Redfield, A. G.: J. Magn. Reson. *7*, 66 (1972)
1032. Stellwagen, E., Shulman, R. G.: J. Mol. Biol. *80*, 559 (1973)
1033. Smith, G. M., Kamen, M. D.: Proc. Nat. Acad. Sci. USA *71*, 4303 (1974)
1034. Jones, W. C., Jr., et al.: J. Biol. Chem. *251*, 7452 (1976)
1035. Batts, B. D., et al.: Aust. J. Chem. *28*, 41 (1975)
1036. Gerig, J. T., Roe, D. C.: J. Am. Chem. Soc. *96*, 233 (1974)
1037. Amstey, J. W., Jindell, S. P., Bender, M. L.: Arch. Biochem. Biophys. *169*, 1 (1975)
1038. Nicholson, B. C., Spotswood, T. M.: Aust. J. Chem. *26*, 135 (1973)
1039. Arnold, L. J., Jr., You, K. S.: Methods Enzymol. *54*, 223 (1978)
1040. Dobson, C. M.: Biochemistry *14*, 4905 (1975)
1041. Arseniev, S., et al.: Eur. J. Biochem. *71*, 595 (1976)
1042. Blicharska, B., et al.: Flavins and flavoproteins: physiocochemical properties and functions. Ostrowski, M. (ed.), p. 51, Warsaw: Panst. Wydawn Nauk 1977
1043. Tanswell, P., Westhead, F. W., Williams, R. J. P.: Biochem. Soc. Trans. *2*, 79 (1974)
1044. Khalifah, R. G.: Biochemistry *16*, 2236 (1977)
1045. Aksenov, S. I., Kharchuk, O. A., Vitvitskii, V. N.: Stud. Biophys. *58*, 223 (1976)
1046. Muller, N., Mead, R. J., Jr.: Biochemistry *12*, 3831 (1973)
1047. Von Dreele, P. H., Rae, I. D., Scheraga, H. A.: Biochemistry *17*, 956 (1978)
1048. Rae, I. D., Scheraga, H. A.: Int. J. Pept. Protein Res. *13*, 304 (1979)
1049. Morrison, C. A., Bradbury, E. M., Garret, R. A.: FEBS Lett. *81*, 435 (1977)
1050. Manuck, B. A., Ho, C.: Biochemistry *18*, 566 (1979)
1051. Rothgeb, T., et al.: Biochemistry *17*, 4564 (1978)
1052. Wagner, G., Wüthrich, K., Tschesche, H.: Eur. J. Biochem. *89*, 367 (1978)
1053. Littlechild, J., et al.: FEBS Lett. *104*, 90 (1979)
1054. Schmidt-Aderjan, I., et al.: Eur. J. Biochem. *96*, 43 (1979)

1055. Granot, J., et al.: Biochemistry *16*, 2339 (1979)
1056. Bartle, K. D., Jones, D. W., L'Amie, R.: Appl. Polym. Symp. *18*, 85 (1971)
1057. Negura, M., Ishikawa, H.: Polym. J. *11*, 139 (1979)
1058. De Wit, J. L., et al.: Biochemistry *18*, 3973 (1979)
1059. Blake, C. C. F., et al.: Proc. Roy. Soc. *B167*, 378 (1967)
1060. Mc Donald, C. C., Phillips, W. D.: J. Am. Chem. Soc. *89*, 6332 (1967)
1061. Deiscuhofer, J., Steigemann, W.: Acta. Cryst. *B31*, 238 (1975)
1062. Antonini, E., Brunori, M.: Haemoglobin and myoglobin in their reactions with ligands. London: Elsevier North Holland 1971
1063. Baldwin, J. M.: Progr. Biophys. Mol. Biol. *29*, 225 (1975)
1064. Geissner-Prettre, C., Pullman, B.: J. Theor. Biol. *31*, 287 (1971)
1065. Tyaglov, B. V., et al.: Mol. Biol. (Moscow) *10*, 347 (1976)
1066. Durand, M., et al.: Biochimie *58*, 395 (1976)
1067. Schimmack, W., Lohmann, W.: Biophys. Struct. Mech. *1*, 319 (1975)
1068. Lilley, D. M. J., et al.: FEBS Lett. *62*, 7 (1976)
1069. Tyaglov, B. V., et al.: Mol. Biol. (Moscow) *9*, 652 (1975)
1070. Bushuev, V. N., et al.: Stud. Biophys. *52*, 173 (1975)
1071. Sloan, D. L., et al.: J. Biol. Chem. *250*, 8913 (1975)
1072. Adawadkar, P., et al.: J. Am. Chem. Soc. *97*, 1959 (1975)
1073. Gabbay, E. J., Sanford, K.: Bioorg. Chem. *3*, 91 (1974)
1074. Wagner, K. G., Lawaczeck, R.: J. Magn. Reson. *8*, 164 (1972)
1075. Lancelot, G., Mayer, R., Helene, C.: J. Am. Chem. Soc. *101*, 1569 (1979)
1076. Hilbers, C. W., et al.: Jerusalem Symp. Quant. Chem. Biochem. *11*, 351 (1978)
1077. Coleman, J. E., Armitage, I. M.: NMR in biology. Dwek, R. A., Campbell, I. D., Richards, R. E., Williams, R. J. P. (eds.), p. 171. London: Academic Press 1977
1078. Tyaglov, B. V., Zenin, S. V.: Nucl. Acids Res. *4*, 2609 (1977)
1079. Tyaglov, B. V., Zenin, S. V.: Mol. Biol. (Moscow) *11*, 801 (1977)
1080. Tyaglov, B. V., et al.: Tezisy Dokl.-Vses. 3rd. Simp. Khim. Pept. Belkov., p. 148 (1974)
1081. Lancelot, G., Helene, C.: Nucl. Acids Res. *6*, 1063 (1979)
1082. Mayer, R., et al.: J. Biol. Chem. *254*, 75 (1979)
1083. Lancelot, G., Helene, C.: Proc. Nat. Acad. Sci. USA *74*, 4872 (1977)
1084. Sellini, H., et al.: FEBS Lett. *30*, 219 (1973)
1085. Niu, C., Black, S.: J. Biol. Chem. *254*, 265 (1979)
1086. Helene, C., et al.: Jerusalem Symp. Quant. Chem. Biochem. *5*, 361 (1973)
1087. Von Hipple, P., Mc Ghec, J. D.: Ann. Rev. Biochem. *41*, 231 (1972)
1088. Toulme, J. J., Helene, C.: J. Biol. Chem. *252*, 244 (1977)
1089. Seeman, N. C., Rosenberg, J. M., Rich, A.: Proc. Nat. Acad. Sci. USA *73*, 804 (1976)
1090. Hosur, R. V., Kumar, N. V., Govil, G.: Int. J. Q. Chem. *20*, 23 (1981)
1091. Crane-Robinson, C.: Biological magnetic resonance Berliner, E. J., Reuben, J. (eds.), p. 33. New York: Plenum Press 1978
1092. Chen, D. M., et al.: Biochem. Biophys. Res. Commun. *92*, 197 (1980)
1093. La Mar, G. N., et al.: J. Biol. Chem. *255*, 66 (1980)
1094. Littlechild, J.: FEBS Lett. *111*, 51 (1980)
1095. Lewis, C. A., Ellis, P. D., Dunlap, R. B.: Biochemistry *19*, 116 (1980)
1096. Florkin, M., Stotz, E. H. (eds.): Comprehensive biochemistry, vol. 5. Amsterdam: Elsevier Publ. 1963
1097. Rao, V. S. R.: J. Ind. Inst. Sci. *56*, 253 (1974)
1098. Haworth, W. N.: The constitution of sugars, p. 90. London: Edward Arnold 1929
1099. Lemieux, R. U., Stevens, J. D.: Can. J. Chem. *43*, 2059 (1965)
1100. Lemieux, R. U., Stevens, J. D.: Can. J. Chem. *44*, 249 (1966)
1101. Durette, P. L., Horton, D.: Org. Magn. Reson. *3*, 417 (1971)
1102. Coxon, B.: Tetrahedron *22*, 2281 (1966)
1103. Rao, V. S. R., et al.: Conformational studies of amylose in conformations of biopolymers. G. N. Ramachandran (ed.), p. 721. London, New York: Academic Press 1967
1104. Sundararajan, P. R., Rao, V. S. R.: Tetrahedron *24*, 289 (1968)

1105. Rao, V. S. R., Vijayalakshmi, K. S., Sundararajan, P. R.: Carbohyd. Res. *17*, 341 (1971)
1106. Vijayalakshmi, K. S., Rao, V. S. R.: Carbohyd. Res. *22*, 413 (1972)
1107. Vijayalakshmi, K. S., Rao, V. S. R.: Carbohyd. Res. *29*, 427 (1973)
1108. Rao, V. S. R., Vijayalakshmi, K. S.: Carbohyd. Res. *33*, 363 (1974)
1109. Koch, H. J., Perlin, A. S.: Carbohyd. Res. *15*, 403 (1970)
1110. Angyal, S. J.: Angew. Chem., Int. Ed. Eng. *8*, 157 (1969)
1111. Coxon, B.: Tetrahedron *21*, 3481 (1965)
1112. Casu, B., et al.: Tetrahedron *22*, 3061 (1966)
1113. Casu, B., et al.: Tetrahedron Lett. *27*, 2253 (1965)
1114. Rao, V. S. R., Foster, J. F.: J. Phys. Chem. *69*, 636 (1965)
1115. Coxon, B.: Carbohyd. Res. *1*, 357 (1966)
1116. Bhacca, N. S., Horton, D., Raulsen, H.: J. Org. Chem. *33*, 2484 (1968)
1117. Lenz, R. W., Hecschen, J. P.: J. Polym. Sec. *51*, 247 (1961)
1118. Jeffrey, G. A., Pople, J. A., Randon, L.: Carbohyd. Res. *25*, 117 (1972)
1119. Perlin, A. S., et al.: Can. J. Chem. *50*, 2437 (1972)
1120. Bundle, D. R., Jennings, H. J., Kenny, C. P.: Carbohyd. Res. *26*, 268 (1973)
1121. Jennings, H. J., Smith, I. C. P.: J. Am. Chem. Soc. *95*, 606 (1973)
1122. Bundle, D. R., Smith, I. C. P., Jennings, H. J.: J. Biol. Chem. *249*, 2275 (1974)
1123. Rao, V. S. R., Foster, J. F.: J. Am. Chem. Soc. *67*, 951 (1963)
1124. Glass, C. A.: Can. J. Chem. *43*, 2652 (1965)
1125. Rees, D. A., Wight, A. W.: J. Chem. Soc. B. 1366 (1971)
1126. Paulsen, H., Friedman, M.: Chem. Ber. *105*, 705 (1972)
1127. Gatti, G., Casu, B., Perlin, A. S.: Biochem. Biophys. Res. Commun. *85*, 14 (1978)
1128. Gatti, G.: Jerusalem Symp. Quant. Chem. Biochem. *11*, 423 (1978)
1129. Pasika, W. M., Cragg, L. H.: Can. J. Chem. *41*, 293 (1963)
1130. Pasika, W. M., Cragg, L. H.: Can. J. Chem. *41*, 777 (1963)
1131. Friebolin, H., Keilich, G., Siefert, E.: Angew. Chem. *81*, 791 (1969)
1132. Friebolin, H., Keilich, G., Siefert, E.: Org. Magn. Reson. *2*, 457 (1970)
1133. Frank, N., et al.: Org. Magn. Reson. *4*, 725 (1972)
1134. Keilich, G., Siefert, E., Friebolin, H.: Org. Magn. Reson. *3*, 31 (1971)
1135. Saito, H., Ohki, T., Sasaki, T.: Carbohyd. Res. *74*, 227 (1979)
1136. Colson, P., et al.: Carbohyd. Res. *71*, 265 (1979)
1137. Seymour, F. R., et al.: Carbohyd. Res. *71*, 231 (1979)
1138. Usui, T., et al.: J. Chem. Soc. Perk. Trans. I. *1973*, 2425
1139. Streefkerk, D. G., De Bie, M. J. A., Vliegenthart, J. F. G.: Tetrahedron *29*, 833 (1973)
1140. Hirano, S.: Agr. Biol. Chem. *37*, 51 (1973)
1141. Jennings, H. J., Rosell, K. G., Kenny, P. C.: Can. J. Chem. *57*, 2902 (1979)
1142. Saito, H., Miyata, E., Sasaki, T.: Macromolecules *11*, 1244 (1978)
1143. Parfondry, A., Cyr, N., Perlin, A. S.: Carbohyd. Res. *59*, 299 (1977)
1144. Excoffier, G., Gangnaire, D. Y., Taravel, F. R.: Carbohyd. Res. *56*, 229 (1977)
1145. Morris, E. R., et al.: J. Mol. Biol. *110*, 1 (1977)
1146. Izumi, K.: Mem. Fac. Ind. Arts. Kyoto. Tech. Univ., Sci. Technol. *24*, 72 (1975)
1147. Inoue, S., Inoue, Y.: Biochem. Biophys. Res. Commun. *23*, 513 (1966)
1148. Jaques, L. B., et al.: Biochem. Biophys. Res. Commun. *24*, 447 (1966)
1149. Dorman, D. E., Roberts, J. D.: J. Am. Chem. Soc. *92*, 1355 (1970)
1150. Dorman, D. E., Roberts, J. D.: J. Am. Chem. Soc. *93*, 4463 (1971)
1151. Breitmaier, E., Voelter, W.: Tetrahedron, *29*, 227 (1973)
1152. Brietmaier, E., Jung, G., Voelter, W.: Chimia, *26*, 136 (1972)
1153. Perlin, A. S., Casu, B., Koch, H. J.: Can. J. Chem. *48*, 2596 (1970)
1154. Voelter, W., et al.: Angew. Chem., Int. Ed. Engl. *9*, 803 (1970)
1155. Dorman, D. E., Angyal, S. J., Roberts, J. D.: J. Am. Chem. Soc. *92*, 1351 (1970)
1156. Breitmaier, E., et al.: Chem. Ber. *104*, 1147 (1971)
1157. Voelter, N., Bilik, V., Breitmaier, E.: Collect. Czech. Chem. Commun. *38*, 2054 (1973)
1158. Voelter, W., Breitmaier, E.: Org. Magn. Reson. *5*, 311 (1973)
1159. Voelter, W., et al.: Chimia *25*, 168 (1971)

1160. Voelter, W., Breitmaier, E., Jung. G.: Angew. Chem., Int. Ed. Engl. *10*, 935 (1971)
1161. Voelter, W., Burrenich, C., Breitmaier, E.: Angew. Chem. Int. Engl. Ed. *11*, 539 (1972)
1162. Bundle, D. R., Jannings, H. J., Smith, I. C. P.: Can. J. Chem. *51*, 3812 (1973)
1163. Colson, P., Jennings, H. J., Smith, I. C. P.: J. Chem. Soc. *96*, 8081 (1974)
1164. Saito, H., et al.: FEBS Lett. *68*, 15 (1976)
1165. Saito, H., Ohki, T., Sasaki, T.: Biochem. *16*, 908 (1977)
1166. Oster, O., Breitmaier, E., Voelter, W.: Nuclear magnetic resonance spectroscopy of nuclei other than proton. Axenrod, T., Webb, C. A. (eds.), p. 233. New York: John Wiley and Sons 1974
1167. Spiro, R. G.: Ann. Rev. Biochem. *39*, 599 (1970)
1168. Marshall, R. O., Neuberger, A.: Adv. Carbohyd. Chem. *25*, 407 (1970)
1169. Rogister, M. G., Degelaen, J., Roberts, G. C. K.: FEBS Lett. *87*, 247 (1978)
1170. Brown, F. F.: J. Biol. Chem. *254*, 4335 (1979)
1171. Dill, K., Allerhand, A.: J. Biol. Chem. *254*, 4524 (1979)
1172. Unkefer, C. J., Gander, J. E.: J. Biol. Chem. *254*, 12131 (1979)
1173. Champman, D. (ed.): Biological membranes, physical fact and function. London, New York: Academic Press 1968
1174. Jain, M. K.: The Bimolecular lipid membrane A system. New York, Cincinnati, Toronto, London, Melbourne: Van Nostrand-Reinhold Publ. 1972
1175. Jain, M. K., White, B. H.: Adv. Lipid. Res. *15*, 1 (1977) Titien, H. Biomolecular lipid membrane: theory and practice. New York: Marcel Dekker 1974
1176. Dalton, A. J., Haguenau, F. (eds.): The membrane. New York, London: Academic Press 1978
1177. Rothfield, L. I. (ed.): Structure and function of biological membranes. New York, London: Academic Press 1971
1178. Leive, L. (ed.): Bacterial membranes and walls. New York: Marcel Dekker 1973
1179. Hatefi, Y., Djavadi – Ohaniance, L. (ed.): The structural basis of membrane function. New York, San Francisco, London: Academic Press 1976
1180. Chapman, D., Wallach, D. F. H. (ed.). Biological membranes. London, New York: Academic Press 1973
1181. Lee, A. G.: Prog. Biophys. Mol. Biol. *29*, 3 (1975)
1182. Singer, S. J., Nicolson, G. L.: Science *175*, 720 (1972)
1183. Chapman, D.: Quart. Rev. Biophys. *8*, 185 (1975)
1184. Hakan, W., Goran, L.: Quart. Rev. Biophys. *10*, 67 (1977)
1185. Kreutz, W.: Angew. Chem. *11*, 551 (1972)
1186. Govil, G., Hosur, R. V.: Int. J. Quant. Chem. *16*, 19 (1979)
1187. Bennet, L. G., et al.: Biomolecular structure and function. Agris, P. F. (ed.), p. 3. New York, San Francisco, London: Academic Press 1978
1188. Seelig, J.: Quart. Rev. Biophys. *10*, 353 (1977)
1189. Rattle, H. W. E.: Prog. Biophys. Mol. Biol. *28*, 1 (1974)
1190. Levine, Y. K.: Prog. Biophys. Mol. Biol. *24*, 1 (1972)
1191. Larson, K.: Ark. Kemi *23*, 1 (1964)
1192. Jensen, L. H., Mabis, A. J.: Acta Cryst. *21*, 770 (1966)
1193. Hitchcock, P. B., et al.: Proc. Nat. Acad. Sci. USA *71*, 3036 (1974)
1194. Pearson, R. H., Pascher, I.: Nature *281*, 499 (1979)
1195. Sundaralingam, M. Ann. N. Y. Acad. Sci. USA *195*, 324 (1972)
1196. Gupta, S. P., Govil, G.: FEBS Lett. *27*, 68 (1972)
1197. Gupta, S. P., Govil, G., Mishra, R. K.: J. Theor. Biol. *51*, 13 (1975)
1198. Pullman, B., Berthod, H.: FEBS Lett. *44*, 266 (1974); Pullman, B., Berthod, H., Gresh, N.: FEBS Lett. *53*, 199 (1975)
1199. Pullman, B., Saran, A.: Int. J. Quant. Chem. Quant. Biol. Symp. No. 2, *1975*, 71
1200. Pullman, B., Courriere, Ph.: Int. J. Quant. Chem, Quant. Biol. Symp. No. 3, *1976*, 29
1201. Hosur, R. V., Saran, A., Govil, G.: Ind. J. Biochem. Biophys. *16*, 165 (1979)
1202. Govil, G., Hosur, R. V., Saran, A.: Chem. Phys. Lipids *21*, 77 (1978)
1203. Hosur, R. V., Govil, G. Proc.: Ind. Nat. Sci. Acad. *45A*, 513 (1979)

1204. Govil, G., Hosur, R. V.: Proc. Int. Conf. on Biomolecular Structure, Conformation, Function and Evolution, Editor R. Srinivasan, 1979, Pergamon Press, p. 507
1205. Birdsall, N. J. M., et al.: J. Chem. Soc. Perk. II *1972*, 1441
1206. Hauser, H., et al.: Biochem. Biophys. Acta. *508*, 450 (1978)
1207. Hauser, H., et al.: Nature *261*, 390 (1976)
1208. Gally, H. U., Niederberger, W., Seelig: J. Biochemistry *14*, 3647 (1975)
1209. Niederberger W., Seelig, J.: J. Am. Chem. Soc. *98*, 3704 (1976)
1210. Seelig, J., Gally, H. U., Woglemuth, R.: Biochim. Biophys. Acta *467*, 109 (1977)
1211. Lee, A. G., Birdsall, N. J. M., Metcalfe, J. C.: Chem. Br. *9*, 116 (1973)
1212. Akutsu, H., Kyogoku, Y.: Chem. Phys. Lipids *18*, 285 (1977)
1213. Hosur, R. V., Govil, G.: J. Ind. Inst. Sci. *57*, 165 (1975)
1214. Levine, Y. K., et al.: Biochemistry *11*, 1416 (1972)
1215. Levine, Y. K., et al.: FEBS Lett. *23*, 203 (1972)
1216. Espersen, W. G., Martin, R. B.: J. Phys. Chem. *80*, 741 (1976)
1217. Marsh, D., Watts, A.: FEBS Lett. *85*, 124 (1978)
1218. Hauser, H., et al.: Biochemistry *19*, 366 (1980)
1219. McAlister, J., Yathindra, N., Sundaralingam, M.: Biochemistry *12*, 1189 (1973)
1220. Dufourcq, J., Lussan, C.: FEBS Lett. *26*, 35 (1972)
1221. Barker, R. W., et al.: Biochim. Biophys. Acta. *260*, 161 (1972)
1222. Yeagle, P. L., et al.: Proc. Nat. Acad. Sci. USA *72*, 3477 (1975)
1223. Kohlar, S. J., Klein, M. P.: Biochemistry *15*, 967 (1976)
1224. Birdsall, N. J. M., et al.: J. Chem. Soc. Chem. Commun. *1971*, 1171
1225. Lee, A. G., Birdsall, N. J. M., Metcalfe, J. C.: Biochim. Biophys. Acta. *255*, 43 (1972)
1226. Arvidson, G., Lindblom, G., Drakenberg, T.: FEBS Lett. *54*, 249 (1975)
1227. Lindblom, G., Persson, N. O., Arvidson, G.: Lyotropic liquid crystals and the structure of biomembranes. Friberg, S. (ed.), chap. 9, Adv. chem. ser. New York: Academic Press 1975
1228. Griffin, R. G., Powers, L., Pershan, P. S.: Biochemistry *17*, 2718 (1978)
1229. Urbina, J., Waugh, J. S.: Proc. Nat. Acad. Sci. USA *71*, 5062 (1974)
1230. Griffin, R. G.: J. Am. Chem. Soc. *98*, 851 (1976)
1231. Pines, A., Gibby, M., Waugh, J. S.: J. Chem. Phys. *59*, 564 (1973)
1232. Schrier, S., Polnaszek, C. F., Smith, I. C. P.: Biochim. Biophys. Acta *515*, 375 (1978)
1233. Butter, K. W., Smith, I. C. P.: Can. J. Biochem. *56*, 117 (1978)
1234. Hosur, R. V.: Ph.D. Thesis, Bombay University 1978
1235. Hosur, R. V., Joshipura, A. S.: Int. J. Quant. Chem. *17*, 983 (1980)
1236. Jendrasiak, G. L.: Chem. Phys. Lipids *9*, 133 (1972)
1237. Dahlquist, F. W., et al.: Proc. Nat. Acad. Sci. USA *74*, 5435 (1977)
1238. Hunt, Q. R., Tipping, L. R. H.: Biochim. Biophys. Acta. *507*, 242 (1978)
1239. Keough, K. M., Oldfield, E., Chapman, D.: Chem. Phys. Lipids *10*, 37 (1973)
1240. Phillips, M. C., Finer, E. G.: Biochim. Biophys. Acta. *356*, 199 (1974)
1241. Ulmius, J., et al.: Biochim. Biophys. Acta. *389*, 197 (1975)
1242. Lee, A. G., Birdsall, N. J. M., Metcalfe, J. C.: Methods in membrane biology, vol. 2. New York, London: Plenum Publish. Corp. 1974
1243. Stockton, G. W., et al.: Biochemistry *15*, 954 (1976)
1244. Stockton, G. W., et al.: Biochem. Biophys. Res. Commun. *60*, 844 (1974)
1245. Joyce, A., Smith, I. C. P.: Microbiology *1979*, 5
1246. Rice, D. M., et al.: Biochemistry *18*, 5885 (1979)
1247. Rice, D. M., et al.: Biochemistry *18*, 5893 (1979)

9 Appendix

Since the manuscript of this book has been submitted to Springer-Verlag, several more papers dealing with the conformational aspects of biological molecules have appeared. We have therefore updated the references. The list now include papers published until the end of 1980. The new references are arranged according to the chapter of the book, and the titles have been retained to help the reader with the contents.

Chapter 1

Section 1.5

1. Rabenstein, D. L., Isab, A. A.: Water elimination by T2 relaxation in proton spin-echo FT-NMR studies of intact human erythrocytes and protein solutions. J. Mag. Reson. *36*, 281 (1979)
2. Chapman, G. E.: Nuclear magnetic resonance. NMR of natural macromolecules. Nucl. Magn. Reson. *8*, 242 (1979)
3. Danyluk, S. S., Schwartz, H. M.: Carbon-13 cross polarization magic-angle NMR studies of biological systems. Energy Res. Abstr. *4*, 20 (1979)
4. Nagayama, K., Wuethrich, K., Ernst, R. R.: Two-dimensional spin echo correlated spectroscopy (SECSY). Biochem. Biophys. Res. Commun. *90*, 305 (1979)

Chapter 2

Section 2.1

Niu, C., et al.: Cross-peptide bond carbon-13-nitrogen-15 coupling constants by carbon-13 and J cross-polarization nitrogen-15 NMR. J. Biochem. Biophys. 1, 315 (1979)

Section 2.2

Perkins, S. J.: Ring current models for the heme ring in cytochrome. J. Mag. Reson. *38*, 297 (1980)

Section 2.5

Burt, C. T., Wyrwicz, A. M.: Phosphorus-31 nuclear magnetic resonance observations in biological systems. II Nucleic acids and proteins. Trends Biochem. Sci. (Pers. Ed.) *4*, 276 (1979)

Section 2.6

Niccolai, N., et al.: Solution structure, dynamics, and proton relaxation mechanisms of natural products and biopolymers. N-acetyl-D-alloisoleucine. J. Am. Chem. Soc. *102*, 1412 (1980)

Section 2.7

Bothner-By, A. A., Johner, P. M.: Spin-diffusion in macromolecules and its effect on nuclear Overhauser efefcts in proteins. Proc. Colloq. Spectrose, Int. Invited Lect. 20th, 77. 1, 355-71. Published by Statni Pedagog, Nakladetelstvi, Prague

Section 2.8

1. Cheng, H. N., Gutowsky, H. S.: Lanthanide shift reagents and their use in dynamic NMR studies. J. Phys. Chem. *84*, 1039 (1980)
2. Geraldes, C.F.G.C.: A case of nonaxial symmetry of lanthanide NMR dipolar shifts as related to the conformational usse of the lanthanide probe method. J. Mol. Struct. *60*, 7 (1980)

Section 2.12

1. Cantow, H. J.: NMR studies on conformation and dyanmics of macromolecules with emphasis on their configuration. Makromol. Chem. Suppl. *3*, 319 (1979)
2. Danyluk, S. S., Schwartz, H. M.: Crabon-13 cross-polarization magic-angle NMR studies of biological systems. Stereodyn. Mol. Syst., Prco. Symp., Pergamon Press *1979*, p. 439
3. Siminovitch, D. J., Rance, M., Jeffrey, K. R.: The use of wide-line nitrogen-14 NMR as a probe in model membranes. FEBS Lett. *112*, 79 (1980)
4. Lee, L., Sykes, B. D.: High resolution NMR. Adv. Inorg Biochem. (Methods Determining Met. Ion Environ, Proteins: Struct. Funct. Metalloproteins) *2*, 183 (1980)
5. Birdsall, B., et al.: The use of saturation transfer NMR experiments to monitor the conformation al selection accompanying ligand-protein interactions. FEBS Lett. *120*, 107 (1980)
6. Kumar, A., et al.: Studies of J-connectivities and selective proton-proton Overhauser effects in aqueous solutions of biological macromolecules by two-dimensional NMR experiments. Biochem. Biophys. Res. Commun. *96*, 1156 (1980)
7. Kumar, A., Ernst, R. R., Wuethrich, K.: A two-dimensional nuclear Overhauser enhancement (2D NOE) experiment for the elucidation of complete proton-proton cross-relaxation networks in biological macromolecules. Biochem. Biophys. Res. Commun. *95*, 1 (1980)
8. Jokisaari, J.: How useful is the NMR spectroscopy of partially oriented molecules in conformational studies? Kem.–Kemi 7, 11 (1980)
9. Dhingra, M. M., Sarma, R. H.: Nuclear magnetic resonance spectroscopy, Chemical Shifts, coupling constants and molecular geometry. Stereodyn. Mol. Syst. Proc. Symp., Pergamon Press *1979*, p. 3
10. Nagarajan, K.: Current trends in Fourer transform NMR spectroscopy. J. Sci. Ind. Res. *39*, 28 (1980)
11. Williamson, K. L., Williams, R. J. P.: Conformational analysis by nuclear magnetic resonance: insulin. Biochemistry *18*, 5965 (1979)

Chapter 3

Section 3.4

1. Gerlt, J. A., Younsblood, A. V.: The solution conformational preferences of the sugar and sugar phosphate constituents of RNA and DNA. J. Am. Chem. Soc. *102*, 7433 (1980)
2. Ferrige, A. G., Lindon, J. C.: Application of resolution enhancement in FT NMR to the structural determination of doxy sugars. Spectrosc. Lett. *13*, 339 (1980)

Section 3.5

1. Klimke, G., Luedemann, H. D., Townsend, L. B.: Further evidence for a S-syn correlation in the purine (beta) ribosides: the solution conformation of two tricyclic analogs of adenosine and guanosine. Z. Naturforsch., C, Biosci. *34C*, 653 (1979)
2. Klimke, G., et al.: Ribose conformations of adenosine analogs modified at the 2′, 3′ or 5′ positions. Z. Naturforsch., C. Biosci. *34C*, 1075 (1979)
3. Uzawa, J., Uramoto, M.: Assignment of indirect carbon-13-proton couplings in the carbon-13 NMR spectra of some purine and pyrimidine nucleosides and their analogs by long-range selective proton decoupling. Org. Mag. Res. *12*, 612 (1979)

4. Blieszner, K. C., Horton, D., Markovs, R. A.: Acyclic-sugar purine nucleosides derived from D-glucose: sterochemical correlations in acyclic-sugar derivatives unequally substituted at C-1. Carbohydr. Res. *80*, 241 (1980)
5. Stolarski, R., Dudycz, L., Shugar, D.: NMR studies on the syn-anti dynamic equilibrium in purine nucleosides and nucleotides. Eur. J. Biochem. *108*, 111 (1980)
6. Stolarski, R., et al.: Comparison of theoretical and experimental approaches to determination of conformation of nucleosides about the glycosidic bond. Biochim. Biophys. Acta *610*, 1 (1980)
7. Lipnick, R. L., Fissekis, J. D.: A comparative proton magnetic resonance conformational study of the tRNA "Wobble" nucleosides 5-carboxymethyl-, 5-carbo-methoxymethyl-, and 5-carbomoylmethyluridine. Can. J. Biochem. *58*, 1355 (1980)
8. Guschlbauer, W.: Conformational analysis of ribonucleosides from proton-proton coupling constants. Biochim, Biophys. Acta *610*, 47 (1980)
9. Kharasch, E. D., Novak, R. F.: Ring current effects in adriamycin-flavin mononucleotide complexation as observed by proton FT NMR spectroscopy. Biochem. Biophys. Res. Commun. *92*, 1320 (1980)
10. Guschlbauer, W., Jankowski, K.: Nucleoside conformation is determined by the electronegativity of the sugar substituent. Nucleic Acids Res. *8*, 1421 (1980)

Section 3.6

1. Bobruskin, I. D. et al.: Nuclear Magnetic Resonance study of the conformation of nucleotides, oligonucleotides, and their analogs in solution. Mol. Biol. (Moscow) *13*, 1407 (1979)
2. Bobruskin, I. D., et al.: NMR study of the conformation state of nucleosides and nucleotides in aqueous solutions. Bioorg. Khim. *6*, 1163 (1980)
3. Cadet, J., et al.: Conformational studies of .alpha.- and .beta.-pyrimidine. Biochim. Biophys. Acta *608*, 435 (1980)
4. Chang, Ching-Jer, Lee, Chi-Gen.: Methylation and conformational analysis of purine ribonucleotides. J. Carbohydr., Nucleosides, Nucleotides *7*, 93 (1980)
5. Hayashi, F., Akasaka, K., Hatano, H.: Conformation of adenosine 3',5'-monophosphate in solution as studied by the NMR-desert method. II. Self-association and temperature-dependent glycosidic isomerization at pH 7. Biochim. Biophys. Acta *588*, 181 (1979)
6. Cohn, M., Hu, A.: Isotopic oxygen-18 shifts in phosphorus-31 NMR of adenine nucleotides synthesized with oxygen-18 in various positions. J. Am. Chem. Soc. *102*, 913 (1980)
7. Yarbrough, L. R., Bock, J. L.: Stacking interactions in fluorescent nucleotide analogs containing 1-aminoaphthalene-5-sulfonate at the phosphoryl terminus. J. Biol. Chem. *225*, 9907 (1980)
8. Zolin, V. F., Koreneva, L. G., Lagodzinskaya, G. V.: Use of rare earth elements as probe for studying biologically active compounds. VIII. Study of the conformation of some nucleotides using a lanthanide shifting reagent and luminescence spectroscopy. Biofizika *24*, 945 (1979)
9. Neidle, S., et al.: 9-beta.-D-Arabinofuranosyl-8-n-butylaminoadenine, a C-8 substituted nucleoside in the anti conformation. Biochim. Biophys. Acta *565*, 379 (1979)
10. Watanabe, K., et al.: CD and NMR studies on the conformational thermostability of 2-thioribothymidine found in the T.PSI.C loop of thermophile tRNA. Biochem. Biophys. Res. Commun. *91*, 671 (1979)
11. Evans, F. E., Wright, J. M.: Proton and phosphorus-31 nuclear magnetic resonance study on the stabilization of the anti conformation about the glycosyl bond of 8-alkylamino adenyl nucleotides. Biochemistry *19*, 2113 (1980)
12. Kan, L., S. et al.: Proton nuclear magnetic resonance studies en dideoxyribonucleoside methylphosphonates. Biochemistry *19*, 2122 (1980)
13. Haasnoot, C.A.G., et al.: Prediction of anti and gauche vicinal proton-proton coupling constants for hexapyranose rings using a generalized Karplus equation. Bull. Soc. Chim. Belg. *89*, 125 (1980)
14. Egert, E., et al.: Influence of substituents at the 5 position on the structure of uridine. J. Am. Chem. Soc. *102*, 3707 (1980)

Section 3.7

1. Lappi, D. A., Evans, F. E., Kaplan, N. O.: Reduced nicotinamide 8-(alkylamino) adenine dinucleotides: enzyme-coenzyme interactions with different adenyl glycogyl bond conformations. Biochemistry *19*, 3841 (1980)
2. Lee, Che-Hung, Tinoco, I. Jr.: Conformation studies of 13 trinucleoside diphosphates by 360 MHz PMR spectroscopy. A bulged base conformation. I. Base protons and H1′ protons. Biophys. Chem. *11*, 283 (1980)
3. Cheng, D. M., et al.: Conformational flexibility of the 3′ acceptor end of transfer ribonucleic acid. Biochemistry *19*, 2491 (1980)
4. Bobruskin, I. D., Kirpichnikov, M. P., Florentev, V. L.: NMR study of the conformation of nucleotides, oligonucleotides and their analogs in solution. II. Syn-anti equilibrium of adenosine, 5′-AMP, 3′-CMP and 3′-CMP in solution. Mol. Biol. (Moscow) *13*, 870 (1979)
5. Bobruskin, I. D., et al.: Nuclear magnetic resonance study of the conformational states of dinucleoside phosphates in aqueous solution. Biofizika *25*, 745 (1980)
6. Sarma, R. H.: Spatial configuration of oligonucleotides in solution as reflected in nuclear resonance study. Nucleic Acid Geom, Dyn., Pergamon Press *1980*, p. 143
7. Bobruskin, I. D., et al.: Nuclear magnetic resonance study of the conformation of nucleotides, oligonucleotides and their analogs in solution. IV. Syn-anti equilibrium in aqueous solution of 2′-deoxynucleosides and nucleotides. Mol. Biol. (Moscow) *14*, 830 (1980)
8. Bobruskin, I. D., et al.: Nuclear magnetic resonance study of the conformation of nucleotides, oligonucleotides and their analogs in solution. III. Syn-anti equilibrium in a solution of cytidine and 2′-deoxycytidine. Mol. Biol. (Moscow) *13*, 1407 (1979)
9. Sarma, R. H., Dhingra, M. M., Feldmann, R. J.: The diverse spatial configurations of DNA. Evidence for a vertically stabilized double helix. Sterodyn. Mol. Syst., Proc. Symp., Pergamon Press *1979*, p. 251
10. Green, M. K., Kotowyez, G.: The Nature of the nicotinamide adenine dinucleotide phosphate (NADP) complex with manganese (II) ions and studied by proton and phosphorus magnetic resonance. Can. J. Chem. *457*, 2434 (1979)
11. Olsthoorn, C. S., et al.: Conformational characteristics of the trinucloside diphosphate dApdApdA and its constituents from nuclear magnetic resonance and circular dichroism studies, Extrapolation to the stacked conformers. Eur. J. Biochem. *112*, 95 (1980)
12. Everett, J. R., et al.: Nearest-neighbour and next-nearest-neighbour effects in the proton NMR spectra of the oligoribonucleotides ApGpX and CpApX. Biopolymers *19*, 557 (1980)
13. Shindo, H. McGhee, J. D., Cohen, J. S.: Phosphorus-31 NMR studies of DNA in nucleosome core particles. Biopolymers *19*, 523 (1980)
14. Ikehara, M., Uesugi, S., Shida, T.: Polynucleotides, LV. Synthesis and properties of dinucleoside monophosphates derived from adening 8,2′-S- and uracil 6,2′-O-cyclonucleosides. Further support for the left-handed stacking of oligonucleotides having high-anti base torision angles. Chem. Pharm. Bull. *28*, 189 (1980)

Section 3.9

1. Shindo, H., et al.: Nonuniform backbone conformation of deoxyribonucleic acid indicated by phosphorus-31 nuclear magnetic resonance chemical shift anisotropy. Biochemistry *19*, 518 (1980)
2. Temussi, P. A., et al.: Temperature dependence of phosphorus-31 NMR spectra of chicken erythrocyte nucleosomes. Physiol. Chem. Phys. *11*, 445 (1979)
3. Hilbers, C. W.: Hydrogen-bonded proton exchange and its effect on NMR spectra of nucleic acids. Biol. Appl. Magn. Reson., Academic Press *1979*, p. 1
4. Robillard, G. D., Reid, B. R.: Elucidation of nucleic acid structure by proton NMR. Biol. Appl. Magn. Reson., Academic Press *1979*, p. 45
5. Patel, D. J.: Structure and dynamics of poly (dG-dC) in solution. Steroid diamine cntdot.-nucleic acid complexes and generation of an "alternating B-DNA" conformation in high salt. Stereodyn. Mol. Syst., Proc. Symp., Pergamon Press *1979*, p. 397

Section 3.10.

1. Hamill, W. D. Jr., Horton, W. J., David, M.: Nuclear magnetic resonance relaxation studies of carbon-13 labeled uracil in transfer ribonucleic acid. J. Am. Chem. Soc. *102*, 5454 (1980)
2. Buckingham, R. H., Kearns, D. R.: Low field nuclear magnetic resonance studies of the structure of UGA suppressor tryptophan tRNA from Escherichia Coli: normal and photo-crosslinked species. Biochimie *62*, 491 (1980)
3. Schweizer, M. P., et al.: Carbon-13 NMR studies on (4-13C)-uracil-labeled Escherichia coli transfer RNA Val. Nucleic Acids Res. *8*, 2075 (1980)
4. Lipnick, R. L., Fissekis, J. D.: Carbon-13 NMR spectra of the tRNA "wobble" nucleosides 5-carboxymethyl-, 5-carbomethoxymethyl-, and 5-carbomoylmethyluridine. J. Heterocyl, Chem. *17*, 195 (1980)
5. Bolton, P. H., James, T. L.: Conformational mobility of deoxyribonucleic acid, transfer ribonucleic acid, and poly (adenylic acid) as monitored by carbon-13 nuclear magnetic resonance relaxation. Biochemistry *19*, 1388 (1980)
6. Luoma, G. A., et al.: Melting of Saccharomyces cerevisiae 5S ribonucleic acid: Ultra absorption, circular dichroism, and 360-MHz proton nuclear magnetic resonance spectroscopy. Biochemistry *19*, 5456 (1980)
7. Reid, B. R., Hurd, R. E.: High-resolution NMR studies on tRNA structure in solution. Cold Spring Harbor Monogr. Ser. 79, 9A (Transfer RNA: Struct. Prop., Recognition), pp. 177–190
8. Schmidt, P. G., Tompson, J. G., Agris, P. F.: Tranfer RNA structure by carbon-13 NMR: C2 of adenine, uracil and cytosine. Nucleic Acids Res. *8*, 643 (1980)
9. Marshall, A. G., Smith, J. L.: Raman and $\{^{19}F\}^{1}H$ nuclear Overhauser evidence for a rigid solution information of Escherichia coli 5-fluorouracil 5S ribonucleic acid. Biochemistry *19*, 5955 (1980)
10. Johnston, P. D., Redfield, A. G.: Proton FT NMR studies of tRNA structure and dynamics. Cold Spring Harbor Monogr. Ser. 79, 9A (Transfer RNA: Struct., Prop. Recognition), p. 191

Section 3.11

1. Chen, D. M., et al.: B1 eomycin-A2 complexes with poly (dA-dT): a proton nuclear magnetic resonance study of the nonexchangeable hydrogens. Biochem. Biophys. Res. Commun. *92*, 197 (1980)
2. Cozzone, P., Toniolo, C., Jardetzky, O.: NMR study of the main components of culpeine and their possible interaction with nucleic acids. FEBS Lett. *110*, 21 (1980)
3. Fritzsche, H., et al.: On the interaction of caffeine with nucleic acids. III and IV. Studies of the caffeine-DNA interaction by infrared and ultraviolet linear dichroism, proton and deuteron nuclear magnetic resonance. Biophys. Chem. *11*, 121 (1980) and *11*, 109 (1980)
4. Krugh, T. R., Nuss, M. E.: Nuclear magnetic resonance studies of drug-nucleic acid complexes. Biol. Appl. Magn. Reson., Academic Press *1979*, p. 113
5. Nuss, M. E., et al.: An NMR study of the interaction of daunomycin with dinucleotides and dinucleoside phosphates. Biochim. Biophys. Acta *609*, 136 (1980)

Chapter 4

Section 4.4

1. Ganapathy, S., Srinivasan, R.: Wide-line NMR studies of organic and biomolecules. Part V. PMR studies of some amino acids and related compounds. Indian J. Biochem. Biophys. *16*, 310 (1979)
2. Nagaraj, R., Venkatachalapathi, Y. V., Balaram P.: Rational isomerism about the C. alpha..CO bond in proline derivatives. Proton and carbon-13 NMR studies of benzyloxycarbonyl-pro-N-methylamide and privaloyl-pro-N-methylamide. Int. J. Pept. Protein Res. *16*, 291 (1980)
3. Mossoyan, J., Asso, M., Berlain, D.: Ligand conformation in lanthanide complexes by NMR paramagnetic shifts: L-proline and L-valine. Org. Magn. Reson. *13*, 287 (1980)
4. Cung, M. T., et al.: Angular dependence of vicinal coupling constants J (13CH3-N-C. alpha.-H-alpha.) in N-methylated peptides. C. R. Seances Acad. Sci., Ser. C *290*, 291 (1980)

Section 4.5

1. Madison, V., Kopple, K. D.: Solvent-dependent conformational distributions of some peptides. J. Am. Chem. Soc. *102*, 4855 (1980)
2. Krishna, N. R., Huang, D. H., Goldstein, G.: Solution conformation of peptide through proton nuclear magnetic double resonance measurement of amide hydrogen exchange rates: study of the active pentapeptide fragment of thymopoietin. Appl. Spectrosc. *34*, 460 (1980)
3. Kobayashi, J., et al.: Conformational analysis of the side-chains of aromatic amino acid residues in Met5-enkephalin. Pept. Chem. 16th, 125 (1979)
4. Marchal, J. P., et al.: Nitrogen-15 chemical shifts and 1J15N1H of some tripeptides measured at the natural abundance level. Biochemistry *19*, 1301 (1980)
5. Piriou, F., et al.: Amino acid side chain conformation in angiotensin II and analogs: Correlated results of circular dischroism and proton nuclear magnetic resonance. Proc. Natl. Acad. Sci. U.S.A. *77*, 82 (1980)
6. London, R. E.: Quantitative evaluation of .gamma.-turn conformation in proline-containing peptides using carbon-13 NMR Int. J. Pept. Protein Res. *14*, 377 (1979)
7. Toma, F., et al.: NMR studies for a type I.beta.-turn in (Pro2)-tetrapeptides and interdependence of cis:trans isomerism, ring flexibility, and backbone conformation. Biopolymers *19*, 781 (1980)
8. Garbay-Jaureguiberry, et al.: Conformational analysis of linear peptides by nitrogen-15 NMR spectroscopy using the enkephalin-related fragment Try-Gly-Phe as a model compound. FEBS Lett. *115*, 315 (1980)
9. Khaled, A. M., Okamoto, K., Urry, D. W.: Nitrogen-15 NMR of repeat peptides of tropoelastin. The tetrapeptide. Biochim. Biophys. Acta *623*, 229 (1980)
10. Premilat, S., Maigret, B.: Statistical molecular models for angiotensin II and enkephal in related to NMR coupling constants. J. Phys. Chem. *84*, 293 (1980)
11. Fermandjian, S., et al.: Arrangement of the amino acid side chains in angiotensin II Pept., Struct. Biol. Funct., Proc. Am. Pept. Symp., 6th 205 (1979)
12. Hallenga, K., et al.: The conformational properties of the peptide hormone somatostatin. III. Assignment and analysis of the proton and carbon-13 high resolution NMR spectra of somatostatin in aqueous solution. FEBS Lett. *119*, 47 (1980)
13. Toniolo, C., et al.: Proton magnetic resonance study of linear sarcosine oligomers. Macromolecules *13*, 1381 (1980)
14. Stevens, E. S., et al.: Conformational analysis of linear peptides. 3. Temperature dependence of NH chemical shifts in chloroform. J. Am. Chem. Soc. *102*, 7048 (1980)
15. Deslauriers, R., Smith, I.C.P.: The multinuclear NMR approach to peptides. Structures. conconformations, and dynamics. Biol. Magn. Reson. *2*, 243 (1980)
16. Higuchi, N., Kyogoku, Y., Yajima, H.: Nuclear magnetic resonance studies of pentapeptides with weak ACTH activity and their solution conformations. Pept. Chem. 17th, 159 (1980)
17. Kobayashi, J.: Conformational analysis of physiologically active peptides by the use of deuterium. A window of NMR analysis. Kagaku to Seibutsu *18*, 106 (1980)
18. Hull, W. E., Kricheldorf, H. R.: Nitrogen-15 NMR spectroscopy, 20. The cis/trans isomerism and neighbouring residue effects of proline-containing peptides. Biopolymers *19*, 1103 (1980)
19. Premilat, S., Maigret, B.: Conformations of angiotensin II and enkephalin related to NMR. Biochem. Biophys. Res. Commun. *91*, 534 (1979)
20. Bleich, H. E., et al.: Study of the rotational diffusion properties of peptides in solution. Biopolymers *18*, 2849 (1979)

Section 4.6

1. Kobayashi, J.: Conformational analysis of physiologically active peptides by the use of deuterium. A window of NMR analysis. Kagaku to Seibutsu *18*, 106 (1980)
2. Hindenlang, D. M., et al.: The carbon-13 NMR spectra of cyclopeptide alkaloids. Liebigs Ann. Chem. *447* (1980)
3. Higashijima, T., et al.: NMR studies on the conformations of gramicidin S analogs. Pept. Chem. 15th, *97* (1978)

4. Kobayashi, J., et al.: Nuclear magnetic resonance study of side-chain conformation of tyrosyl residue in (Met 5).enkephalin. Solvent and temperature dependenc. Biochim. Biophys. Acta. *621*, 190 (1980)
5. Anteunis, M. J. O., et al.: Solution conformation by proton NMR of the diastereometric cyclopeptides cyclo-4-thiapipecolic acid-phenylalanine. Bull. Soc. Chim. Belg. *88*, 683 (1979)
6. Kobayashi, Y., et al.: Structures of oxytocin analogs without disulfide bond in solution, II. Pept. Chem. 16th, 109 (1979)
7. Vicar, J., et al.: Carbon-13 nuclear magnetic resonance study of cyclodipeptides containing carbon-13 enriched hydrophobic amino acids. Collect. Czech. Chem. Commun. *45*, 482 (1980)
8. Wyssbroad, H. R., et al.: A comparison of the conformations of arginine vasopressin, arginine vasotocin, oxytocin, and oxypressin in aqueous solution by proton nuclear magnetic resonance spectroscopy. Pept., struct. Biol. Funct., Proc. Am. Pept. Symp., 6th, 213 (1979)
9. Philson, S. B., Bothner-By, A. A.: NMR studies of 1H-2H exchange in gramcidin S. Pept., Struct. Biol. Funct., Proc. Am. Pept. Symp., 6th, 209 (1979)
10. Prasad, K. U., Trapane, T. L., Urry, D. W.: Conformation studies on cyclic peptide analogs of the repeat tetrapeptide of tropoelastin. Pept., Struct. Biol. Funct., Proc. Am. Pept. Symp., 6th, 193 (1979)
11. Fischman, A. J., et al.: Torsion angles in the cystine bridge of oxytocin in aqueous solution. Measurements of circumjacent vicinal couplings between proton, carbon-13, and nitrogen-15. J. Am. Chem. Soc. *102*, 2533 (1980)
12. Khaled, Md. A., Sugano, H., Urry, D. W.: Conformational study of the cyclic hexapeptide cyclo (L-Ala-L-Pro-Gly-L-Val-Gly-L-Val), by nuclear magnetic resonance spectroscopy. J. Chem. Soc. Perkin Trans. *2*, 206 (1980)
13. Kricheldrof, H. R., Hull, W. E.: Nitrogen-15 NMR spectroscopy. 12. Steric effects in diastereomeric oligopeptides of alanine, phenylalanine, and valine. Org. Magn. Reson. *12*, 607 (1979)
14. Haslinger, E., Robien, K.: Proton-spin-lattice relaxation and internal movement of molecules. Studies on the conformation of frangulanine, a cyclic peptide alkaloid. Monatsh. Chem. *110*, 1011 (1979)
15. Antenuis, M. J. O., et al.: Structure assignments and conformations of the four diastereometric cyclic dipeptides of phenylalanine, S-methyl-4-thiazolidinium-2carboxylic acid. Bull. Soc. Chim. Belg. *88*, 817 (1979)
16. Tori, K., et al.: Proton NMR spectral evidence for the structure and conformation of peptide antibiotic siomycin-A. J. Antibiot. *32*, 1072 (1979)
17. Khaled, M. A., et al.: Nuclear magnetic resonance and conformational energy calculations of repeat peptides of tropoelastin: conformational characterization of the cyclododecapaptide. J. Chem. Soc. Perkin Trans *2*, 1119 (1980)
18. Kondor, P., Kessler, H.: Conformation of cyclic 5-peptides. Pept., Struct. Biol. Funct., Proc. Am. Pept. Symp., 6th, 181 (1979)
19. Smolders, R. R., et al.: Carbon-13 nuclear magnetic resonance conformational study of peptides with analgesic activity. Can. J. Biochem. *58*, 1241 (1980)
20. Pease, L. G.: Preferred hydrogen-bond conformations of cyclic pentapeptides. Pept., Struct. Biol. Funct., Proc. Am. Pept., Symp., 6th, 197 (1979)
21. Kessler, H., et al.: Peptide conformations. VI. Determination of intramolecular hydrogen bonds in cyclo(-Phe-Gly-Xxx-Val-Ala-) by proton NMR spectroscopy. Chem.Ber. *112*, 3528 (1979)
22. Kopple, K. D., Sarkar, S. K.: Solvent and structural effects on cyclic hexapeptide conformation. Pept., Struct. Biol. Funct., Proc. Am. Pept. Symp., 6th, 201 (1979)
23. Fonia, L. A., et al.: Structure-activity relation for cyclodepsipeptides of the valinomycin series, VIII. Isoleucinomycin and its cyclopolymeric homologs. Bioorg. Khim. *6*, 1285 (1980)
24. Rich, D. H., Jasensky, R. D.: Observation of 3 .fwdraw. 1 intramolecular hydrogen bonds (.gamma. turns) in the cyclic tetrapetides, (Ala4)-desimethylchlamydocin and cylco-(D-Phe-Pro-D-Phe-Pro), by by NMR spectrometry. Effect of solvent on solution conformation. J. Am. Chem. Soc. *102*, 1112 (1980)

Section 4.7

1. Suzuki, Y., Inoue, Y., Chujo, R.: Induced carbon-13 shifts by coil-supporting solvents observed in oligopeptides as model compounds of polypeptides Makromol. Chem. *181*, 177 (1980)
2. Ribeiro, A. A., Goodman, M., Naider, F.: The preferred conformations of protected homodito homoheptamethionine peptides. A proton NMR study in deuterochloroform medium. Int. J. Pept. Protein Res. *14*, 414 (1979)
3. Wittebort, R. J., Szabo, A., Gurd, F. R. N.: Rotational motions of side chains of poly-L-lysine. J. Am. Chem. Soc. *102*, 5723 (1980)
4. Toniolo, C., et al.: Nitrogen-15 magnetic resonance spectroscopy. Natural-abundance spectra of isomeric homooligopeptides of L-norvaline and L-valine to the tetramers. J. Org. Chem. *45*, 288 (1980)
5. Kricheldorf, H. R.: Nitrogen-15 NMR spectroscopy, 26, Coil-helix transition of poly-L-ornithine. Polym. Bull. *2*, 177 (1980)
6. Ribeiro, A., Saltman, R. P., Goodman, M.: Preferred conformations of protected homo-oligo-L-glutamate peptides in CDC13 and CDC13/TFA mixtures. Biopolymers *19*, 1771 (1980)
7. Naider, F., Ribeiro, A. A., Goodman, M.: Proton NMR study of protected methionine homo-oligopeptides in helix-supporting environment. Biopolymers *19*, 1791 (1980)
8. Zambelli, A., et al.: Correlation between carbon-13 NMR chemical shifts and conformational of polymers. 3. Hexad sequence assignments of methylene spectra of polypropylene. Macromolecules *13*, 267 (1980)
9. Perly, B., Chachaty, C., Tsutsumi, A.: A nuclear magnetic resonance study of the local conformation and molecular motions of poly(N5-3-hydroxypropyl)-L-glutamine) in aqueous solution. J. Am. Chem. Soc. *102*, 1521 (1980)
10. Suzuki, Y., Inoue, Y, Chujo, R.: Helix-coil transition in polypeptides having linear alkyl side-chain and ester group, and some conformational aspects observed by carbon-13 NMR. Makromol. Chem. *181*, 165 (1980)
11. Khaled, M. A.,Urry, D. W., Bradley, R. J.: pH and solvent titrations of enkephalins by carbon-13 nuclear magnetic resonance spectroscopy: complete assignments of resonances. J. Chem. Soc. Perkin Trans. 1693 (1979)
12. Lader, H.J., Mandelkern, L.: Origin of multipeak behaviour in the NMR spectra of poly (L-glutamic acid). Biopolymers *18*, 2607 (1979)

Section 4.8

1. Niccolai, N., Garsky, V., Gibbons, W. A.: Proton spin-lattice relaxation studies of (D-Ala2-Met5) enkephalin. J. Am. Chem. Soc. *102*, 1517 (1980)
2. Morishima, I., Ogawa, S., Yamada, H.: High-pressure proton nuclear magnetic resonance studies of hemoproteins. Pressure-induced structural change in heme environments of myoglobin, hemoglobin and horseradish peroxidase. Biochemistry *19*, 1569 (1980)
3. Sekacis, I., et al.: Carbon-13 NMR studies of the three-dimensional structure of tuftsin. Biorg. Khim. *5*, 1617 (1979)
4. Levine, B., A., et al.: Nuclear magnetic resonance studies of the solution structure of proteins. Int. Rev. Biochim. (Chem. Macromol. 2A – Simple Macromol.) *24*, 77 (1979)
5. Houghten, R. A., Yamashiro, D., Li, C. H.: Beta-Endorphin and Analogs: 360 MHz proton NMR spectroscopy Pept., Struct. Biol. Funct., Proc. Am. Pept. Symp., 6th, 869 (1979)
6. Wuethrich, K.: Complementation of protein crystal structures by NMR studies in solutions. Front Bioorg. Chem. Mol. Biol. Pengamon Press *1980*, p. 161
7. Tonelli, A. E.: Carbon-13 nuclear magnetic resonance chemical chifts and polypeptide structure. J. Am. Chem. Soc. *102*, 7635 (1980)
8. Llinas, M., De Marco, A., Lecomte, J. T. J.: Proton magnetic resonance study of Canambin, a hyperstable hydrophobic protein, at 250 and 600 MHz. Biochemistry *19*, 1140 (1980)
9. Wuethrich, K., et al.: Correlations between internal mobility and stability of globular proteins. J. Biophys. *32*, 549 (1980)
10. Ribeiro, A. A., et al.: An approach to the mapping of internal motions in proteins. Analysis of carbon-13 NMR relaxation in the bovine pancreatic trypsin inhibitor. J. Am. Chem. Soc. *102*, 4040 (1980)

11. Calvert, R., Unge Wickel, E., Gratzer, W.: A conformational study of human spectrin. Eur. J. Biochem. *107*, 363 (1980)
12. Higashijima, T., Tsumi, M., Miyazawa, T.: Proton NMR studies on molecular conformations of synthetic substrates of collagenase in solution. Pept. Chem. 14th, 89 (1977)
13. Okhanov, V. V., et al.: Three-dimensional structure of apamin in solution. Conformational NMR analysis. Bioorg. Khim. *6*, 840 (1980)
14. Jelinski, L. W., Torchia, D. A.: Investigation of labeled amino acid side-chain motion in collagen using carbon-13 nuclear magnetic resonance. J. Mol. Biol. *138*, 255 (1980)
15. Littlechild, J.: Proton magnetic resonance studies of Escherichia coliribosomal proteins S16. FEBS Lett. *111*, 51 (1980)
16. Morishima, I., Neya, S., Yone Zawa, T.: Proton NMR study of hemoproteins. Ionization and orientation of iron-bound imidazole in methemoglobin and metmyoglobin. Biochim. Biophys. Acta *621*, 218 (1980)
17. Sederel W. L., et al.: Solid-state conformation of copolymers of .beta.-benzyl-L-aspartate with L-alanine, L-leucine, L-valine, .gamma.-benzyl-L-glutamate, .epsilon.-carbobenzoxy-L-lysine. Biopolymers *19*, 1603 (1980)
18. De Marco A., Manegatti, E., Guarneri, M.: Proton nuclear magnetic resonance studies of the procine pancreatic secretary trypsin inhibitor at 270 MHz. Eur. J. Biochem. *102*, 185 (1979)
19. Buttlaire, D. H.: Magnetic resonance and optical spectroscopic studies of substrate-and anion-induced conformational changes of arginine kinase. NMR Biochem. Symp., Published by Dekkar, N.Y. 1979, p. 389
20. Brnjas-Kraljevic, J., Pifat, G., Maricic, S.: Quaternary structure, hydration, and self-associatiion of hemoglobin. A proton magnetic relaxation study. Physiol. Chem. Phys. *11*, 371 (1979)
21. Konishi, Y., Scheraga, H. A.: Regeneration of ribonuclease A from the reduced protein. 2. Conformational analysis of the intermediates by nuclear magnetic resonance spectroscopy. Biochemistry *19*, 1316 (1980)
22. Hill, H. A. O., et al.: Investigation of the structure of the blue copper protein from Rhus vernicifera stellacyanin by proton nuclear magnetic resonance spectroscopy. J. Inorg. Biochem. *11*, 101 (1979)
23. Morishima, I., Ogawa, S., Yamada, H.: Nuclear magnetic resonance studies of the effects of pressure on the heme environmental structure of hemoproteins. J. Am. Chem. Soc. *101*, 7074 (1979)
24. Perkins, S. J., Wuethrich, K.: Conformational transition from trypsinogen to trypsin. Proton nuclear magnetic resonance at 360 MHz and ring current calculations. J. Mol. Biol. *138*, 43 (1980)
25. Jardetzky, O.: NMR relaxation studies of protein dynamics. NMR Biochem., Symp. Dekker *1979*, p. 141
26. Garbay-Jaureguiberry, C., et al.: X-ray and NMR studies of L-4-hydroxyproline conformation in oligopeptides related to collagen. J. Am. Chem. Soc. *102*, 1827 (1980)
27. Higuchi, N,. Kyogoku, Y., Yajima, H.: Nitrogen-15 nuclear magnetic resonance of (Met5)-enkephalin: signal assignments and the effect of pH on the chemical shifts. Chem. Lett. 171 (1980)
28. Harada, K., Ito, S., Suzukii, M.: Structural investigation of an antibiotic sporaviridin. II. Application of carbon-13 NMR to the structural elucidation of viridopentaose B. Tetrahedron Lett. 3965 (1980)
29. Lee, A. G.: The study of protein structure and conformation using magnetic resonance. Tech. Life Sci. Sect. Biochem. *B107*, 1 (1978)
30. James, T. L., et al.: Unfolding of ribonuclease A by guanidinium chloride. Protein internal motions studied by nuclear magnetic resonance spin-lattice relaxation in an off-resonance rotating frame. J. Am. Chem. Soc. *101*, 7050 (1979)
31. Jansen, E. H. J. M., et al.: A 360-MHz photo-CIDNP study of bovine pancreatic phospholipase A2. Observation of a pH-dependent conformational change. J. Am. Chem. Soc. *101*, 7397 (1979)
32. Jardetzky, O., Wade-Jardetzky, N. G.: Comparison of protein structures by high resolution solid state and solution NMR. FEBS Lett. *110*, 133 (1980)

33. Miyazawa, T.: Structures of neurotoxic proteins and their nmr spectra. Kagaku (Kyoto) *34*, 920 (1979)
34. La Mar, G. N., Budd, D. L.: Proton NMR study of model substrate binding in hemoproteins. Intercalation of mecuric triiodide in sperm whale met-aquo myoglobin. Biochim. Biophys. Acta *581*, 201 (1979)
35. Neya, S., Morishima, F.: Proton magnetic resonance study of p-mercuribenzoate binding and structural changes in methemoglobin. Biochemistry *19*, 258 (1980)
36. Bolton, P. H., James, T. L.: Fast and slow conformational fluctuations of RNA & DNA. Subnanosecond internal motion correlation times determined by phosphrus-31 NMR. J. Am. Chem. Soc. *102*, 25 (1980)
37. Evans, J. S., et al.: Proton magnetic resonance studies on proteolytic fragments of troponin C. Structural homology with the native molecule. Biochim. Biophys. Acta *623*, 10 (1980)
38. Feeney, J., et al.: Proton nuclear magnetic resonance studies of the effects of ligand binding on tryptophan residues of selectively deuterated dihydrofolate reductase from Lactobacillus casel. Biochemistry *19*, 2316 (1980)
39. Stollery, J. G., et al.: Direct observation by carbon-13 nuclear magnetic resonance of membrane-bound human myelin basic protein. Biochemistry *19*, 2391 (1980)
40. Hershberger, M. V., et al.: Phosphrescence and optically detected magnetic resonance studies of a class of anomalous tryptophan residues in globular proteins. Biochemistry *19*, 2204 (1980)
41. Edwards, B. F. P., Sykes, B. D.: Nuclear magnetic resonance evidence for the coexistence of several conformational states of rabbit cardiac and skeletal tropomyosins. Biochemistry *19*, 2577 (1980)
42. Moore, G. R., Williams, R. J. P.: Nuclear magnetic resonance studies of ferrocytochrome c. pH and temperature dependence. Eur. J. Biochem. *103*, 513 (1980)

Section 4.9

1. Granot, J., et al.: NMR studies of the nucleotides conformation and the arrangement of substrates and activators on phosphoribosylpyrophosphate synthetase. J. Biol. Chem. *255*, 10931 (1980)
2. Ishikawa, T., et al.: Functional monomers and polymers. LXXI. Conformation and interaction studies of poly-L-lysine derivatives with pendant nuclei acid bases. J. Polym. Sci., Polym. Chem. Ed. *18*, 1847 (1980)
3. Davanloo, P., et al.: Nuclear magnetic resonance investigation of lysine oligopeptides and a complex with d(pA) 3 pGpC(pT)3. Biopolymers *18*, 2213 (1979)

Chapter 5

Section 5.2

1. Jaques, L. W., Glant, S., Weltner, W.: Spin-lattice relaxation times for two isomers of N-acetylneuraminyllactose. Carbohydr. Res. *80*, 207 (1980)
2. Angyal, S. J., Le Fur, R.: Conformations of acyclic sugar derivatives, Part V. The carbon-13 NMR spectra of alditols. Carbohydr. Res. *84*, 201 (1980)

Section 5.3

1. Saito, H.: NMR studies on carbohydrates. Application of 13C NMR to determine composition, conformation and dynamics of polysaccharides. Yuki Gosei Kagaku Kyokaishi *37*, 935 (1979)
2. Coxon, B.: Carbon-13 nuclear magnetic resonance spectroscopy of food-related disaccharides and trisaccharides. Dev. Food Carbohydr. *2*, 351 (1980)
3. Jennings, H. J., Rosell, K. J., Kasper, D. L.: Structural determination and serology of the native polysaccharide antigen of type-III group B Streptococcus. Can. J. Biochem. *58*, 112 (1980)
4. Matsuhiro, B., Zanlungo, A. B.: Polysaccharides from Chilean seaweeds, Part VI. Structural studies on the soluble polysaccharide of Chondrus canaliculatus. Proton NMR study of hexa-o-acetylcarrabiose dimethyl acetal. Carbohydr. Res. *81*, 335 (1980)
5. Vliegenthart, J. F. G.: High resolution proteon NMR spectroscopy of carbohydrate structures. Adv. Exp. Med. Biol. *125* (Struct. Funct. Gangliosides), 77 (1980)

6. Nuez, H. A., Barker, R.: Enzymic synthesis and carbon-13 nuclear magnetic resonance conformational studies of disaccharides containing .beta.-D-galactopyranosyl and .beta.-D-(1.13C)galactopyranosyl residues. Biochemistry *19*, 489 (1980)
7. Koto, S., et al.: Upfield shift of carbon-13 NMR of anomeric carbons of 1,1- linked glycopyranosyl glycopyranosides. Chem. Lett. *4*, 403 (1980)
8. Smith, I. C. P., Saito, H.: Composition, conformation, sequence and dynamics of complex carbohydrates as revealed by carbon-13 NMR spectroscopy. Int. Congr. Pure Appl. Chem. (Proc.) *27*, 213 (1980)
9. Lemieux, R. U., Bock, K.: The conformations of the Lewis blood group determinants. Jpn. J. Antibiot. *32* (Supl.), 163 (1979)
10. Brown, L. R., Lauterwein, J., Wuethrich, K.: High-resolution proton NMR studies of self-aggregation of melittin in aqueous solution. Biochim. Biophys. Acta *622*, 231 (1980)
11. Lauterwein, J., Brown, L. R., Wuethrich, K.: High resolution proton NMR studies of monomeric melittin in aqueous solution. Biochim. Biophys. Acta *622*, 219 (1980)
12. Meredith, P., et al.: The developing starch granule. Part IX. Saccharides of developing wheat grain determined by carbon-13 NMR spectroscopy. Starch/Stärke *32*, 198 (1980)
13. Dorland, L., et al.: Investigation by 360-MHz proton nuclear magnetic resonance spectroscopy and methylation analysis of the single-glycan chain of chicken evotransferrin. E. J. Biochem. *100*, 569 (1979)
14. Lemieux, R. U., et al.: The conformations of oligosaccharides related to the ABH and Lewis human blood group determinants. Can J. Chem. *58*, 631 (1980)
15. Riess-Maurer, I., Wagner, H., Liptak, A.: Synthesis and carbon-13 NMR study of flavonol 3-0-(cirhamnosyl) glactosides of the xanthorhamin, catharticin and alaterin series. Tetrahedron Lett. *39*, 3695 (1979)
16. Nicolaisen, F. M., Meyland, I., Schaumburg, K.: Carbon-13 NMR spectra at 67.9 MHz of aqueous solutions of agarose and partly 6-0-methylated agarose at 95, degree C. Acta Chem. Scand. *B34*, 103 (1980)
17. Hall, L. D., Morris, G. A.: Measurement of carbon-13 proton coupling constants in oligosaccharides by two-dimensional carbon-13 NMR spectroscopy. Carbohydr. Res. *82*, 175 (1980)
18. Atalla, R. H., et al.: Carbon-13 NMR spectra of cellulose polymorphs. J. Am. Chem. Soc. *102*, 3249 (1980)
19. Usov, A. I., Yarotskii, S. V., Shashkor, A. S.: Carbon-13 NMR spectroscopy of red algal galactans. Biopolymers *19*, 977 (1980)

Section 5.4

1. Bush, C. A., Duben, A., Ralapati, S.: Conformation of the glycopeptide linkage in asparagine-linked glycoproteins. Biochemistry *19*, 501 (1980)
2. Koto, S., Inada, S., Zen, S.: Upfield shift of carbon-13 NMR of anomeric carbons of 1,1'-linked glycopyranosyl glycopyranosides. Chem. Lett. 403 (1980)
3. Berman, E., Allerhand, A., DeVries, A. L.: Natural abundance carbon-13 nuclear magnetic resonance spectroscopy of antifreeze glycoproteins. J. Biol. Chem. *255*, 4407 (1980)
4. Cohen, R. E., Ballou, C. E.: Linkage and sequence analysis of mannose-rich glycoprotein core oligosaccharides by proton nuclear magnetic resonance spectroscopy. Biochemistry *19*, 4345 (1980)
5. Van Halbeek, H., et al.: The application of 500-MHz proton-NMR spectroscopy for the structure elucidation of N-acetyllactosamine type asparagine-bound carbohydrate chains of glycoproteins. FEBS Lett. *114*, 11 (1980)
6. Dill, K., Allerhand, A.: Natural abundance carbon-13 NMR spectra of some glycosylated amino acids. FEBS Lett. *107*, 26 (1979)
7. Cramer, J. A., Marchesi, V. T., Armitage, M.: Reversible trifluoroacetic acid-induced conformational changes in glycophorin A as detected by proton nuclear magnetic resonance spectroscopy. Biochim. Biophys. Acta *595*, 235 (1980)
8. Maggio, J. E.: Structure of a myco bacterial polysaccharidefatty acyl-Co A complex. Proc. Nat. Aca. Sci. U.S.A. *77*, 2582 (1980)

Chapter 6

Section 6.3

1. Seelig, J., Dijkman, R., de Haas, G. H.: Thermodynamic and conformational studies on sn-2-phosphatidyl-cholines in monolayers and bilayers. Biochemistry *19*, 2215 (1980)
2. London, R. E., et al.: Application of doubly decoupled carbon-13 protons, nitrogen-14-NMR spectroscopy to studies of the conformation and dynamics of the choline headgroup of phospholipids. Chem. Phys. Lipids *25*, 7 (1979)
3. Hauser, H., et al.: The polar group conformation of a lysophosphatidylcholine analogue in solution. A high-resolution nuclear magnetic resonance study. J. Mol. Biol. *137*, 265 (1980)

Section 6.4.

1. Hauser, H., et al.: Polar group conformation of phosphatidylcholine. Effect of solvent and aggregation. Biochemistry *19*, 366 (1980)
2. Burnell, E., et al.: Phosphorus-31 nuclear magnetic resonance and freeze-fracture electron microscopy studies on Escherichia coli. I. Cytoplasmic membrane and total phospholipids. Biochim. Biophys. Acta *597*, 492 (1980)
3. Burnell, E., et al.: Phosphorus-31 nuclear magnetic resonance and freeze-fracture electron microscopy studies on Escherichia coli. III. The outer membrane. Biochim. Biophys. Acta *597*, 518 (1980)
4. Rance, K., et al.: Orientational order of unsaturated lipids in the membranes of Acholoplasma laidlawii as observed by deuterium NMR. Biochim. Biophys. Acta *600*, 245 (1980)
5. Cornell, B. A.: The dynamics of the carbonyl groups in phospholipid bilayers from a study of their carbon-13 chemical shift anisotropy. Chem. Phys. Lett. *72*, 462 (1980)
6. Rance, M., et al.: Orientational order of unsaturated lipids in the membranes of Acholeslasma laidlawii as observed by deuterium NMR. Biochim. Biophys. Acta *600*, 245 (1980)
7. Van Meer, G., et al.: Preservation of bilayer structure in human erythrocytes and erythrocyte ghosts after phospholipase treatment. A phorphorus-31 NMR study. Biochim. Biophys. Acta *596*, 1 (1980)
8. Burn, R. A. Jr., Roberts, M. F.: Carbon-13 nuclear magnetic resonance studies of short-chain lectithins. Motional and conformational characteristics of micellar and monomeric phospholipid. Biochemistry *19*, 3100 (1980)
9. Neuringer, L. J., et al.: Deuterium NMR studies of cerebroside-phospholipid bilayers. Biochim. Biophys. Acta *558*, 325 (1979)
10. Kuo, An-Li, et al.: Lipid lateral diffusion in oriented lipid/deuterium oxide multilayers by pulsed NMR. Chem. Phys. Lipids *25*, 135 (1979)
11. Cushley, R. J., et al.: Carbon-13 and phosphorus-31 nuclear magnetic resonance of lipid dispersions from human aorta. Can. J. Biochem. *58*, 206 (1980)
12. Basosi, R., et al.: Electron paramagnetic resonance and nuclear magnetic resonance studies of lipid-water systems. J. Chem. Soc., Faraday Trans. 2, *76*, 91 (1980)
13. Wohlgemuth, R., et al.: Bilayers of phosphatidylglycerol. A deuterium and phosphorus nuclear magnetic resonance study of the head-group region. Biochemistry *19*, 3315 (1980)
14. Barsukov, L. I., et al.: Investigation of the inside-outside distribution, intermembrane exchange and transbilayer movement of phospholipids in sonicated vesicles by shift reagent NMR. Biochim. Biophys. *598*, 153 (1980)
15. Nichol, C. P., et al.: Quantitative study of the fluidity of Escherichia coli membranes using deuterium magnetic resonance. Biochemistry *19*, 451 (1980)
16. Crawford, M. S., et al.: Diffusion in rigid bilayer membranes. Use of combined multiple pulse and multiple pulse gradient techniques in nuclear magnetic resonance. J. Am. Chem. Soc. *102*, 3728 (1980)
17. Arnold, K., Gawrisch, K., Volke, F.: Phosphorus-31 NMR investigations of phospholipids. I. Dipolar interactions and the Phosphorus-31 NMR lineshape of oriented phospholipid/water dispersions. Stud. Biophys. *75*, 189 (1979)

18. Skarjune, R., Oldfield, E.: Physical studies of cell surface and cell membrane structure. Determination of phospholipid head group organization by deuterium and phosphorus nuclear magnetic resonance spectroscopy. Biochemistry *18*, 5902 (1979)

Section 6.5

1. Rice, D., et al.: Protein-lipid interactions. High-field deuterium and Phosphorus nuclear magnetic resonance spectroscopic investigation of the cytochrome oxidase-phospholipid interaction and the effects of cholate. Biochemistry *18*, 5885 (1979)
2. Van Alphen, L., et al.: Phosphorus-31 nuclear magnetic resonance and freeze-fracture electron microscopy studies on Escherichia Coli. II. Lipopolysaccharide and lipopolysaccharide-phospholipid complexes. Biochim. Biophys. Acta. *597*, 502 (1980)
3. Neya, S., et al.: Interaction of human adult methemoglobin in low-spin state with inositol hexaphosphate. A proton magnetic resonance study. Biochem. Biophys. Res. Commun. *92*, 825 (1980)
4. Pogliani, L., et al.: Proton and carbon-13 NMR study of phosphopeptides. I. Acetylphosphoserine and acetylphosphothreonine. Tetrahedron *35*, 2867 (1979)
5. Rice, D. M., et al.: Protein-lipid interactions. A nuclear magnetic resonance study of sarcoplasmic reticulum (calcium(2+).magnetium(2+)ion)-activated ATPase, lipophilin, and proteolipid apoprotein-lecithin systems and a comparison with the effects of cholesterol. Biochemistry *18*, 5892 (1979)
6. Feigenson, G. W., Meers, P. R.: Proton NMR study of valinomycin conformation in a phospholipid bilayer. Nature (London) *283*, 313 (1980)
7. Ptak, M., et al.: A 400 MHz proton NMR study of peptidolipin NA, a natural cyclic popeptide. Biochem. Biophys. Res. Commun. *94*, 1311 (1980)
8. Wüthrich, K., Bosch, C., Brown, L. R.: Conformational studies of lipid-bound polypeptides by elucidation of proton-proton cross-relaxation networks. Biochem. Biophys. Res. Commun. *95*, 1504 (1980)
9. Veatch, W., et al.: Conformational of the gramicidin A channel in lipid vesicles: a carbon-13 and fluorine-19 nuclear magnetic resonance study. Pept., Struct. Biol. Funct., Proc. Am. Pept. Symp. 6th, 635 (1979)
10. Chen, Tsung-Chang, et al.: Magnetic resonance studies on apolipoprotein C-I nitroxide- labeled or (13C) methyl enriched at methionine-38. Biochemistry *19*, 5140 (1980)
11. Dabrowkki J., Egge, H., Hanfland, P.: High resolution nuclear magnetic resonance spectroscopy of glycosphingolipids. I. 360 MHz proton and 90.5 MHz carbon-13 NMR analysis of galactosyceramides. Chem. Phys. Lipids *26*, 187 (1980)
12. Cullis, P. R., Hope, M. J.: The bilayer stabilizing role of sphingomyelin in the presence of choles cholesterol. A phosphorus-31 NMR study. Biochem. Biophys. Acta *597*, 533 (1980)
13. Jacobs, R. E., Oldfield, E.: NMR in membranes. Progress in NMR Spectroscopy *14*, 113 (1981)

10 Subject Index

Actinomycin D 80, 117
Adenosine diphosphate (ADP) 15, 62
Adenosine triphosphate 15, 62
Alamethicin 109
Allose 147
Altrose 147
Alumichrome 118
Amino acids
– table, 82, 84
– conformation 101
Angiotensin 103, 108
Antibody combining site 131,137
Arabinose 147

Backbone 8
– torsion angles in nucleotides 40
– torsion angles in peptides 92–96
Bohr's effect 140
Butane 4–6

^{13}C chemical shifts 22, 25
– of polynucleotides 68
– of polypeptides 100
– of polysaccharides 151
^{13}C NMR 12
– cis: trans ration in Pro peptides 25, 92
– molecular symmetry 99
- of carbohydrates 151
– relative sensitivity 10, 12
– spin-lattice relaxation time (T_1) 28, 99, 120, 127
– X–Pro bonds 92, 93
^{13}C NMR spectra
– of 5′-AMP 68
– of 5′-8 BrAMP 68
– of collagen 13
– of cyanoferrimyoglobin 12
– of cyclo(Pro-Gly)$_3$ 98
– of dipalmitoyl lacithin 162
– of glycoamylase 153
– of poly(A) 68
– of poly(8BrA) 68
– of poly(L-Glutamic acid) 126
– of poly(U) 72
– of proline peptides 94
– of sarcoplasmic reticulum membranes 173
C1, 1C: Conformations of saccharides 144–145
Calcium binding proteins 131, 137

Carbohydrates 144–154
Chemical shifts 10, 22
(see also individual muclei)
Chromatin 143
Chymotrypsin 140
^{35}Cl NMR 16
Clover-leaf model of t-RNA 75
Collagen 13, 122, 140
Configuration 3
Conformation
(see also individual compounds)
– classical states and torsion angles 3
– definition 2
– dependent long range couplings 22, 94
– dependent NMR parameters 10
– dependent parameters in valinomycin 122
– dependent parameters of peptides 92–101
– dependent parameters of polynucleotides 69
– dependent parameters of saccharides 146, 149
– dependent vicinal couplings 21
– use of nematic liquid crystals 35
– use of paramagnetic probes 30–34
Conformational mobility 6, 28, 99
– in membranes 163, 170
– in oxytocin 119
– in proline 102
– in proteins 130
Contact interaction 32
Convolution difference spectroscopy 12, 13
Correlation time 10, 28, 99
Coupling constant 10, 18
(see also *vicinal* couplings)
Cyclonucleosides 46
Cyclonucleotides 46, 61
Cyclo(Pro-Gly)$_3$ 98, 99
Cyclic peptides 110–123
– dipeptides 110, 111
– hexa peptides 112, 117
– pentapeptides 111, 116
– tetrapeptides 111, 116
– tripeptides 111, 115
Cyanoferrimyoglobin 12
Cytochrome c 140, 172

Deoxyribonucleic acid (DNA) 39, 56, 70, 71, 73, 74, 142

DESERT 28
Deuterium NMR (see ^{2}H-NMR)
Dextrans 150
Difference spectroscopy 12, 13, 33
Dihedral angles 2–4
Dihydrofolate reductase 131, 136
Diketopiperazine 110
Dinucleoside monophosphate 63
– table 63
Dipeptides 103, 105
Dipeptide maps 88, 89
Dipole-dipole interaction 35, 99
Dipole-dipole relaxation 27
Drug-nucleic acid interactions 79
– literature table 79

Elastin 132, 137
Elastin peptides 23
Electric quadrupole relaxation 31
Electronegativity
– effect on J 20, 146
Electron-nuclear interactions 31
Enkephalin 108
Enzyme-substrate binding 130
ESR signal quenching 34
ESR spinlabel 169

^{19}F NMR 16, 140
^{19}F NMR spectra
– of fluorotyrosine alkaline phosphate 16
Ferricytochrome 118
Ferrimyoglobin 31
Fourier transform (FT) NMR 10, 13, 58
– 2D 11

g-tensor 31
Galactose 147
Gd(III)
– as an NMR probe 32, 134
Gene-5 143
Globular proteins 128, 140
Glucose 147
Glycoproteins 152
Glycosyl torsion angle 40, 41, 43–46
Gramicidin S 113, 121

^{1}H chemical shifts
– of nucleotides 68
– of peptides 100
– of saccharides 149, 150
^{1}H NMR 9
– relative sensitivity 10
^{1}H NMR spectra
– of BPTI 129
– of d(TpTp) 66
– of dipalmitoyl lecithin 160
– of horse ferrimyoglobin 31
– of N-methyl acetamide in nematic phase 35
– of poly(A) 67
– of poly(BrA) 67
– of proline peptides 93
^{2}H NMR 11
– quadrupole couplings 36
– in membranes 164, 167–169
^{2}H NMR spectra
– of cytochrome c-lipids 172
– of dipalmitoyl lecithin 168
Helix 6
– α 89, 134
– 3_{10} 91
– DNA, RNA, polynucleotides 70–73
Helix-coil transitions
– in polynucleotides 71
– in polypeptides 124–128
Heme 138
Hemeproteins 132, 138–140
Hemoglobins 138–140
Hexapeptides 104, 108
Histones 143
Ho(III) as NMR probe 32
Human parathyroid hormone 105, 110
Hydrogen bonds 22
– in α-helix 89
– in β-sheet structures 90
– in DNA 71
– in protein-nucleic acid interactions 142
– in t-RNA 76, 77
Hydrogen bonded
– base pairs 71, 72, 76
– imino proton chemicals shifts 74
– in t-RNA 76, 77

Idose 147
INDOR 11, 95
Insulin 105, 110
Interaction, of drugs with nucleic acids 79
Isoenergy maps
– for n-pentane 7
– for nucleotides 64, 65
– for peptides 87–89, 91
Isosheilding contours 25
Isotope labelling methods
– for NMR studies 11, 16

J-couplings (see coupling constants)

^{39}K-NMR 16
Karplus relations 18–22, 43, 47, 96
– for ^{1}H–^{13}C couplings 43

– for $^{13}C-^{31}P$ couplings 51
– for $^{1}H-^{31}P$ couplings 51
– for peptides 95
– for sacharides 146
– parameters, for 19
– substituent electronegativity effects 19, 20, 146

Larmor frequency 26
Lande g-factor 32
Lanthanides 32, 101
Lentinan 152
Linear peptides 102–110
– tabulated literature 103–105
Lipid-bilayer structures 157
Lipid-protein interactions 172–173
Lipids 156
Liquid crystal 35
Luteinizing Hormone-release factor 105, 109
Lysozyme 131, 133

Magnetic anisotropy 23, 45
Magnetic susceptibility 23
Mannose 147
Melanocyte-stimulating hormone release inhibiting factor 107
Membranes 1, 11, 155–173
^{25}Mg NMR 16
Mn(II)
– as NMR probe 143
Molecular symmetry 25, 99
Myelin basic proteins 131, 137
Myoglobin 138

N-acetylglucosamine 134
^{14}N NMR 14
^{15}N NMR 14
– chemical shift 14
– relative sensitivity 14
^{15}N NMR spectra
– of oxytocin 121
– of poly(-Ala-Ser-Gly) 14
^{23}Na NMR 16
N-Methyl acetamide 35
N-Methyl formamide 36
Newmann projection diagrams 2, 5, 47
– for amino acid side chains 96
– for peptides 85
– for phospholipids 159, 161
for nucleotides 41, 54
Nicotinamide adenine dinucleotide (NADH) 34
Nuclear Overhauser Effect (NOE) 14, 29, 43, 44, 100, 128, 137, 164, 167

Nucleic acids 39
(see also under DNA, RNA and t-RNA)
– double helices 70–73
– rigidity in 70
– short segments 63–66, 71
– t-RNA 70–73
Nucleosides 39, 58
– pyridine 62
– literature table 59–60
Nucleotides 39, 58
– cyclic 61
– mono 61
– pyridine 62
– literature table 60–61

^{17}O NMR 16
Oligonucleotides 67
Oligopeptides
– reviews on 86
Oligo(Pro-Pro-Gly) 128
Ordered structure 6
Order parameter 36, 167–169

^{31}P chemical shifts 26, 57
– T_1, T_2 173
^{31}P NMR 15
– chemical shift anisotropy in membranes 164
– of intact cells 15
– of lipoproteins 323
– relative intensity 10
^{31}P NMR spectra of
– phosphate group 166
– rabbit heart 15
– t-RNA 15
Paramagnetic probes 26–34
Pentapeptide 23, 104, 108
Peptide groups, cis and trans 92–93, 128
Peptide hormones 104, 108
Peptide planarity 36
Phospholipids 157–163
Photochemical crosslinking in t-RNA 79
Poly(A) 67–69
Poly(β-Ala-Ser-Gly) 14
Poly(γ-benzyl-L-glutamate) 128
Poly(8BrA) 67–69
Poly(Gly-Gly-Pro-Gly) 128
Poly(L-Glutamic acid) 124
Poly(L-Pro) 128
Poly(Pro-Gly) 128
Poly(L-Tyr) 124
Poly(U) 72
Polydispersity 124, 127
Polypeptide
– review table 86

– helix-coil transitions 124–128
– dynamic character in solution 128
Polyribnucleotides 67–69
Polysaccharides 144–154
Porphyrin 138
Potential energy diagrams
– for H_2O_2 and H_2S_2 4
– for butane 5
– for ribose 49
– for angle ϕ' in nucleotides 52
– orf angle ω in peptidse 85
Pr(III)
– as a NMR probe 32, 134
Proflavin-nucleic acid interactions 80
Proline 89
– cis-trans isomerisation 92
– ring conformation 101
Propidium diiodide 80
Protein conformation, reviews 86
– structure (literature table) 131
Protein-nuclei acid interactions 74, 141
– literature table 141
Proton relaxation enhancement (PRE) 34, 37
Protoporphyrin 138
Pseudocontact shifts 32
Pseudorotation of furanose ring 46–50
Pyridine nucleosides 62
– nucleotides 62
Pyranose ring 144–148

Quadrupole coupling 11, 36, 167
Quadrupole nuclei
– as NMR probes 17
Quadrupole relaxation 28
Quaternary structure 3, 140
Quenching of ESR signal 34

Random coil 9, 17
– polynucleotides 66
– polypeptides 89
– proteins 129
Relaxation times (T_1 & T_2) 9, 26, 45, 99, 102
Ribonuclease 11, 132, 138
Ribonucleic Acid (RNA) 39, 56, 68–70, 71, 73, 74, 142
Ribose puckering 40, 46, 147
Ribosomal Proteins 74, 133
Ring current shift 23, 56, 120, 164
Rotamers 2
Rotational correlation time 29
Rotational isomeric state model 6, 21
Rotational isomers 2

Saturation transfer 37, 107
Scalar coupling 26
Schizophyllan 152
Secondary structure 2
– β-sheet structures 90, 117, 121, 134
Shielding in nucleic acid base pairs 73
Solvent accessibility 36, 97
Solvent titration 23, 97
Spatial structure 37, 129
Spin label 33
Spin rotation relaxation 26
Suzukacillin 109

t-RNA 40, 57, 58, 70, 71, 74–79, 143
– clover-leaf model 75
– conformational changes in 77–79
– ^{31}P NMR spectra 15
– ^{1}H NMR spectra 75, 78
Talose 147
Temperature dependence
– amide proton chemical shifts 97
Tetrapeptides 104, 107
Tertiary structure 2
– of t-RNA 76
– of hemoglobin 140
Thyrotropin-releasing factor (TRF) 106
Time averages 5, 102
Titration shifts 25, 100
Torsion angles 2–4
– in nucleotides 39–43
– in peptides 81, 84, 94
– in phospholipids 158
Tripeptides 103, 106
Tropoelastin 108, 132, 137
Trypsin inhibitor, basic pancreatic 129, 131, 134
β-turns 91, 107, 109, 118

Valinomycin 114, 122
Vasopressin 113, 119
Vicinal couplings 18–22
– and electronegativity 19
– in amino acids 101
– in nucleotides 43, 47, 51, 54, 55
– in peptides 94–97
– in saccharides 149, 150

Watson-Crick, base pairs 71

X-ray crystallography 1, 45
– t-RNA 75, 76
– in relation to NMR 37, 77, 95, 123, 129, 134
Xylose 147

Yb(III)- as NMR probe 32

A. Hebeish, J. T. Guthrie

The Chemistry and Technology of Cellulosic Copolymers

1980. 91 figures. XII, 351 pages
(Polymers/Properties and Applications, Volume 4)
ISBN 3-540-10164-0

The driving force behind the great scientific interest in copolymer science and technology is the search for products with useful, new or interesting properties. This monograph provides an informative account of new, improved cellulosic materials and the chemistry and technology involved in their production, as well as the first detailed description of grafted and modified celluloses. The topics it covers include:
Grafting reaction and the factors which influence the grafting process. – Mechanism of free radical formation on cellulose. – Radical decay processes. – Charge transfer mechanisms of grafting. – Grafting by chain transfer. – Uncatalyzed grafting. – Grafting by chemical activation of cellulose. – High energy radiation induced grafting. – Grafting by photochemical activiation. – Grafting on chemically modified celluloses.

NMR in Medicine

Editor: R. Damadian

1981. Approx. 77 figures. approx. 44 tables.
Approx. 200 pages. (NMR, Volume 19)
ISBN 3-540-10460-7

Contents:

R. Damadian, M. Goldsmith, L. Minkoff: NMR Scanning
I. D. Weisman, L. H. Bennet, S. Maxwell, D. E. Henson: Cancer Detection by NMR in the Living Animal
P. T. Beall, D. Medina, C. F. Hazlewood: The "Systematic Effect" of Elevated Tissue and Serum Relaxation Times for Water in Animals and Humans with Cancers
M. Goldsmith, R. Damadian: Proton Magnetic Resonance of Human Tissues. Further Development as a Method of Cancer Diagnosis
G.-J. Béné, B. Borcard, E. Hiltbrand, P. Magnin: Medical Diagnosis by Nuclear Magnetism in the Earth Field Range
J. A. Koutcher, K. Zaner, R. Damadian: ^{31}P as a Nuclear Probe for the Diagnosis and Treatment of Malignant Tissue
T. Glonek, C. T. Burt, M. Bárány: ^{31}P NMR Analysis of Intact Tissue Including Several Examples of Normal and Diseased Human Muscle Determinations
G. N. Ling, M. Tucker: NMR Relaxation and Water Contents in Normal Tissues and Cancer Cells

Springer-Verlag Berlin Heidelberg New York